The Enzymes

VOLUME XXIX

PROTEIN PRENYLATION

PART A

THE ENZYMES

Edited by

Fuyuhiko Tamanoi

Department of Microbiology,
Immunology and Molecular Genetics
University of California, Los Angeles
Los Angeles, CA 90095-1489, USA

Christine A. Hrycyna

Department of Chemistry
Purdue University
560 Oval Dr.West Lafayette
IN 47907-2084, USA

Martin O. Bergo

Wallenberg Laboratory
Institute of Medicine
Sahlgrenska University Hospital
S-413 45 Gothenburg, SWEDEN

Volume XXIX

PROTEIN PRENYLATION

PART A

AMSTERDAM • BOSTON • HEIDELBERG • LONDON
NEW YORK • OXFORD • PARIS • SAN DIEGO
SAN FRANCISCO • SINGAPORE • SYDNEY • TOKYO
Academic Press is an imprint of Elsevier

ELSEVIER

Academic Press is an imprint of Elsevier
32 Jamestown Road, London NW1 7BY, UK
Radarweg 29, PO Box 211, 1000 AE Amsterdam, The Netherlands
Linacre House, Jordan Hill, Oxford OX2 8DP, UK
225 Wyman Street, Waltham, MA 02451, USA
525 B Street, Suite 1900, San Diego, CA 92101-4495, USA

First edition 2011

ISBN: 978-0-12-381339-8
ISSN: 1874-6047
1006717614

For information on all Academic Press publications
visit our website at elsevierdirect.com

Printed and bound in USA

11   12   13   14   10 9 8 7 6 5 4 3 2 1

Contents

8. Organization and Function of the Rab Prenylation and Recycling Machinery

KIRILL ALEXANDROV, YAOWEN WU, WULF BLANKENFELDT, HERBERT WALDMANN, AND ROGER S. GOODY

9. Protein Prenylation CaaX Processing in Plants

SHAUL YALOVSKY

10. Posttranslational Isoprenylation of Tryptophan Residues in Bacillus subtilis

MASAHIRO OKADA, FUMITADA TSUJI, AND YOUJI SAKAGAMI

11. Global Analysis of Prenylated Proteins by the Use of a Tagging via Substrate Approach

LAI N. CHAN AND FUYUHIKO TAMANOI

12. Global Identification of Protein Prenyltransferase Substrates: Defining the Prenylated Proteome

CORISSA L. LAMPHEAR, ELAINA A. ZVERINA, JAMES L. HOUGLAND,
AND CAROL A. FIERKE

13. Structural Biochemistry of CaaX Protein Prenyltransferases

MICHAEL A. HAST AND LORENA S. BEESE

14. Genetic Analyses of the *CAAX* Protein Prenyltransferases in Mice

MOHAMED X. IBRAHIM, OMAR M. KHAN, AND MARTIN O. BERGO

15. Farnesyl Transferase Inhibitors: From Targeted Cancer Therapeutic to a Potential Treatment for Progeria

W. ROBERT BISHOP, RONALD DOLL, AND PAUL KIRSCHMEIER

Preface

The posttranslational modification of eukaryotic proteins by the addition of isoprenyl lipids at their C-termini was first observed in the 1970s and 1980s. These discoveries emerged from two different lines of investigation. First, studies of cells treated with cholesterol biosynthesis inhibitors led to the identification of prenylated proteins such as the nuclear lamins. Second, studies on yeasts, such as *Rhodosporidium toluroides*, revealed the presence of isoprenoids in peptide mating pheromones. Since then, more than a hundred proteins have been shown to be modified by C15 farnesyl or C20 geranylgeranyl groups, including most of the members of the Ras, Rho, and Rab families of G proteins. The precise number of prenylated proteins is not yet known, but novel proteomics approaches have been developed to define the entire prenylated proteome.

Soon after these modifications were discovered, the search for prenyltransferase enzymes intensified. The enzymes that catalyze protein prenylation, including protein farnesyltransferase, protein geranylgeranyltransferase type I, and protein geranylgeranyltransferase type II (Rab geranylgeranyltransferase) were identified and characterized in the 1990s. Subsequent structural studies identified subunit composition and catalytic properties of these enzymes and resulted in a more complete understanding of the mechanism of action at the atomic level.

In the late 1980s, it became clear that for some proteins that contain a C-terminal CaaX sequence (C is cysteine, "a" is a generally aliphatic residue, and X is one of a number of amino acids), prenylation is followed by proteolytic cleavage of C-terminal aaX residues and carboxylmethylation of the newly exposed C-terminal cysteine. The enzymes responsible for these modifications, Rce1 and Icmt, respectively, were subsequently identified and their biological and biochemical characterization is ongoing.

Since the discovery of these modifications, the biological significance of protein prenylation and the subsequent modifications have been investigated extensively. The modifications are thought to be important for membrane association, protein–protein interactions, protein stability, and receptor–ligand interactions. In particular, the demonstration that the transforming activity of mutant Ras depends on prenylation and methylation pointed to the significance of these modifications in oncogenesis. More recently, important insights into the biological role of these modifications have been obtained by a series of studies using knockout mice.

Finally, small molecule inhibitors of protein farnesyltransferase (FTIs), protein geranylgeranyltransferase (GGTIs), and Rab geranylgeranyltransferase have been developed and both FTIs and GGTIs have been evaluated as anticancer drugs. In addition, small molecule inhibitors of Rce1 and Icmt have been developed.

To encompass past and recent developments in the study of protein prenylation, we designed a two-part miniseries. This volume is the first part (Volume 29, Part A). We believe that this miniseries captures the current knowledge and advances in the field and hope that they will be of interest to a wide range of researchers including biochemists, molecular and cell biologists, and cancer researchers.

We thank the authors for their excellent and informative chapters, Mary Ann Zimmerman and Malathi Samayan of Elsevier for their guidance, and Gloria Lee at UCLA for her outstanding assistance in communication and in preparing and editing chapters.

Fuyuhiko Tamanoi
Christine A. Hrycyna
Martin O. Bergo
May 2011

1

Protein Prenylation: A Perspective on Initial Discoveries

JOHN A. GLOMSET

Departments of Medicine and Biochemistry and Regional Primate Research Center
University of Washington, Seattle
Washington, USA

I.   Abstract

The discovery of protein prenylation by members of our group was the indirect result of our early studies of the effects of plasma lipoproteins on arterial smooth muscle cells. We had discovered that platelets contain PDGF, a growth factor that can stimulate the replication of cells in culture, and had begun to use partially purified PDGF to examine the effects of cholesterol on different types of replicating cells. For example, we added high concentrations of compactin (an inhibitor of 3-hydroxy-3-methylglutaryl coenzyme A reductase) to the cells to block the synthesis of cholesterol, showed that the compactin also blocked the synthesis of DNA and induced cell rounding, and sought to determine whether added unesterified choles-terol or plasma lipoproteins could prevent these effects. However, neither of these additives was effective, while added exogenous mevalonic acid pre-vented both effects. So we added radioactive mevalonic acid to compactin-treated cells in amounts that could promote shape change reversal, searched for products of the mevalonic acid that could have been involved, and found that much of the radioactivity was insoluble in lipid solvents. Further, a study of Swiss 3T3 cells that had been cultured in the presence of mevinolin (an inhibitor of mevalonic acid synthesis) and then labeled with radioactive exogenous mevalonic acid showed that much of the material that became

1
ISSN NO: 1874-6047
DOI: 10.1016/B978-0-12-381339-8.00001-9

labeled was present in proteins which had apparent molecular masses of 13,000–58,000 Da. And subsequent studies of other cells identified specific proteins that were modified by farnesyl groups or geranylgeranyl groups.

II.   Steps in the Trail of Research

Many years ago, Russell Ross and I decided to do a collaborative study of the effects of plasma lipoproteins on cultures of arterial smooth muscle cells, and to use plasma lipoproteins and arterial smooth muscle cells from non-human primates as models. So, I obtained blood from caged nonhuman primates in the Regional Primate Research Center at the University of Washington and used a centrifugation approach to separate the plasma lipoproteins from the blood cells, while members of the Ross laboratory obtained fresh arterial tissue from the Primate Center's tissue distribution program, and used this tissue to prepare cultures of the arterial tissue smooth muscle cells. However, when the lipoproteins that I had isolated were added to the cultures of smooth muscle cells, the smooth muscle cells failed to replicate, and we later discovered that the reason for this was that the centrifugation procedure which I had used to prepare the plasma lipoproteins had effectively separated the lipoproteins from platelets and a previously unknown growth factor that they contained [1]. Further, subsequent experiments showed that a partially purified form of this growth factor (PDGF) could not only stimulate the replication of arterial smooth muscle cells in culture but also stimulate the synthesis of DNA and cell division in Swiss 3T3 cells in culture [2,3].

We then studied cultures of PDGF-stimulated nonhuman primate arterial smooth muscle cells or Swiss 3T3 cells to investigate the relation between cholesterol synthesis and DNA synthesis within a single cell cycle. In each case, we used high concentrations of compactin (ML-236B), an inhibitor of 3-hydroxy-3-methylglutaryl coenzyme A reductase, to block the synthesis of cholesterol, examined the effect of this block on the synthesis of DNA, and then sought to determine whether added exogenous cholesterol could prevent the effect of the compactin on DNA synthesis. The results of these experiments showed that compactin could indeed block the synthesis of DNA in each type of cell, but we were unable to prevent or reverse this effect by adding exogenous unesterified cholesterol or low- density lipoproteins to the cells. However, we found that we could prevent the effect of compactin on DNA synthesis by adding exogenous mevalonic acid to the cells, and this raised the possibility that either mevalonic acid itself or an unknown product of mevalonic acid metabolism might have promoted the growth of the cells. Note that other investigators had been considering this possibility as well [4,5].

We next examined the effects of compactin on Swiss 3T3 cells that were either quiescent or traversing through a single cell cycle in response to added PDGF, and found that compactin could not only inhibit the synthesis of DNA in the PDGF-stimulated cells but also cause both the quiescent cells and the replicating cells to acquire a characteristic rounded shape [6]. To explore the molecular basis of these effects, we added radioactive mevalonic acid to cultures of Swiss 3T3 cells and sought to identify a product of mevalonic acid metabolism that could promote the synthesis of DNA and regulate the shape of the cells. The initial results of these experiments showed that the cells incorporated much of the radioactive mevalonic acid into cell proteins [7], and subsequent studies showed that a few specific proteins were involved [8]. For example, studies of HeLa cells and Chinese hamster ovary cells provided evidence that the lamin B in these cells can be modified by a derivative of mevalonic acid [9] and that the corresponding derivative in human HeLa cells is a farnesyl group which modifies a cysteine residue toward the end of lamin B [10]. Moreover, a study of *Saccharomyces cerevisiae* identified mutants that were defective in Ras protein farnesylation and showed that at least two genes were involved [11].

In addition to this, a study of HeLa cells that had been incubated in the presence of radioactive mevalonic acid identified labeled fragments of cell proteins that contained an all-*trans* geranylgeranyl group [12]. Brain G protein gamma subunits were found to contain an all-*trans* geranylgeranyl-cysteine methyl ester at their carboxyl termini [13]. Evidence was obtained that the human platelet protein, smg p21B is geranylgeranylated and carboxylmethylated at its C-terminal cysteine residue [14]. The properties of a protein geranylgeranyl transferase from bovine brain were studied [15]. The C-terminus of smg p25A was shown to contain two geranylgeranylated cysteine residues and a methyl ester [16]. A Rab geranylgeranyl transferase was shown to catalyze the geranylgeranylation of adjacent cysteines in the small GTPases Rab1A, Rab3A, and Rab5A [17]. And several reviews related to protein prenylation were published including one that described the methodology that we had been using to characterize protein-bound prenyl groups [18–20].

References

1. Ross, R., Glomset, J.A., Kariya, B., and Harker, L. (1974). A platelet-dependent serum factor that stimulates the proliferation of arterial smooth muscle cells in vitro. *Proc Natl Acad Sci USA* 71:1207–1210.
2. Rutherford, R.B., and Ross, R. (1976). Platelet factors stimulate fibroblasts and smooth muscle cells quiescent in plasma serum to proliferate. *J Cell Biol* 69:196–203.

3. Vogel, A., Raines, E., Kariya, B., Rivest, M.J., and Ross, R. (1978). Coordinate control of 3T3 cell proliferation by platelet-derived growth factor and plasma components. *Proc Natl Acad Sci USA* 75:2810–2814.

4. Kaneko, I., Hazama-Shimada, Y., and Endo, A. (1978). Inhibitory effects on lipid metabolism in cultured cells of ML-236B, a potent inhibitor of 3-hydroxy-3-methylglutaryl-coenzyme-A reductase. *Eur J Biochem* 87:313–321.

5. Quesney-Huneeus, V., Wiley, M.H., and Siperstein, M.D. (1979). Essential role for mevalonate synthesis in DNA replication. *Proc Natl Acad Sci USA* 76:5056–5060.

6. Habenicht, A.J., Glomset, J.A., and Ross, R. (1980). Relation of cholesterol and mevalonic acid to the cell cycle in smooth muscle and swiss 3T3 cells stimulated to divide by platelet-derived growth factor. *J Biol Chem* 255:5134–5140.

7. Schmidt, R.A., Glomset, J.A., Wight, T.N., Habenicht, A.J., and Ross, R. (1982). A study of the Influence of mevalonic acid and its metabolites on the morphology of swiss 3T3 cells. *J Cell Biol* 95:144–153.

8. Schmidt, R.A., Schneider, C.J., and Glomset, J.A. (1984). Evidence for post-translational incorporation of a product of mevalonic acid into Swiss 3T3 cell proteins. *J Biol Chem* 259:10175–10180.

9. Wolda, S.L., and Glomset, J.A. (1988). Evidence for modification of lamin B by a product of mevalonic acid. *J Biol Chem* 263:5997–6000.

10. Farnsworth, C.C., Wolda, S.L., Gelb, M.H., and Glomset, J.A. (1989). Human lamin B contains a farnesylated cysteine residue. *J Biol Chem* 264:20422–20429.

11. Goodman, L.E., Judd, S.R., Farnsworth, C.C., Powers, S., and Gelb, M.H. (1990). Mutants of Saccharomyces cerevisiae defective in the farnesylation of Ras proteins. *Proc Natl Acad Sci USA* 87:9665–9669.

12. Farnsworth, C.C., Gelb, M.H., and Glomset, J.A. (1990). Identification of geranylgeranyl-modified proteins in HeLa cells. *Science* 247:320–322.

13. Yamane, H.K., Farnsworth, C.C., Xie, H., Howald, W., Fung, B.K.K., Clarke, S., Gelb, M.H., and Glomset, J.A. (1990). Brain G protein gamma subunits contain an all-trans-geranylgeranylcysteine methyl ester at their carboxyl termini. *Proc Natl Acad Sci USA* 87:5868–5872.

14. Kawata, M., Farnsworth, C.C., Yoshida, Y., Gelb, M.H., Glomset, J.A., and Takai, Y. (1990). Posttranslationally processed structure of the human platelet protein smg p21B: evidence for geranylgeranylation and carboxyl methylation of the C-terminal cysteine. *Proc Natl Acad Sci USA* 87:8960–8964.

15. Yokoyama, K., Goodwin, G.W., Ghomashchi, F., Glomset, J.A., and Gelb, M.H. (1991). A protein geranylgeranyltransferase from bovine brain: implications for protein prenylation specificity. *Proc Natl Acad Sci USA* 88:5302–5306.

16. Farnsworth, C.C., Kawata, M., Yoshida, Y., Takai, Y., Gelb, M.H., and Glomset, J.A. (1991). C terminus of the small GTP-binding protein smg p25A contains two geranylgeranylated cysteine residues and a methyl ester. *Proc Natl Acad Sci USA* 88:6196–6200.

17. Farnsworth, C.C., Seabra, M.C., Ericsson, L.H., Gelb, M.H., and Glomset, J.A. (1994). Rab geranylgeranyl transferase catalyzes the geranylgeranylation of adjacent cysteines in the small GTPases Rab1A, Rab3A, and Rab5A. *Proc Natl Acad Sci USA* 91:11963–11967.

18. Glomset, J.A., Gelb, M.H., and Farnsworth, C.C. (1990). Prenyl proteins in eukaryotic cells: a new type of membrane anchor. *Trends Biochem Sci* 15:139–142.

19. Farnsworth, C.C., Casey, P.J., Howald, W.N., Glomset, J.A., and Gelb, M.H. (1990). Structural characterization of prenyl groups attached to proteins. *Methods* 1:231–240.

20. Glomset, J.A., and Farnsworth, C.C. (1994). Role of protein modification reactions in programming interactions between ras-related GTPases and cell membranes. *Annu Rev Cell Biol* 10:181–205.

2

Insights into the Function of Prenylation from Nuclear Lamin Farnesylation

MICHAEL SINENSKY

Department of Biochemistry and Molecular Biology
East Tennessee State University, Johnson City
Tennessee, USA

I. Abstract

The discovery of mammalian protein prenylation was originally motivated by an effort to identify a nonsterol isoprenoid which indirect evidence suggested was a coregulator of isoprenoid biosynthesis and played a critical role in cellular proliferation. The first prenylated proteins to be identified were the nuclear lamin proteins—B lamins and prelamin A—which were subsequently shown to be farnesylated at a carboxyl-terminal CAAX motif. In both types of lamin, the farnesylation and carboxymethylation play a role in targeting these proteins to the nuclear envelope. The nucleus can be demonstrated to be a CAAX processing compartment for the lamins. In the case of prelamin A, there is removal of a carboxyl-terminal polypeptide which is specifically catalyzed by the enzyme Zmpste24. This processing event is necessary for assembly of lamin A into the lamina and may play a role in cell cycle control. Because the nucleus contains only one target membrane, lamin farnesylation and carboxymethylation may be sufficient to allow association with this membrane. This stands in contrast to farnesylated proteins expressed in the cytoplasm.

ISSN NO: 1874-6047
DOI: 10.1016/B978-0-12-381339-8.00002-0

II. Introduction

The discovery of protein prenylation, as a general phenomenon, and the more particular discovery of lamin farnesylation began with studies directed at understanding the regulation of isoprenoid metabolism. Early efforts to understand the regulation of cholesterol biosynthesis had by the late 1970s led to the confirmation of the hypothesis that regulation was centered on the enzyme 3-hydroxy-3-methyl glutaryl coenzyme A reductase (HMGR). Driven by the importance of cholesterol metabolism in atherosclerosis, pharmaceutical research directed at the discovery of drugs useful in lowering serum cholesterol levels became focused on the search for drugs that could inhibit HMGR. The most successful class of such compounds are the statins (e.g., lovastatin), the first of which, compactin, was reported by Endo in 1976 [1].

Subsequent studies with compactin on the regulation of HMGR led to the concept that not only was HMGR regulated in response to sterols but, as the rate limiting step of isoprenoid biosynthesis, was also regulated by other isoprenoids (reviewed in Ref. [2]). This was most clearly exhibited as a loss of the sterol regulatory response of HMGR in cells starved for mevalonate by compactin treatment. Other studies on compactin effects on the cycling of cultured cells revealed that a mevalonate metabolite, other than cholesterol, was required for reentry of quiescent cells into S-phase [3,4]. This observation suggested that it would be possible to isolate a somatic cell mutant auxotrophic for mevalonate via the bromodeoxyuridine (BrdU)-visible light technique. This procedure selects for mutants that undergo a cell cycle arrest in response to starvation for a required nutrient preventing the incorporation of light-activated toxic BrdU into DNA [5]. The resultant mevalonate auxotroph (Mev-1) was expected to be defective in one of the enzymes of mevalonate synthesis which turned out to be cytosolic 3-hydroxy-3-methyl glutaryl coenzyme A synthase based on enzyme assay [6,7]. As was the case with cultured cells treated with compactin, Mev-1 exhibited defective regulation of HMGR by sterols during mevalonate starvation. These observations were generally taken as consistent with the view that there was a nonsterol isoprenoid critical for cell cycle progression and, hence, also regulatory for mevalonate biosynthesis—a concept dubbed "multivalent regulation" [2].

A. DISCOVERY OF LAMIN FARNESYLATION

The discovery by the Glomset laboratory [8] of the posttranslational incorporation of labeled mevalonate into proteins in cultured mammalian cells suggested the hypothesis that a protein modified with an isoprenoid substituent was this coregulatory molecule. This hypothesis was shown to be supported by the kinetic relationship between isoprenylated protein turnover and cell cycle arrest in mevalonate starved Mev-1 cells [9]. At this point, there had not yet been any identification of a mammalian protein that was prenylated, a metabolic process that (as discussed in Volume 30 Protein Prenylation Part B and in [10]) was known only for fungi. To explore for a possible candidate protein, it seemed plausible to examine mammalian cell nuclei for proteins which incorporated label from mevalonate in the hope that whatever protein(s) might be involved in cell cycle control were affecting DNA synthesis directly. Exploration of mevalonate-labeled nuclear proteins [11] led to the finding that there were three such proteins which, eventually, could be demonstrated to be lamin B [11,12], prelamin A [11,13], and lamin B_2 (unpublished results). Mevalonate incorporation into the respective lamin proteins can most readily be illustrated by 2D-NEPHGE electrophoresis of nuclear matrix intermediate filament preparations [13,14] (Figure 2.1). Verification of the identity of lamin B, as indicated by 2D-gel migration, was performed by immunoprecipitation and was initially reported simultaneously by our laboratory and that of Glomset [12]. The identity of the prenyl group as farnesyl was later reported by the Glomset lab as well [15], and lamin B thus became the first example of a prenylated mammalian protein.

Our laboratory chose to pursue the biochemistry and cell physiology of prelamin A prenylation. This was motivated by our original interest in a possible role for one or more prenylated proteins in cell cycle regulation.

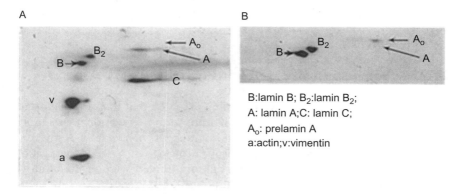

B:lamin B; B_2:lamin B_2;
A: lamin A; C: lamin C;
A_o: prelamin A
a:actin;v:vimentin

FIG. 2.1. Coomassie stained (A) or R-[2-^{14}C] mevalonate-labeled proteins (B) found in nuclear matrix intermediate filament preparations from cultured CHO cells. Such preparations [13] are enriched for nuclear lamin proteins. Radioactivity is visualized by fluorography. The pH increases from left to right.

We observed in our original work on lamin prenylation that the 74-kDa protein, we found to be mevalonate labeled, corresponded to the reported size and charge of a previously identified lamin A precursor [16,17]. We confirmed the identity of this protein as prelamin A by peptide mapping [11]. Suggestive of a functional role of this protein in the regulatory responses to mevalonate starvation, we noted that it accumulated to very high levels in Mev-1 cells starved for mevalonate but was virtually undetectable in cells growing normally.

B. ENDOPROTEOLYTIC PROCESSING OF PRELAMIN A

Understanding the role of prelamin A prenylation required further, more detailed, chemical characterization of the processing reactions which was technically difficult due to the small amounts of precursor that normally accumulate in proliferating cells. However, employing a variety of genetic and biochemical methods, not to mention considerable patience, we were able to demonstrate that, as is the case for lamin B and many other proteins, prelamin A is farnesylated by thioether linkage to a carboxyl-terminal CAAX-box cysteine [18,19]. It then undergoes a canonical CAAX-box endoproteolytic cleavage removing the carboxyl-terminal tripeptide followed by carboxymethylation of the now carboxyl-terminal cysteine [20]. A second farnesylation-dependent internal endoproteolytic cleavage [21], which is unique to mammalian proteins but observable in the maturation of fungal mating pheromones (Volume 30 Protein Prenylation Part B), results in the formation of mature lamin A. Further, we could show that these processing reactions were obligatorily sequential, that is, they constituted a biochemical pathway. A diagram of this pathway is shown in Figure 2.2.

III. Subcellullar Trafficking and Processing of Prelamin A

Early studies of prelamin A processing revealed that it was transported into the nucleus and processed to the mature form there prior to assembly into the lamina [22]. As a nuclear protein, transport of prelamin A into the nucleus is dependent on its nuclear localization signal (NLS). Mutation of the NLS results in accumulation of immunofluorescently detected lamin A in the cytoplasm [23]. Mutation of the CAAX-box cysteine or treatment with lovastatin blocks cysteine prenylation. In cells treated with lovastatin, wild-type prelamin A enters the nucleus but accumulates in the nucleoplasm rather than assembling into the lamina [18,19]. Pulse–chase and epitope-tagged studies of nonprenylated wild-type prelamin A

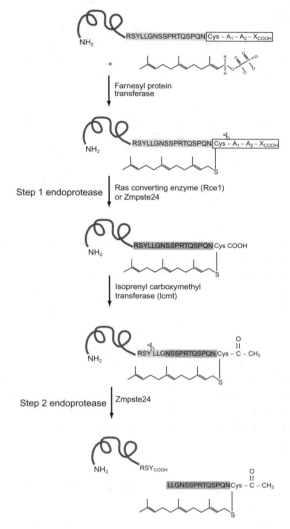

FIG. 2.2. The processing pathway of prelamin A. The step 2 endoproteolytic cleavage, catalyzed by Zmpste24, is unique to this pathway in mammalian cells.

accumulated in the nucleus by inhibition of polyisoprenoid synthesis with lovastatin indicates that relief of the block by mevalonate supplementation results in the rapid conversion of prelamin A to mature lamin A prior to incorporation of the mature lamin A into the lamina [13,19,24].

These observations suggest that prelamin A processing pathway enzymes are expressed in the nucleus. Consistent with this hypothesis, visual tracking of microinjected prelamin A into the nucleus of mammalian cells indicated that it assembled into the lamina without exiting the nucleus [25], which requires that its maturation be in the nucleus. In further support of this hypothesis, it was possible to detect the presence of farnesyl transferase, an enzyme mostly associated with the cytoplasm [26], in the nucleus of mammalian cells, as well, by immunofluorescence [20]. Dual localization to cytoplasmic and nuclear membranes of the CAAX-box enzymes Icmt [27] and Zmpste24 [27,28] has been reported. Cells appear to have the functional capability for prelamin A processing in either the cytoplasm or the nucleus. Various chimeric constructs of prelamin A, possessing the prelamin A carboxyl-terminal tail, but lacking an NLS, were shown some time ago to undergo maturation [29]. More recent studies with similar constructs have shown that in the absence of an NLS, maturation occurs in the cytoplasm, while prelamin A tail constructs possessing an NLS undergo maturation in the nucleus [27]. Kinetic analysis of cells expressing these constructs, but reversibly blocked in farnesylation by treatment with a farnesyl transferase inhibitor, shows similar kinetics of prelamin A processing after removal of the prenylation block in either cytosolic or nuclear compartments [27]. Nonetheless, under physiological circumstances, the rapid entry of prelamin A into the nucleus would argue for a primarily nucleoplasmic processing which is in agreement with pulse–chase studies [13,22]. In support of this view, immunofluorescent visualization of prelamin A, with prelamin A-specific antibody, is only detectable in the nucleus [24,30] in normally growing cells. In contrast, studies on B-lamin (lamin B3) processing in *Xenopus* oocytes suggest that although lamin B processing can also occur in either cytoplasm or nucleus, it is kinetically favored in the cytoplasm [31].

IV.   Step 2 Endoprotease Activity and Zmpste 24

As described elsewhere in this volume, CAAX-box prenylation and the subsequent posttranslational modifications of step 1 endoprotease cleavage of the –AAX tripeptide and carboxymethylation are common features to all CAAX-box proteins. In all reported cases, including lamin B [32], step 1 endoprotease activity can be ascribed to the enzyme Rce1 [33–35]. In unpublished studies, we have found that, *in vitro*, Rce1 can utilize the prelamin A CAAX-box as a substrate which we have also reported for Zmpste24 [28]. In cells cultured from patients with restrictive dermopathy (a genetic total loss of expression of Zmpste24), prelamin A is seen to accumulate [36,37]. We have found that this prelamin A is carboxymethy-lated and farnesylated based on a combination of base-release assay (see Volume 30 Protein Prenylation Part B) and mass spectral evidence (unpublished results). These findings support the concept that Rce1 can catalyze the step 1 endopeptidase reaction in addition to Zmpste24. The bifunction-ality of Zmpste24 and redundancy in CAAX endoproteolysis with Rce1 also occurs and, indeed, was first reported [33,38,39], in the processing of the farnesylated *Saccharomyces cerevisiae* a-mating factor.

Thus, the unique process catalyzed by Zmpste24 is the step 2 endopro-tease activity shown in Figure 2.2 as has been verified *in vitro* [28]. The cleavage site was shown to be after Y^{646} by Weber et al. [40]. The mecha-nism of this reaction has only been partially elucidated, as both the pre-lamin A substrate and Zmpste24 share an incomplete characterization of their higher-order structure and interaction. Our laboratory defined some characteristics of this interaction using a synthetic substrate, shown below that is identical in primary structure to the prelamin A carboxyl-terminal tail.

$$Y* = [^{125} I]Tyr$$

H$_2$NRSY*LLGNSSPRTQSPQNC$_{COCH_3}^{S}$

When this substrate is cleaved by the step 2 endoprotease activity, it releases the iodinated RSY tripeptide allowing the reaction to be assayed. The RSYLLG sequence containing the cleavage site is highly conserved among vertebrate lamins [21] and occurs uniquely in prelamin A among vertebrate proteins suggesting that it is specifically recognized by the step 2 endoprotease activity of Zmpste24. Kinetic studies with this substrate revealed the following properties of step 2 endoproteolysis: (1) as expected, prenylation is required for cleavage, as is carboxymethylation; (2) there is specificity for farnesyl as the substituent, compared to geranylgeranyl, and farnesylated analogues are noncompetitive inhibitors. These observations are consistent with binding of the prenyl group to a different site than the proteolytic cleavage site on the enzyme; (3) the simple hexapeptide RSYLLG is both a substrate and a competitive inhibitor of the prenylated substrate consistent with the active site recognizing this hexapeptide as the substrate.

Zmpste24 possesses the canonical HEXXH domain that is the signature of zinc metalloprotease active sites [41,42]. In unpublished studies, we have found that mutation of the HEXXH domain results in a loss of the step 1 endoprotease activity but not the step 2 activity. We speculate that the prenyl-binding site for prelamin A on Zmpste24 is adjacent to the step 1

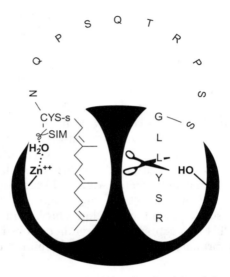

FIG. 2.3. A possible mechanism for the bifunctional activity of the protease Zmpste24, in which the step 1 endoprotease activity is associated with the Zn^{2+} site and the step 2 activity is associated with a ser OH. The reactions are sequential, suggesting a conformational change in the enzyme or substrate or both between steps 1 and 2.

endoprotease active site, but that there is a second active site on the enzyme that catalyzes step 2 endoproteolysis—a model shown in Figure 2.3. We have reported that inhibitors of serine proteases inhibit step 2 endoprotease activity *in vitro* [21].

V.   Functional Aspects of Lamin Farnesylation

As mentioned above, lamin B undergoes CAAX processing via the Rce1 endoprotease. The fate of lamin B is assembly into the nuclear envelope as a farnesylated protein. Modification with the farnesyl group is essential for lamin B assembly into the lamina and for its association with the nuclear membrane during mitosis, as demonstrated originally for the B_2 isoform [43] and later for lamin B [44]. It is possible that the farnesylation of lamin B mediates its nuclear membrane association via insertion into the lipid bilayer. However, there is evidence that lamin B binding to the nuclear envelope is mediated by a specific binding partner, the lamin B receptor (LBR) [45–47] which has been shown to bind farnesylated lamin B both *in vivo* and *in vitro*. The loss of binding of lamin B to the nuclear envelope upon elimination of prenylation raises the possibility that the farnesyl group is recognized by this protein, but there have been no reports comparing the binding of farnesylated and nonfarnesylated lamin B to the LBR *in vitro*. It is also possible that stabilization of association of lamin B with the nuclear envelope occurs through a combination of protein–protein and lipid–lipid interaction [48].

The functional significance of prelamin A farnesylation has attracted considerable interest both because of the uniqueness of its farnesylation-dependent endoproteolytic processing in vertebrate tissue and because of the remarkable progeria, Hutchinson–Gilford progeria syndrome (HGPS) [49,50] associated with a deficit in that processing resulting in a permanently farnesylated mutant protein (progerin or LAΔ50). The functional significance of this complex pathway does not seem to be due to a requirement for it in targeting lamin A to the lamina. Preexisting mature lamin A becomes cytoplasmic during the course of mitosis and is reassembled into the lamina directly on chromatin [51–54]. However, our laboratory has also demonstrated that mature lamin A expressed in interphase cells is also competent for assembly into the lamina [19]. This observation has recently been confirmed in whole animals by the transgenic expression of mature lamin A in lamin A null mice [55].

So what is the function of the complex processing pathway of lamin A? The answer may lie in the subnuclear localization of prelamin A. Prelamin A mutants defective in farnesylation [18,19] show a nucleoplasmic localization of the protein. So do cells treated with lovastatin so that prenylation of prelamin A is blocked [19]. Immunofluorescent localization of prelamin A in proliferating cells with prelamin A-specific antibody also suggests that this precursor is initially localized to the nucleoplasm [24,56]. This led us to propose some time ago [19] that the biochemical function of the prelamin A tail is to prevent its incorporation into the lamina. Nucleoplasmic farnesylation would then lead to translocation of the farnesylated prelamin A to the nuclear membrane where it undergoes proteolytic conversion to mature lamin A. In agreement with this assertion, farnesylated prelamin A mutants defective in the cleavage domain localize to the nuclear membrane [29] as does farnesylated prelamin A in cells knocked down for Zmpste24 [56]. Interestingly, the permanently farnesylated prelamin A HGPS mutant, LAΔ50, remains associated with the nuclear membrane during mitosis, which is in common with lamin B, but in contrast to mature lamin A [57].

These observations still leave unresolved the question of the biological significance of prelamin A processing. There is some evidence that prelamin A accumulation may play a role in cell cycle control. The tumor suppressor and cell cycle regulatory protein pRb has been shown to associate with lamin A through specific binding domains on each protein [58]. Interaction of pRb with lamin A is at intranuclear foci, based on immunofluorescent colocalization [59]. It is noteworthy that prelamin A accumulation is also at intranuclear foci [24]. This colocalization of lamin A and pRb epitopes is cell cycle dependent and is seen [59] to promote cell cycle arrest by preventing proteolytic degradation of pRb. In agreement with this possible role for lamin A–pRb interaction in cell cycle arrest, it has been reported that inhibition of cell cycling by TGFβ is also mediated by the A-lamins through facilitation of the PP2A catalyzed dephosphorylation of pRb [60]. Expression of the permanently farnesylated mutant prelamin A, LAΔ50, in HeLa cells also results in a G1 cell cycle arrest [57] and in a hypophosphorylated pRb. However, cells expressing a processing defective mutant of prelamin A have been reported to exhibit a dominant negative response to induction of cell cycle arrest by p16^{ink4a} [61]. Taken together, these results suggest that the processing of prelamin A may play a role in cell cycle control, at least in some cells. A possible mechanism would be for nucleoplasmic localization of prelamin A to participate in cell cycle arrest though mediating sequestration, stabilization, or hypophosphorylation of pRb. Release from prelamin A modulation of pRb activity would be through maturation and relocalization of lamin A to the lamina. The mechanism of regulation of pRb by both mature and prelamin A would appear to be a fruitful area for future investigation.

VI. Relationship of Lamin Prenylation to the Functional Role of Prenylation in Other Proteins

Posttranslational modifications of proteins by lipids include cysteine thioacylation [62,63], amino acylation by myristate on N-terminal glycines [62] or (less commonly) internal lysines [64], carboxyl-terminal derivitization with glycosyl phosphatidyl inositol [65], and, in the case of the hedgehog proteins [66,67], carboxy terminal esterification of glycine by cholesterol. Functionally, all the lipid posttranslational modifications have the general property of increasing the hydrophobicity of the proteins they modify facilitating their attachment to lipid bilayers and further into specific lipid domains such as rafts [68]. Membrane attachment facilitated by lipid modification, as opposed to structural features resulting from primary amino acid sequence, has intrinsic reversibility.

In the instances of GPI-anchored and S-acylated proteins, hydrolytic reversal plays an important role in their functionality [63,69]. However, unlike these modifications, the thioether linkage of prenyl substituents [70] cannot be removed by enzymatic hydrolysis. Reversibility of membrane association of prenylated proteins is achieved through recognition of the prenyl group by partner proteins, and partner protein-prenylated protein heterodimer formation involving prenyl group recognition is an important feature of prenylated protein function [71].

Obviously, specificity of such heterodimerization must also involve recognition by the heterodimeric prenylated protein partner of a specific structural domain within the target prenylated protein. There are now a number of well-characterized examples of how the formation of protein heterodimers between prenylated proteins and their partners plays a critical role in the function of the many prenylated proteins that facilitate trafficking and signaling as has been reviewed previously [71,72]. In the case of prelamin A processing, this concept is consistent with the mechanism for Zmpste24 proposed in Figure 2.3. Although this model awaits structural confirmation, the existing evidence suggests that the enzyme has the ability to recognize both the prenyl group and a hexapeptide domain at what appears to be two distinct binding sites. Such a two-site model could also be the mechanism that produces the interaction between the LBR and lamin B.

The farnesylation and carboxymethylation of CAAX proteins have been shown to be adequate to produce kinetically stable association with a lipid bilayer [73]. In the case of cytosolic farnesylated proteins, the multiple membrane compartments require a second signal to assure specificity of subcellular localization [74]. This is clearly not an issue for the farnesylated nuclear lamins whose association with the nuclear envelope may be mediated solely by posttranslational modification, thus the association of lamin B and LAΔ50 with the nuclear membrane during mitosis.

Another feature of lamin farnesylation, that is common to other farnesylated proteins, is that geranylgeranylation can substitute when farnesylation is blocked, for example, by treatment with a farnesyl transferase inhibitor [75,76]. This observation probably explains why farnesyl transferase inhibition has no effect on lamin structure or ras function [77]. In prelamin A processing, substitution of geranylgeranyl for farnesyl by CAAX-box mutation results in a slower rate of processing but normal assembly into the lamina [21].

REFERENCES

1. Endo, A., Kuroda, M., and Tanzawa, K. (1976). Competitive inhibition of 3-hydroxy-3-methylglutaryl coenzyme A reductase by ML-236A and ML-236B fungal metabolites, having hypocholesterolemic activity. *FEBS Lett* 72:323–326.
2. Brown, M.S., and Goldstein, J.L. (1980). Multivalent feedback regulation of HMG CoA reductase, a control mechanism coordinating isoprenoid synthesis and cell growth. *J Lipid Res* 21:505–517.

3. Quesney-Huneeus, V., Wiley, M.H., and Siperstein, M.D. (1979). Essential role for mevalonate synthesis in DNA replication. *Proc Natl Acad Sci USA* 76:5056–5060.
4. Habenicht, A.J., Glomset, J.A., and Ross, R. (1980). Relation of cholesterol and mevalonic acid to the cell cycle in smooth muscle and swiss 3T3 cells stimulated to divide by platelet-derived growth factor. *J Biol Chem* 255:5134–5140.
5. Kao, F.T., and Puck, T.T. (1969). Genetics of somatic mammalian cells. IX. Quantitation of mutagenesis by physical and chemical agents. *J Cell Physiol* 74:245–258.
6. Schnitzer-Polokoff, R., von Gunten, C., Logel, J., Torget, R., and Sinensky, M. (1982). Isolation and characterization of a mammalian cell mutant defective in 3-hydroxy-3-methylglutaryl coenzyme A synthase. *J Biol Chem* 257:472–476.
7. Ortiz, J.A., Gil-Gomez, G., Casaroli-Marano, R.P., Vilaro, S., Hegardt, F.G., and Haro, D. (1994). Transfection of the ketogenic mitochondrial 3-hydroxy-3-methylglutaryl-coenzyme A synthase cDNA into Mev-1 cells corrects their auxotrophy for mevalonate. *J Biol Chem* 269:28523–28526.
8. Schmidt, R.A., Schneider, C.J., and Glomset, J.A. (1984). Evidence for post-translational incorporation of a product of mevalonic acid into Swiss 3T3 cell proteins. *J Biol Chem* 259:10175–10180.
9. Sinensky, M., and Logel, J. (1985). Defective macromolecule biosynthesis and cell-cycle progression in a mammalian cell starved for mevalonate. *Proc Natl Acad Sci USA* 82:3257–3261.
10. Omer, C.A., and Gibbs, J.B. (1994). Protein prenylation in eukaryotic microorganisms: genetics, biology and biochemistry. *Mol Microbiol* 11:219–225.
11. Beck, L.A., Hosick, T.J., and Sinensky, M. (1988). Incorporation of a product of mevalonic acid metabolism into proteins of Chinese hamster ovary cell nuclei. *J Cell Biol* 107:1307–1316.
12. Wolda, S.L., and Glomset, J.A. (1988). Evidence for modification of lamin B by a product of mevalonic acid. *J Biol Chem* 263:5997–6000.
13. Beck, L.A., Hosick, T.J., and Sinensky, M. (1990). Isoprenylation is required for the processing of the lamin A precursor. *J Cell Biol* 110:1489–1499.
14. Dalton, M., and Sinensky, M. (1995). Expression systems for nuclear lamin proteins: farnesylation in assembly of nuclear lamina. *Methods Enzymol* 250:134–148.
15. Farnsworth, C.C., Wolda, S.L., Gelb, M.H., and Glomset, J.A. (1989). Human lamin B contains a farnesylated cysteine residue. *J Biol Chem* 264:20422–20429.
16. Dagenais, A., Bibor-Hardy, V., Laliberte, J.F., Royal, A., and Simard, R. (1985). Detection in BHK cells of a precursor form for lamin A. *Exp Cell Res* 161:269–276.
17. Laliberte, J.F., Dagenais, A., Filion, M., Bibor-Hardy, V., Simard, R., and Royal, A. (1984). Identification of distinct messenger RNAs for nuclear lamin C and a putative precursor of nuclear lamin A. *J Cell Biol* 98:980–985.
18. Holtz, D., Tanaka, R.A., Hartwig, J., and McKeon, F. (1989). The CaaX motif of lamin A functions in conjunction with the nuclear localization signal to target assembly to the nuclear envelope. *Cell* 59:969–977.
19. Lutz, R.J., Trujillo, M.A., Denham, K.S., Wenger, L., and Sinensky, M. (1992). Nucleoplasmic localization of prelamin A: implications for prenylation-dependent lamin A assembly into the nuclear lamina. *Proc Natl Acad Sci USA* 89:3000–3004.
20. Sinensky, M., Fantle, K., Trujillo, M., McLain, T., Kupfer, A., and Dalton, M. (1994). The processing pathway of prelamin A. *J Cell Sci* 107(Pt 1):61–67.
21. Kilic, F., Dalton, M.B., Burrell, S.K., Mayer, J.P., Patterson, S.D., and Sinensky, M. (1997). In vitro assay and characterization of the farnesylation-dependent prelamin A endoprotease. *J Biol Chem* 272:5298–5304.

22. Lehner, C.F., Furstenberger, G., Eppenberger, H.M., and Nigg, E.A. (1986). Biogenesis of the nuclear lamina: in vivo synthesis and processing of nuclear protein precursors. *Proc Natl Acad Sci USA* 83:2096–2099.
23. Loewinger, L., and McKeon, F. (1988). Mutations in the nuclear lamin proteins resulting in their aberrant assembly in the cytoplasm. *EMBO J* 7:2301–2309.
24. Sasseville, A.M., and Raymond, Y. (1995). Lamin A precursor is localized to intranuclear foci. *J Cell Sci* 108(Pt 1):273–285.
25. Goldman, A.E., Moir, R.D., Montag-Lowy, M., Stewart, M., and Goldman, R.D. (1992). Pathway of incorporation of microinjected lamin A into the nuclear envelope. *J Cell Biol* 119:725–735.
26. Reiss, Y., Goldstein, J.L., Seabra, M.C., Casey, P.J., and Brown, M.S. (1990). Inhibition of purified p21ras farnesyl:protein transferase by Cys-AAX tetrapeptides. *Cell* 62:81–88.
27. Barrowman, J., Hamblet, C., George, C.M., and Michaelis, S. (2008). Analysis of prelamin A biogenesis reveals the nucleus to be a CaaX processing compartment. *Mol Biol Cell* 19:5398–5408.
28. Corrigan, D.P., Kuszczak, D., Rusinol, A.E., Thewke, D.P., Hrycyna, C.A., Michaelis, S., and Sinensky, M.S. (2005). Prelamin A endoproteolytic processing in vitro by recombinant Zmpste24. *Biochem J* 387:129–138.
29. Hennekes, H., and Nigg, E.A. (1994). The role of isoprenylation in membrane attachment of nuclear lamins. A single point mutation prevents proteolytic cleavage of the lamin A precursor and confers membrane binding properties. *J Cell Sci* 107(Pt 4):1019–1029.
30. Sinensky, M., Fantle, K., and Dalton, M. (1994). An antibody which specifically recognizes prelamin A but not mature lamin A: application to detection of blocks in farnesylation-dependent protein processing. *Cancer Res* 54:3229–3232.
31. Firmbach-Kraft, I., and Stick, R. (1995). Analysis of nuclear lamin isoprenylation in Xenopus oocytes: isoprenylation of lamin B3 precedes its uptake into the nucleus. *J Cell Biol* 129:17–24.
32. Maske, C.P., Hollinshead, M.S., Higbee, N.C., Bergo, M.O., Young, S.G., and Vaux, D.J. (2003). A carboxyl-terminal interaction of lamin B1 is dependent on the CAAX endoprotease Rce1 and carboxymethylation. *J Cell Biol* 162:1223–1232.
33. Ashby, M.N. (1998). CaaX converting enzymes. *Curr Opin Lipidol* 9:99–102.
34. Winter-Vann, A.M., and Casey, P.J. (2005). Post-prenylation-processing enzymes as new targets in oncogenesis. *Nat Rev Cancer* 5:405–412.
35. Bergo, M.O., Wahlstrom, A.M., Fong, L.G., and Young, S.G. (2008). Genetic analyses of the role of RCE1 in RAS membrane association and transformation. *Methods Enzymol* 438:367–389.
36. Navarro, C.L., Cadinanos, J., De Sandre-Giovannoli, A., Bernard, R., Courrier, S., Boccaccio, I., Boyer, A., Kleijer, W.J., Wagner, A., Giuliano, F., et al. (2005). Loss of ZMPSTE24 (FACE-1) causes autosomal recessive restrictive dermopathy and accumulation of Lamin A precursors. *Hum Mol Genet* 14:1503–1513.
37. Moulson, C.L., Go, G., Gardner, J.M., van der Wal, A.C., Smitt, J.H., van Hagen, J.M., and Miner, J.H. (2005). Homozygous and compound heterozygous mutations in ZMPSTE24 cause the laminopathy restrictive dermopathy. *J Invest Dermatol* 125:913–919.
38. Chen, P., Sapperstein, S.K., Choi, J.D., and Michaelis, S. (1997). Biogenesis of the Saccharomyces cerevisiae mating pheromone a-factor. *J Cell Biol* 136:251–269.
39. Tam, A., Nouvet, F.J., Fujimura-Kamada, K., Slunt, H., Sisodia, S.S., and Michaelis, S. (1998). Dual roles for Ste24p in yeast a-factor maturation: NH2-terminal proteolysis and COOH-terminal CAAX processing. *J Cell Biol* 142:635–649.

40. Weber, K., Plessmann, U., and Traub, P. (1989). Maturation of nuclear lamin A involves a specific carboxy-terminal trimming, which removes the polyisoprenylation site from the precursor; implications for the structure of the nuclear lamina. *FEBS Lett* 257:411–414.

41. Freije, J.M., Blay, P., Pendas, A.M., Cadinanos, J., Crespo, P., and Lopez-Otin, C. (1999). Identification and chromosomal location of two human genes encoding enzymes potentially involved in proteolytic maturation of farnesylated proteins. *Genomics* 58:270–280.

42. Pendas, A.M., Zhou, Z., Cadinanos, J., Freije, J.M., Wang, J., Hultenby, K., Astudillo, A., Wernerson, A., Rodriguez, F., Tryggvason, K., *et al.* (2002). Defective prelamin A processing and muscular and adipocyte alterations in Zmpste24 metalloproteinase-deficient mice. *Nat Genet* 31:94–99.

43. Kitten, G.T., and Nigg, E.A. (1991). The CaaX motif is required for isoprenylation, carboxyl methylation, and nuclear membrane association of lamin B2. *J Cell Biol* 113:13–23.

44. Mical, T.I., and Monteiro, M.J. (1998). The role of sequences unique to nuclear intermediate filaments in the targeting and assembly of human lamin B: evidence for lack of interaction of lamin B with its putative receptor. *J Cell Sci* 111(Pt 23):3471–3485.

45. Worman, H.J., Evans, C.D., and Blobel, G. (1990). The lamin B receptor of the nuclear envelope inner membrane: a polytopic protein with eight potential transmembrane domains. *J Cell Biol* 111:1535–1542.

46. Worman, H.J., Yuan, J., Blobel, G., and Georgatos, S.D. (1988). A lamin B receptor in the nuclear envelope. *Proc Natl Acad Sci USA* 85:8531–8534.

47. Ye, Q., and Worman, H.J. (1994). Primary structure analysis and lamin B and DNA binding of human LBR, an integral protein of the nuclear envelope inner membrane. *J Biol Chem* 269:11306–11311.

48. Nigg, E.A., Kitten, G.T., and Vorburger, K. (1992). Targeting lamin proteins to the nuclear envelope: the role of CaaX box modifications. *Biochem Soc Trans* 20:500–504.

49. Davies, B.S., Fong, L.G., Yang, S.H., Coffinier, C., and Young, S.G. (2009). The post-translational processing of prelamin A and disease. *Annu Rev Genomics Hum Genet* 10:153–174.

50. Rusinol, A.E., and Sinensky, M.S. (2006). Farnesylated lamins, progeroid syndromes and farnesyl transferase inhibitors. *J Cell Sci* 119:3265–3272.

51. Glass, J.R., and Gerace, L. (1990). Lamins A and C bind and assemble at the surface of mitotic chromosomes. *J Cell Biol* 111:1047–1057.

52. Ottaviano, Y., and Gerace, L. (1985). Phosphorylation of the nuclear lamins during interphase and mitosis. *J Biol Chem* 260:624–632.

53. Gerace, L., and Blobel, G. (1980). The nuclear envelope lamina is reversibly depolymerized during mitosis. *Cell* 19:277–287.

54. Gerace, L., Blum, A., and Blobel, G. (1978). Immunocytochemical localization of the major polypeptides of the nuclear pore complex-lamina fraction. Interphase and mitotic distribution. *J Cell Biol* 79:546–566.

55. Coffinier, C., Jung, H.J., Li, Z., Nobumori, C., Yun, U.J., Farber, E.A., Davies, B.S., Weinstein, M.M., Yang, S.H., Lammerding, J., *et al.* (2010). Direct synthesis of lamin A, bypassing prelamin a processing, causes misshapen nuclei in fibroblasts but no detectable pathology in mice. *J Biol Chem* 285:20818–20826.

56. Gruber, J., Lampe, T., Osborn, M., and Weber, K. (2005). RNAi of FACE1 protease results in growth inhibition of human cells expressing lamin A: implications for Hutchinson-Gilford progeria syndrome. *J Cell Sci* 118:689–696.

57. Dechat, T., Shimi, T., Adam, S.A., Rusinol, A.E., Andres, D.A., Spielmann, H.P., Sinensky, M.S., and Goldman, R.D. (2007). Alterations in mitosis and cell cycle progression caused by a mutant lamin A known to accelerate human aging. *Proc Natl Acad Sci USA* 104:4955–4960.
58. Ozaki, T., Saijo, M., Murakami, K., Enomoto, H., Taya, Y., and Sakiyama, S. (1994). Complex formation between lamin A and the retinoblastoma gene product: identification of the domain on lamin A required for its interaction. *Oncogene* 9:2649–2653.
59. Johnson, B.R., Nitta, R.T., Frock, R.L., Mounkes, L., Barbie, D.A., Stewart, C.L., Harlow, E., and Kennedy, B.K. (2004). A-type lamins regulate retinoblastoma protein function by promoting subnuclear localization and preventing proteasomal degradation. *Proc Natl Acad Sci USA* 101:9677–9682.
60. Van Berlo, J.H., Voncken, J.W., Kubben, N., Broers, J.L., Duisters, R., van Leeuwen, R.E., Crijns, H.J., Ramaekers, F.C., Hutchison, C.J., and Pinto, Y.M. (2005). A-type lamins are essential for TGF-beta1 induced PP2A to dephosphorylate transcription factors. *Hum Mol Genet* 14:2839–2849.
61. Nitta, R.T., Jameson, S.A., Kudlow, B.A., Conlan, L.A., and Kennedy, B.K. (2006). Stabilization of the retinoblastoma protein by A-type nuclear lamins is required for INK4A-mediated cell cycle arrest. *Mol Cell Biol* 26:5360–5372.
62. James, G., and Olson, E.N. (1990). Fatty acylated proteins as components of intracellular signaling pathways. *Biochemistry* 29:2623–2634.
63. Magee, T., and Seabra, M.C. (2005). Fatty acylation and prenylation of proteins: what's hot in fat. *Curr Opin Cell Biol* 17:190–196.
64. Stevenson, F.T., Bursten, S.L., Fanton, C., Locksley, R.M., and Lovett, D.H. (1993). The 31-kDa precursor of interleukin 1 alpha is myristoylated on specific lysines within the 16-kDa N-terminal propiece. *Proc Natl Acad Sci USA* 90:7245–7249.
65. Eisenhaber, B., Maurer-Stroh, S., Novatchkova, M., Schneider, G., and Eisenhaber, F. (2003). Enzymes and auxiliary factors for GPI lipid anchor biosynthesis and post-translational transfer to proteins. *Bioessays* 25:367–385.
66. Breitling, R. (2007). Greased hedgehogs: new links between hedgehog signaling and cholesterol metabolism. *Bioessays* 29:1085–1094.
67. Jeong, J., and McMahon, A.P. (2002). Cholesterol modification of Hedgehog family proteins. *J Clin Invest* 110:591–596.
68. Levental, I., Grzybek, M., and Simons, K. (2010). Greasing their way: lipid modifications determine protein association with membrane rafts. *Biochemistry* 49:6305–6316.
69. Sharom, F.J., and Radeva, G. (2004). GPI-anchored protein cleavage in the regulation of transmembrane signals. *Subcell Biochem* 37:285–315.
70. Sinensky, M., and Lutz, R.J. (1992). The prenylation of proteins. *Bioessays* 14:25–31.
71. Sinensky, M. (2000). Functional aspects of polyisoprenoid protein substituents: roles in protein-protein interaction and trafficking. *Biochim Biophys Acta* 1529:203–209.
72. Resh, M.D. (2006). Trafficking and signaling by fatty-acylated and prenylated proteins. *Nat Chem Biol* 2:584–590.
73. Silvius, J.R., and l'Heureux, F. (1994). Fluorimetric evaluation of the affinities of iso-prenylated peptides for lipid bilayers. *Biochemistry* 33:3014–3022.
74. Hancock, J.F., Cadwallader, K., Paterson, H., and Marshall, C.J. (1991). A CAAX or a CAAL motif and a second signal are sufficient for plasma membrane targeting of ras proteins. *EMBO J* 10:4033–4039.
75. Whyte, D.B., Kirschmeier, P., Hockenberry, T.N., Nunez-Oliva, I., James, L., Catino, J.J., Bishop, W.R., and Pai, J.K. (1997). K- and N-Ras are geranylgeranylated in cells treated with farnesyl protein transferase inhibitors. *J Biol Chem* 272:14459–14464.

76. Varela, I., Pereira, S., Ugalde, A.P., Navarro, C.L., Suarez, M.F., Cau, P., Cadinanos, J., Osorio, F.G., Foray, N., Cobo, J., *et al.* (2008). Combined treatment with statins and aminobisphosphonates extends longevity in a mouse model of human premature aging. *Nat Med* 14:767–772.
77. Dalton, M.B., Fantle, K.S., Bechtold, H.A., DeMaio, L., Evans, R.M., Krystosek, A., and Sinensky, M. (1995). The farnesyl protein transferase inhibitor BZA-5B blocks farnesylation of nuclear lamins and p21ras but does not affect their function or localization. *Cancer Res* 55:3295–3304.

3

Posttranslational Processing of Nuclear Lamins

BRANDON S.J. DAVIES[a] • CATHERINE COFFINIER[a] •
SHAO H. YANG[a] • HEA-JIN JUNG[a] • LOREN G. FONG[a] •
STEPHEN G. YOUNG[a,b]

[a]*Department of Medicine, David Geffen School of Medicine*
University of California
Los Angeles, California, USA

[b]*Department of Human Genetics, David Geffen School of Medicine*
University of California
Los Angeles, California, USA

I. Abstract

The nuclear lamina, a meshwork formed from intermediate filament proteins, provides a structural scaffolding for the cell nucleus. The principal proteins of the nuclear lamina are lamin A, lamin B1, lamin B2, and lamin C. Prelamin A (the precursor to mature lamin A), lamin B1, and lamin B2 contain a carboxyl-terminal *CaaX* motif that triggers farnesylation and methylation of a carboxyl-terminal cysteine. Prelamin A undergoes one additional processing step: the last 15 amino acids of the protein, including the farnesylcysteine methyl ester, are clipped off, releasing mature lamin A. Here, we review the posttranslational processing of the nuclear lamins and discuss its importance to health and disease.

THE ENZYMES, Vol. XXIX
21
ISSN NO: 1874-6047
DOI: 10.1016/B978-0-12-381339-8.00003-2

II. Introduction

The nuclear lamina is an intermediate filament meshwork that lies beneath the inner nuclear membrane, providing a scaffolding for the cell nucleus and also participating in diverse cellular processes, including DNA replication, heterochromatin organization, cell division, protein trafficking, and gene transcription [1–5]. In mammals, the major proteins of the nuclear lamina are lamin A, lamin B1, lamin B2, and lamin C (Figure 3.1). Lamins A and C, generally called the A-type lamins, are produced from

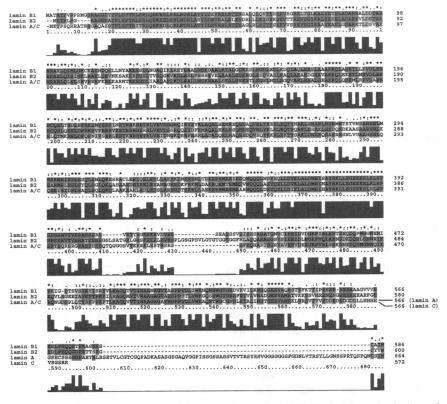

FIG. 3.1. Alignment of amino acid sequences for lamin B1, lamin B2, prelamin A, and lamin C. The divergent carboxyl-terminal domains of prelamin A and lamin C are shown on the bottom row. Identical residues are highlighted green and similar residues are highlighted yellow. (See color plate section in the back of the book.)

a single gene, *LMNA*, by alternative splicing [1,4,6,7]. The B-type lamins, lamins B1 and B2, are encoded by separate genes, *LMNB1* and *LMNB2*, respectively.

Lamin B1, lamin B2, and prelamin A (the precursor to mature lamin A) terminate with a *CaaX* motif (where the "*C*" is a cysteine, the *a* residues are usually aliphatic amino acids, and the *X* can be one of many different residues) [8,9]. The *CaaX* motif triggers three sequential modifications [9,10]. First, the cysteine of the *CaaX* motif is farnesylated by a cytosolic enzyme, protein farnesyltransferase (FTase) [10,11]. Second, the last three amino acids of the protein (i.e., the *–aaX* of the *CaaX* motif) are clipped off by a prenylprotein-specific endoprotease. For most *CaaX* proteins, including lamin B1, this cleavage is carried out by RCE1 [12–14], but for prelamin A, it appears that this cleavage can be carried out by either RCE1 or ZMPSTE24 [15,16]. Third, the newly exposed isoprenylcysteine is methylated by isoprenylcysteine carboxyl methyltransferase (ICMT) [17–20]. Once these "*CaaX* modifications" are complete, prelamin A undergoes an additional processing step. The last 15 amino acids of the protein, including the farnesylcysteine methyl ester, are clipped off by ZMPSTE24, releasing mature lamin A.

III. B-Type Lamins

Lamin B1, often called "lamin B" in early studies, was the first mammalian protein recognized as being isoprenylated [21,22]. Lamin B1 is widely assumed to be expressed by all cell types; expression is detectable in embryonic stem (ES) cells as well as in immune, hematopoietic, and neuroendocrine stem cells [23–26]. Thus far, no human disease has been linked to missense or nonsense mutations in *LMNB1*, but a *LMNB1* duplication causes autosomal-dominant leukodystrophy [27].

Lmnb1 deficiency in mice causes severe disease phenotypes. Vergnes *et al.* [28] used a BayGenomics gene-trap clone with an insertional mutation in intron 5 of *Lmnb1* to create *Lmnb1* knockout mice ($Lmnb1^{\Delta/\Delta}$). $Lmnb1^{\Delta/\Delta}$ embryos grew slowly and died very shortly after birth [28]. The $Lmnb1^{\Delta/\Delta}$ embryos exhibited retarded lung development and had abnormally shaped craniums and reduced ossification of long bones. $Lmnb1^{\Delta/\Delta}$ fibroblasts exhibited aneuploidy, premature senescence, and an impaired ability to differentiate into adipocytes. $Lmnb1^{\Delta/\Delta}$ fibroblasts also exhibited misshapen cell nuclei, with numerous blebs and abnormal chromatin distribution [28].

The physiologic importance of lamin B1 farnesylation has never been thoroughly evaluated, but it seems likely that the farnesyl anchor is crucial for tethering the protein to the inner nuclear membrane. Mutagenesis studies have suggested that farnesylation of lamin B1 is crucial for proper localization of the protein [29,30]. However, Dalton *et al.* [31] found no abnormality in lamin B1 distribution in CHO-K1 cells treated with a protein farnesyltransferase inhibitor (FTI). In our opinion, the physiologic importance of lamin B1 farnesylation will remain in doubt until knock-in mice expressing a nonfarnesylated version of lamin B1 are created and evaluated.

Mammals were initially thought to express only a single B-type lamin. However, following the identification of a second B-type lamin in chickens [32], another B-type lamin, lamin B2, was identified in mice and subsequently in other mammals [33]. Lamins B1 and B2 are highly homologous with ~60% identity at the amino acid level (Figure 3.1). Like lamin B1, lamin B2 terminates with a *CaaX* motif and is farnesylated. Thus far, no human disease has been linked conclusively to mutations in *LMNB2*, but recent studies of *Lmnb2* knockout mice ($Lmnb2^{-/-}$) have shed light on the physiologic importance of lamin B2 in mammals [34].

Coffinier *et al.* [34] found that $Lmnb2^{-/-}$ embryos are normal in size during development, but the pups die shortly after birth. Histopathologic analysis of $Lmnb2^{-/-}$ mice revealed abnormal layering of neurons in the cerebral cortex, and additional experiments, including neuronal birthdating studies, showed that this abnormality was due to defective neuronal migration [34]. Developmental abnormalities were not restricted to the cerebral cortex; the cerebellum in $Lmnb2^{-/-}$ embryos was small and devoid of foliation, and the morphology of the hippocampus was abnormal.

The migration of neurons from the ventricular zone to the cortical plate depends on the coupled movement of the centrosome and the nucleus (nuclear translocation). During neuronal migration, the centrosome moves toward the leading edge of the cell, and the nucleus is then pulled toward the centrosome by dynein motors on microtubules [35]. This nuclear translocation process is regulated by a number of cytosolic proteins, including LIS1, NDE1, and NDEL1 [36,37], and defects in those regulators cause defective neuronal migration [38–40]. The neuronal migration abnormalities in $Lmnb2^{-/-}$ mice suggest that lamin B2 is important for anchoring the nucleus to the microtubule network in the cytoplasm. However, as lamin B2 is located *inside* the nucleus (i.e., beneath the inner nuclear membrane), it seems likely that membrane-spanning proteins are involved in linking the nuclear lamina to the microtubule network in the cytoplasm.

No one knows whether protein farnesylation is critical for lamin B2 function, but it is reasonable to speculate that protein farnesylation could be important for neuronal migration. Nuclear translocation subjects the nucleus to significant deformation, and we hypothesize that farnesyl lipid anchors function to keep the nuclear lamina attached to the inner nuclear membrane during this process.

No one has yet identified clinically significant *LMNB2* mutations in humans, but we suspect that, sooner or later, *LMNB2* mutations will be identified in human patients with neurodevelopmental abnormalities.

IV.   A-Type Lamins

As noted earlier, lamins A and C are generated from *LMNA* by alternative splicing. The first 566 amino acids of lamins A and C are identical, but the carboxyl-terminal domains are unique [1,4,6,7] (Figure 3.1). Lamin C (572 amino acids), which terminates with exon 10 sequences, contains six unique amino acids at its carboxyl terminus. Unlike prelamin A, lamin C does not contain a *CaaX* motif and is not farnesylated. Prelamin A (664 amino acids) contains a 98-amino acid extension not found in lamin C (encoded by exons 11 and 12). As noted earlier, prelamin A has a *CaaX* motif and undergoes the typical *CaaX* modifications—farnesylation, release of the last three amino acids, and carboxyl methylation. The endoproteolytic cleavage and methylation steps are utterly dependent on protein farnesylation; when farnesylation is blocked, those steps do not occur. After the *CaaX* modifications are complete, the last 15 amino acids of the protein (including the farnesylcysteine methyl ester) are clipped off, releasing mature lamin A (646 amino acids) [41]. That final cleavage, which is carried out by ZMPSTE24 [15,16,42,43], is also dependent on protein prenylation [44]. Aside from farnesyl-prelamin A, no other ZMPSTE24 substrates have been identified in mammals.

Lamins A and C appear to be expressed in roughly equal amounts in most tissues, although no systematic study of the relative amounts of these proteins in different tissues has yet been published. The "physiologic rationale" for the existence of two A-type lamins is not clear. Earlier studies had concluded that lamin C targeting to the nuclear rim requires lamin A [45], but recent studies have cast doubt on that conclusion. Fong *et al.* [46] used gene targeting to create a *Lmna* knock-in allele, *Lmna*^{LCO}, that yields exclusively lamin C (and no prelamin A or lamin A) (Figure 3.2). Of note, *Lmna*^{LCO/LCO} mice had no detectable disease phenotypes, and there was no apparent defect in the delivery of lamin C to the nuclear rim in *Lmna*^{LCO/LCO} fibroblasts. Thus, at least in mice, the synthesis of prelamin A and lamin A appears to be dispensable.

While an absence of lamin A synthesis appears to have no consequences, the loss of *both* lamin A and lamin C causes severe disease [54]. At birth, *Lmna* knockout mice (*Lmna*^{−/−}) appear normal, but they grow slowly and

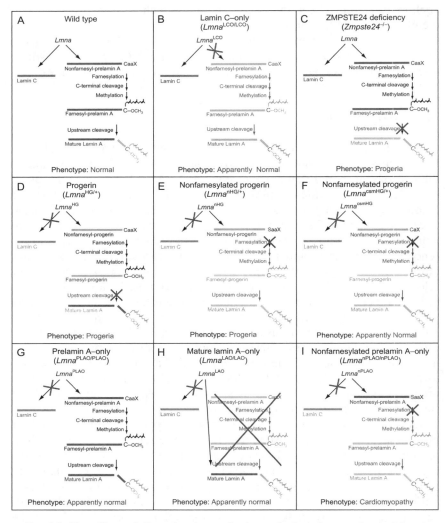

FIG. 3.2. The effects of disrupting A-type lamin synthesis and processing in laboratory mice. (A) Wild-type mice produce both lamin C and prelamin A. Prelamin A then undergoes four sequential posttranslational processing steps to generate mature lamin A. (B) *Lmna*^LCO/LCO mice produce only lamin C. These mice appear to be phenotypically normal [46]. (C) *Zmpste24*-deficient mice produce both lamin C and prelamin A, but the final endoproteolytic step does not occur, resulting in the accumulation of farnesylated prelamin A. These mice display multiple progeria-like disease phenotypes [15,43]. (D) *Lmna*^HG/+ mice make normal lamin C and prelamin A from the wild-type allele, and prelamin A with a 50-amino acid internal deletion from the *Lmna*^HG allele. This deletion prevents the final,

succumb to muscular dystrophy by 5–6 weeks of age [54]. The absence of both lamin A and lamin C disrupts the structural integrity of the nuclear envelope, resulting in grossly misshapen cell nuclei [54] and defective nuclear mechanics [55]. Heterozygous knockout mice ($Lmna^{+/-}$) appear grossly normal but develop late-onset cardiomyopathy and conduction system disease [56]. In humans, heterozygous loss of lamin A/C from nonsense mutations results in Emery-Dreifuss muscular dystrophy [57], dilated cardiomyopathy with conduction defects [58], or autosomal-dominant limb-girdle muscular dystrophy with cardiac conduction disturbances [59]. Homozygous loss of lamin A/C in humans causes severe contractures in both upper and lower limbs, severe generalized muscular dystrophy, and lethality at birth [59].

V.   Defects in Prelamin A Processing and Disease

Defects in prelamin A processing lead to severe disease phenotypes in both humans and mice (Figure 3.2). Homozygosity for a *Zmpste24* knockout allele ($Zmpste24^{-/-}$) abolishes the endoproteolytic release of mature lamin A from prelamin A, leading to an accumulation of prelamin A [15,43,60,61]. Using metabolic labeling experiments, Coffinier *et al.* [62] demonstrated that the prelamin A in ZMPSTE24-deficient cells is farnesylated. $Zmpste24^{-/-}$ mice appear normal at birth, but they gain weight slowly and manifest several other obvious phenotypes, including alopecia, kyphosis, abnormal incisors, and an arthritic gait [15]. They also display a grip abnormality and develop osteolytic lesions in bones, most prominently in

---

ZMPSTE24-mediated, endoproteolytic step in lamin A biogenesis, leading to the accumulation of a farnesylated and truncated form of prelamin A (progerin). $Lmna^{HG/+}$ mice develop progeria-like disease phenotypes [47,48]. (E) $Lmna^{nHG/+}$ mice are identical to $Lmna^{HG/+}$ mice except that the truncated prelamin A generated from the $Lmna^{nHG}$ allele cannot be farnesylated (due to a cysteine-to-serine substitution in the *CaaX* motif), leading to the accumulation of nonfarnesylated progerin. These mice also develop progeria, though it is somewhat less severe than in $Lmna^{HG/+}$ mice [49]. (F) The internally truncated prelamin A from the $Lmna^{csmHG}$ allele also cannot be farnesylated (due to an isoleucine deletion in the *CaaX* motif), leading to an accumulation of nonfarnesylated progerin. In this case, the mice appear normal [50]. (G) $Lmna^{PLAO/PLAO}$ mice produce exclusively wild-type prelamin A but do not produce lamin C. The prelamin A produced by these mice is efficiently converted to mature lamin A. These mice appear to be normal [51]. (H) $Lmna^{nPLAO/nPLAO}$ mice produce a mutant prelamin A with a cysteine-to-serine substitution in prelamin A's *CaaX* motif, leading to an accumulation of nonfarnesylated prelamin A. These mice develop dilated cardiomyopathy and exhibit reduced survival [51]. (I) $Lmna^{LAO/LAO}$ mice produce mature lamin A *directly*, bypassing prelamin A synthesis and processing. These mice cannot make lamin C. Like $Lmna^{LCO/LCO}$ and $Lmna^{PLAO/PLAO}$ mice, $Lmna^{LAO/LAO}$ mice appear to be normal [52]. Adapted from Davies *et al.* [53]. (See color plate section in the back of the book.)

the ribs and zygomatic arch. The rib lesions predispose to fractures, which become surrounded by fracture callus but heal poorly [15]. At ~16 weeks of age, the mice begin to lose weight, and they invariably die by ~30 weeks of age. The cause of death is uncertain, but one possibility is that the bone lesions impair nutrition, leading to progressive inanition.

ZMPSTE24 deficiency in humans causes restrictive dermopathy (RD), a perinatal-lethal progeroid syndrome [63,64]. Hallmarks of RD include sclerodermatous skin, loss of fat, prominent superficial veins, dysplastic clavicles, alopecia, and joint contractures. Like $Zmpste24$-deficient mouse fibroblasts, human RD fibroblasts accumulate farnesyl-prelamin A and have a high frequency of misshapen nuclei [61].

Hutchinson-Gilford progeria syndrome (HGPS), the classic progeroid disorder of children, is caused by a $LMNA$ point mutation that leads to the accumulation of an internally truncated prelamin A that retains a carboxyl-terminal farnesylcysteine methyl ester. Children with HGPS appear normal at birth, but within a few months manifest failure to thrive. Later, they develop dental abnormalities, micrognathia, stiff joints, alopecia, scleroder-matous skin changes, loss of subcutaneous fat, osteoporosis, and osteolytic lesions in bones. Typically, children with HGPS die during their teenage years from atherosclerotic lesions in the coronary and cerebral arteries [65–71].

HGPS is caused by a single-nucleotide substitution in exon 11 of $LMNA$. The mutation does not change an amino acid residue but optimizes a cryptic splice donor site. Use of the alternative splice site deletes the last 150 bp of exon 11, leading to an in-frame deletion of 50 amino acids in the carboxyl-terminal portion of prelamin A [67]. Prelamin A's $CaaX$ motif is unaffected; however, the site for the final ZMPSTE24-mediated cleavage step is eliminated. Thus, the mutant prelamin A, generally called progerin, undergoes the usual $CaaX$ processing steps but retains the carboxyl-terminal farnesylcysteine methyl ester and cannot be further processed to mature lamin A [49,72].

An HGPS knock-in allele, $Lmna^{HG}$, was created by Yang et $al.$ [47]. The sole product of the $Lmna^{HG}$ allele is progerin (Figure 3.2). Like $Zmpste24^{-/-}$ mice, $Lmna^{HG/+}$ mice appear normal at birth. However, $Lmna^{HG/+}$ mice lose weight after 6–8 weeks and develop osteolytic lesions and bone fractures similar to those found in $Zmpste24^{-/-}$ mice [48]. $Lmna^{HG/+}$ mice become progressively malnourished and die by 4–6 months of age. Homozygous knock-in mice ($Lmna^{HG/HG}$) are small, grow very slowly, and die by 3–4 weeks of age [48].

The accumulation of farnesylated progerin (in HGPS) or uncleaved farnesyl-prelamin A (in RD) disrupts the integrity of the nuclear lamina, and both ZMPSTE24-deficient fibroblasts and HGPS fibroblasts (from either humans or mice) exhibit misshapen cell nuclei [42,43,67,73]. But the mechanism by which those farnesylated proteins elicit multisystem disease phenotypes is not clear. Genetic studies have suggested that the farnesylated forms of prelamin A and progerin are toxic and elicit disease in a dose-dependent fashion. For example, mice with two copies of the $Lmna^{HG}$ allele ($Lmna^{HG/HG}$) develop more severe disease than mice with one $Lmna^{HG}$ allele and one knockout allele ($Lmna^{HG/-}$) [74]. Also, increasing the production of farnesyl-prelamin A in $Zmpste24^{-/-}$ mice (by introducing a $Lmna$ knock-in allele that yields exclusively prelamin A) *increases* the severity of disease [51] (Figure 3.3). Conversely, when prelamin A synthesis in $Zmpste24^{-/-}$ mice is reduced (by introducing a $Lmna$ knockout allele or a $Lmna^{LCO}$ allele), disease phenotypes are eliminated [42,46].

In 2004, Fong et al. [42] proposed that prelamin A's farnesyl anchor might be crucial for the pathogenesis of progeroid disorders, and since that time, a variety of experiments have supported that view. Treatment of ZMPSTE24-deficient and HGPS fibroblasts (both mouse and human) with an FTI reduces the frequency of misshapen nuclei [47,61,75–77]. Even more importantly, an FTI ameliorates the progeria-like phenotypes in $Zmpste24^{-/-}$ and $Lmna^{HG/+}$ mice [48,78–80]. However, while the beneficial effects of an FTI in those two mouse models were highly significant, they were far from complete. FTI-treated mice still developed progeria-like disease phenotypes and ultimately succumbed to the disease. In part, the less-than-complete response to FTI therapy was likely due to incomplete inhibition of protein farnesylation [48,80]. However, in the case of $Lmna^{HG/+}$ mice, Yang et al. [48,49] considered another potential explanation—that *nonfarnesylated* progerin might retain intrinsic toxicity. To assess this possibility, Yang et al. [49] created knock-in mice expressing a nonfarnesylated version of progerin ($Lmna^{nHG/+}$ mice, in which the cysteine of the $CaaX$ motif was changed to serine) (Figure 3.2). Remarkably, $Lmna^{nHG/+}$ mice developed all of the progeria-like disease phenotypes found in $Lmna^{HG/+}$ mice (which express farnesylated progerin). The disease phenotypes in $Lmna^{nHG/+}$ mice tended to be milder than in $Lmna^{HG/+}$ mice, but this was likely due to lower steady-state levels of progerin in the tissues of $Lmna^{nHG/+}$ mice [49].

At face value, the persistence of disease in $Lmna^{nHG/+}$ mice suggested that nonfarnesylated progerin retains substantial toxicity, sounding a distinctly pessimistic note for the concept of treating progeria with FTIs. However, this pessimistic conclusion depends on the premise that nonfarnesylated progerin terminating with –SSIM (as in $Lmna^{nHG/+}$ mice) is functionally equivalent to nonfarnesylated progerin that would accumulate in FTI-treated $Lmna^{HG/+}$ mice (which would terminate in –CSIM). To determine whether this premise is valid, Yang et al. [50] generated a second

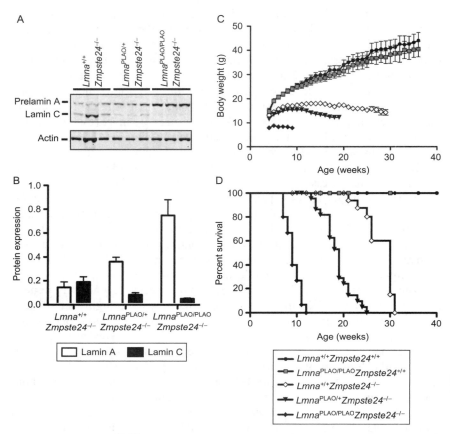

FIG. 3.3. The $Lmna^{PLAO}$ allele worsens the progeria-like disease phenotypes in $Zmpste24^{-/-}$ mice in a dose-dependent manner. (A) Western blots (with an antibody against lamin A/C) showing prelamin A and lamin C in the kidneys of $Lmna^{+/+}Zmpste24^{-/-}$, $Lmna^{PLAO/+}$-$Zmpste24^{-/-}$, and $Lmna^{PLAO/PLAO}Zmpste24^{-/-}$ mice ($n=3$/genotype). (B) Quantitative analysis of prelamin A and lamin C expression in the kidney in $Lmna^{+/+}Zmpste24^{-/-}$, $Lmna^{PLAO/+}Zmpste24^{-/-}$, and $Lmna^{PLAO/PLAO}Zmpste24^{-/-}$ mice (bar graphs depict prelamin A:actin and lamin C:actin ratios). (C) Body weight curves for $Lmna^{+/+}Zmpste24^{+/+}$ ($n=25$), $Lmna^{PLAO/PLAO}Zmpste24^{+/+}$ ($n=17$), $Lmna^{+/+}Zmpste24^{-/-}$ ($n=24$), $Lmna^{PLAO/+}Zmpste24^{-/-}$ ($n=24$), and $Lmna^{PLAO/PLAO}Zmpste24^{-/-}$ ($n=15$) mice. Body weight curves for $Lmna^{+/+}$-$Zmpste24^{+/+}$ and $Lmna^{PLAO/PLAO}Zmpste24^{+/+}$ mice were not significantly different. Body weight curves for all other genotypes were significantly different from wild-type mice and from each other ($P<0.01$ between $Lmna^{+/+}Zmpste24^{-/-}$ and $Lmna^{PLAO/+}Zmpste24^{-/-}$ mice; $P<0.001$ for all other comparisons). (D) Kaplan–Meier survival plots for $Lmna^{+/+}Zmpste24^{+/+}$ ($n=26$), $Lmna^{PLAO/PLAO}Zmpste24^{+/+}$ ($n=16$), $Lmna^{+/+}Zmpste24^{-/-}$ ($n=23$), $Lmna^{PLAO/+}Zmpste24^{-/-}$ ($n=24$), and $Lmna^{PLAO/PLAO}Zmpste24^{-/-}$ ($n=15$) mice. Survival times for $Lmna^{+/+}Zmpste24^{+/+}$ and $Lmna^{PLAO/PLAO}Zmpste24^{+/+}$ mice were not statistically different. Survival times for all other genotypes were significantly different from wild-type mice and from each other ($P<0.001$ for all comparisons). From Davies *et al.* [51].

knock-in allele for nonfarnesylated progerin, $Lmna^{csmHG}$, in which the isoleucine of the *CaaX* motif was deleted (Figure 3.2). Interestingly, $Lmna^{csmHG/+}$ mice were entirely free of disease, even though progerin levels in $Lmna^{csmHG/+}$ mice were virtually identical to those in $Lmna^{nHG/+}$ mice. Homozygous knock-in mice ($Lmna^{csmHG/csmHG}$) were also free of progeria-like disease, in striking contrast to $Lmna^{nHG/nHG}$ mice, which developed severe disease and died within a few months. The findings with the $Lmna^{csmHG}$ mice are more reassuring to those interested in using an FTI to treat progeria.

Davies *et al.* [51] recently assessed the physiologic effects of a nonfarnesylated version of full-length prelamin A. They created a knock-in allele ($Lmna^{nPLAO}$) that yields exclusively nonfarnesylated prelamin A, and a second knock-in allele ($Lmna^{PLAO}$) that yields exclusively wild-type prelamin A [51] (Figure 3.2). Neither allele produces lamin C. As expected, wild-type prelamin A from the $Lmna^{PLAO}$ allele was quickly converted to mature lamin A. In contrast, none of the nonfarnesylated prelamin A from the $Lmna^{nPLAO}$ allele was converted to lamin A [51]. Despite similar steady-state levels of prelamin A, the $Lmna^{nPLAO/nPLAO}$ mice developed none of the progeria-like disease phenotypes found in $Zmpste24^{-/-}$ mice (which accumulate farnesyl-prelamin A). These studies indicate that the farnesyl lipid is critical for the emergence of progeria in the setting of full-length prelamin A.

Although free from progeria-like disease phenotypes, $Lmna^{nPLAO/nPLAO}$ mice were not healthy. The mice developed dilated cardiomyopathy and died prematurely [51]. This phenotype could not be explained by the absence of lamin C synthesis; $Lmna^{PLAO/PLAO}$ mice also lacked the ability to synthesize lamin C and were free of disease. We suspect that the cardiomyopathy in $Lmna^{nPLAO/nPLAO}$ mice is due to the fact that nonfarnesylated prelamin A is a functionally hypoactive lamin protein—as opposed to a toxic protein [53]. This hypothesis is based on a genetic observation— that reducing synthesis of nonfarnesylated prelamin A *worsens* disease. $Lmna^{nPLAO/nPLAO}$ mice survived for $\sim$39 weeks, whereas $Lmna^{nPLAO/-}$ mice survived for only $\sim$15 weeks [53]. This finding—*worsening* of disease by lowering expression of the mutant protein—is quite different from results with the $Lmna^{HG}$ allele. Disease phenotypes in $Lmna^{HG/-}$ mice were significantly *milder* than those in $Lmna^{HG/HG}$ mice [74]. These observations suggest that *Lmna* mutations might be divided into two classes—those yielding a toxic protein (e.g., farnesylated progerin leading to progeria) and those yielding a functionally hypoactive lamin (e.g., nonfarnesylated prelamin A leading to cardiomyopathy).

Cardiomyopathy resulting from a *Lmna* missense mutation is not a peculiarity of the $Lmna^{nPLAO}$ allele. Dozens of different amino acid substitutions in lamins A and C cause cardiomyopathy and/or muscular dystrophy in humans [57,81,82]. Some of these very same mutations have been introduced into the mouse *Lmna* gene and caused cardiomyopathy and premature death in mice [83,84].

VI.   Alternate Prenylation of Prelamin A

Alternate prenylation (geranylgeranylation of proteins that are normally farnesylated) has been documented for some *CaaX* proteins. For example, N-Ras and K-Ras (both of which have *CaaX* motifs that terminate with methionine) normally undergo farnesylation but are efficiently geranylgeranylated by GGTase-I when farnesylation is blocked with an FTI [85]. Because prelamin A also terminates with methionine, it is plausible that it would be geranylgeranylated when FTase activity is inhibited [86]. Alternate prenylation would be expected to blunt any beneficial effects of FTI therapy and could help explain the incomplete therapeutic response of $Lmna^{HG/+}$ mice to FTI treatment.

Several studies have suggested that alternate prenylation of prelamin A is inefficient. Toth *et al.* [61] found that the electrophoretic migration of prelamin A was retarded in FTI-treated ZMPSTE24-deficient cells, consistent with a *nonprenylated* rather than a geranylgeranylated protein. Capell *et al.* [75] reported similar findings in FTI-treated cells expressing a green fluorescent protein-tagged progerin. Also, Lee *et al.* [87] examined FTase-deficient keratinocytes and found substantial accumulation of prelamin A (which appeared to be nonprenylated based on electrophoretic mobility). If alternate prenylation of prelamin A in FTase-deficient keratinocytes had been robust, one might have expected to find little or no prelamin A accumulation.

However, Varela *et al.* [88] reported that alternate prenylation of prelamin A is efficient and robust. Their first line of studies involved pharmacologic inhibitors of FTase and GGTase-I in HeLa cells. They found little accumulation of prelamin A in FTI-treated cells, but significant amounts of prelamin A accumulation when a GGTI (a GGTase-I inhibitor) was added to the FTI. The authors interpreted these studies as supporting alternate prenylation [88]. However, their studies were rather unusual in that their FTI, when administered alone, did not cause large amounts of prelamin A accumulation. Also, it would be interesting to investigate whether the exaggerated accumulation of prelamin A in GGTI/FTI-treated cells may have been due to off-target effects of the GGTI on other enzymes (e.g., ZMPSTE24).

In addition to pharmacologic studies, Varela *et al.* [88] reported mass spectrometry data supporting alternate prenylation. They found that the carboxyl terminus of prelamin A in *Zmpste24*$^{-/-}$ cells is farnesylated; however, a new spectrum emerged after FTI treatment—one suggesting geranylgeranylation of the carboxyl terminus. Similarly, the carboxyl terminus of progerin appeared to be geranylgeranylated in the setting of FTI treatment. These intriguing findings should to be pursued with direct biochemical assays [85] and metabolic labeling studies [89]. Also, additional studies with tissue-specific FTase and GGTase-I knockout mice [87,90] would be useful.

Understanding the extent of alternate prenylation of prelamin A is clearly important for guiding therapeutic strategies. Davies *et al.* [89] showed that a geranylgeranylated version of progerin, while no worse than farnesylated progerin, still causes severe progeria. Thus, if FTI treatment led to geranylgeranylation of most of the progerin (or prelamin A) in tissues, the treatment strategy would be compromised. For this reason, Varela *et al.* [88] proposed that it would be useful to consider drug treatments designed to inhibit both protein farnesylation and protein geranylgeranylation. In particular, they suggested using a combination of statins and bisphosphonates to treat progeroid syndromes. (Using bisphosphonates to treat progeria had been proposed years earlier by Toth *et al.* [61].) Varela *et al.* [88] found that a combination of statins and bisphosphonates improved disease phenotypes in *Zmpste24*$^{-/-}$ mice. However, it was not clear whether the beneficial effects of the combination were superior to the effects of an FTI alone [78]. Also, unlike the situation with FTI treatment, which clearly inhibited protein farnesylation *in vivo* [48,78], it was unclear whether the statin/bisphosphonate combination affected the prenylation of prelamin A (or any other prenylated protein) in mouse tissues [88].

VII.   The Purpose of Prelamin A Processing

The conservation of the prelamin A processing pathway in vertebrate evolution [91–94] suggests that this processing pathway is important. If it were unimportant, it is likely that evolution would have quickly truncated prelamin A to the size of mature lamin A, avoiding the metabolic costs of posttranslational processing. Early *in vitro* studies suggested that prelamin A's farnesyl lipid anchor might be important for the targeting of prelamin A to the nuclear rim [21,44,92,95]. But more recently, several lines of evidence have cast doubt on the notion that farnesylation is critical for targeting of lamins to the nuclear rim. For example, the targeting of nonfarnesylated prelamin A to the nuclear rim in tissues of *Lmna*$^{\text{nPLAO/nPLAO}}$ mice appears to be identical to that of mature lamin A in $Lmna^{PLAO/PLAO}$ tissues, as judged by immunofluorescence microscopy [51] (Figure 3.4). Also, in FTase-deficient keratinocytes, the targeting of prelamin A to the nuclear rim is indistinguishable from that of mature lamin A in wild-type keratinocytes [87]. Of course, these findings do not exclude the possibility that the absence of protein prenylation affects prelamin A targeting in a very subtle fashion—so subtle that it cannot be detected by immunofluorescence microscopy.

To directly assess the importance of prelamin A synthesis and processing, Coffinier et al. [52] generated a Lmna knock-in allele ($Lmna^{LAO}$, or "mature lamin A-only" allele) that yields mature lamin A directly, bypassing prelamin A synthesis and processing. In the $Lmna^{LAO}$ allele, the sequences encoding the last 18 amino acids of prelamin A were deleted (along with introns 10 and 11); thus, translation ends after the last amino acid residue of mature lamin A. Homozygous "mature lamin A-only" mice ($Lmna^{LAO/LAO}$) produce exclusively mature lamin A, and no lamin C or prelamin A (Figure 3.2). Of note, the levels of mature lamin A in $Lmna^{LAO/LAO}$ cells and tissues are identical to those in $Lmna^{PLAO/PLAO}$ mice; thus, bypassing prelamin A processing has no impact on the steady-state levels of mature lamin A in cells. Remarkably, $Lmna^{LAO/LAO}$ mice had no detectable pathology, and body weights and survival curves were entirely normal. Even in $Lmna^{LAO/LAO}$ mice > 18 months of age, there were no obvious phenotypes [52]. Like the prelamin A in $Lmna^{nPLAO/nPLAO}$ mice, the mature lamin A in $Lmna^{LAO/LAO}$ tissues was positioned normally at the nuclear rim, indistinguishable from mature lamin A in wild-type and $Lmna^{PLAO/PLAO}$ mice (Figure 3.4). $Lmna^{LAO/LAO}$ fibroblasts did exhibit a higher frequency of misshapen nuclei than wild-type fibroblasts, with a greater percentage of the mature lamin A appearing in the nucleoplasm, but the physiologic importance of this finding is not clear given that $Lmna^{LAO/LAO}$ mice were phenotypically normal [52].

The absence of disease phenotypes in $Lmna^{LAO/LAO}$ mice suggests that prelamin A processing is largely dispensable, at least in laboratory mice. Ultimately, it is possible that a subtle phenotype will be uncovered in $Lmna^{LAO/LAO}$ mice—one that would begin to illuminate the physiological importance of the prelamin A processing pathway.

VIII. Concluding Thoughts

Although it has long been recognized that lamin B1 and lamin B2 undergo farnesylation and methylation, the purpose of these modifications remains unclear. For lamin B2, we have hypothesized that protein

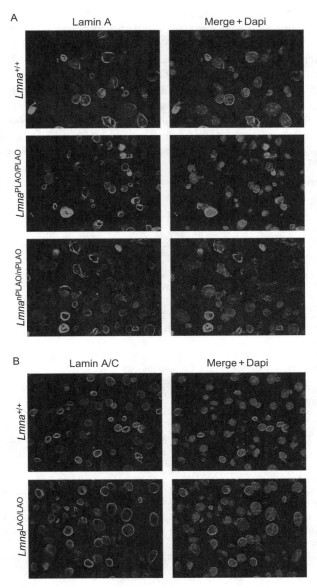

Fig. 3.4. Nuclear rim localization of lamin A and prelamin A in tissues of $Lmna^{PLAO/PLAO}$, $Lmna^{nPLAO/nPLAO}$, and $Lmna^{nPLAO/nPLAO}$ mice. (A) Immunostaining of liver from $Lmna^{+/+}$, $Lmna^{PLAO/PLAO}$, $Lmna^{nPLAO/nPLAO}$, and $Zmpste24^{-/-}$ mice with a lamin A-specific antibody (red). Nuclei were counterstained with DAPI (blue). (B) Immunostaining of liver from $Lmna^{+/+}$ and $Lmna^{LAO/LAO}$ mice with a lamin A/C-specific antibody (red). Nuclei were counterstained with DAPI (blue). Images taken with a 63× objective. Adapted from Davies *et al*. [51] and Coffinier *et al*. [52]. (See color plate section in the back of the book.)

farnesylation might be required for maintaining the integrity of the nuclear envelope during neuronal migration, but this hypothesis has not yet been tested.

The purpose of prelamin A processing remains a mystery. The carboxyl terminus of prelamin A is farnesylated and methylated, but that segment is clipped off and degraded during the biogenesis of mature lamin A. Interfering with the conversion of farnesyl-prelamin A to mature lamin A leads to severe disease (e.g., HGPS, RD). However, eliminating the production of prelamin A and lamin A, as in the $Lmna^{LCO/LCO}$ mice, does not lead to any obvious abnormality in mice. Nor was there detectable pathology in $Lmna^{LAO/LAO}$ mice, where mature lamin A is synthesized directly, bypassing prelamin A synthesis and processing. Nuclear rim localization of lamin A, long assumed to require prelamin A processing, was seemingly normal in tissues of $Lmna^{LAO/LAO}$ mice. Thus, the purpose of prelamin A processing remains elusive, and our ongoing challenge is to understand why this pathway has been preserved in vertebrate evolution.

ACKNOWLEDGMENTS

This work was supported by the National Institutes of Health Grants (HL76839, HL86683, HL089781, GM66152, AG035626); a March of Dimes Grant (6-FY2007-1012); the Ellison Medical Foundation Senior Scholar Program; a postdoctoral fellowship award from the American Heart Association, Western States Affiliate; and a Scientist Development Grant from the American Heart Association (0835489N).

REFERENCES

1. Burke, B., and Stewart, C.L. (2002). Life at the edge: the nuclear envelope and human disease. *Nat Rev Mol Cell Biol* 3:575–585.
2. Muchir, A., and Worman, H.J. (2004). The nuclear envelope and human disease. *Physiology (Bethesda)* 19:309–314.
3. Worman, H.J., and Bonne, G. (2007). "Laminopathies": a wide spectrum of human diseases *Exp Cell Res* 313:2121–2133.
4. Mounkes, L.C., Burke, B., and Stewart, C.L. (2001). The A-Type lamins: nuclear structural proteins as a focus for muscular dystrophy and cardiovascular diseases. *Trends Cardiovasc Med* 11:280–285.
5. Wilson, K.L. (2000). The nuclear envelope, muscular dystrophy and gene expression. *Trends Cell Biol* 10:125–129.
6. Fisher, D.Z., Chaudhary, N., and Blobel, G. (1986). cDNA sequencing of nuclear lamins A and C reveals primary and secondary structural homology to intermediate filament proteins. *Proc Natl Acad Sci USA* 83:6450–6454.
7. Lin, F., and Worman, H.J. (1993). Structural organization of the human gene encoding nuclear lamin A and nuclear lamin C. *J Biol Chem* 268:16321–16326.
8. Casey, P.J. (1992). Biochemistry of protein prenylation. *J Lipid Res* 33:1731–1740.

9. Casey, P.J., and Seabra, M.C. (1996). Protein prenyltransferases. *J Biol Chem* 271:5289–5292.
10. Zhang, F.L., and Casey, P.J. (1996). Protein prenylation: molecular mechanisms and functional consequences. *Annu Rev Biochem* 65:241–269.
11. Schafer, W.R., and Rine, J. (1992). Protein prenylation: genes, enzymes, targets, and functions. *Annu Rev Genet* 26:209–237.
12. Kim, E., *et al.* (1999). Disruption of the mouse Rce1 gene results in defective Ras processing and mislocalization of Ras within cells. *J Biol Chem* 274:8383–8390.
13. Maske, C.P., Hollinshead, M.S., Higbee, N.C., Bergo, M.O., Young, S.G., and Vaux, D.J. (2003). A carboxyl-terminal interaction of lamin B1 is dependent on the CAAX endoprotease Rce1 and carboxymethylation. *J Cell Biol* 162:1223–1232.
14. Otto, J.C., Kim, E., Young, S.G., and Casey, P.J. (1999). Cloning and characterization of a mammalian prenyl protein-specific protease. *J Biol Chem* 274:8379–8382.
15. Bergo, M.O., *et al.* (2002). Zmpste24 deficiency in mice causes spontaneous bone fractures, muscle weakness, and a prelamin A processing defect. *Proc Natl Acad Sci USA* 99:13049–13054.
16. Corrigan, D.P., *et al.* (2005). Prelamin A endoproteolytic processing in vitro by recombinant Zmpste24. *Biochem J* 387:129–138.
17. Clarke, S., Vogel, J.P., Deschenes, R.J., and Stock, J. (1988). Posttranslational modification of the Ha-ras oncogene protein: evidence for a third class of protein carboxyl methyltransferases. *Proc Natl Acad Sci USA* 85:4643–4647.
18. Dai, Q., *et al.* (1998). Mammalian prenylcysteine carboxyl methyltransferase is in the endoplasmic reticulum. *J Biol Chem* 273:15030–15034.
19. Bergo, M.O., *et al.* (2001). Isoprenylcysteine carboxyl methyltransferase deficiency in mice. *J Biol Chem* 276:5841–5845.
20. Bergo, M.O., Leung, G.K., Ambroziak, P., Otto, J.C., Casey, P.J., and Young, S.G. (2000). Targeted inactivation of the isoprenylcysteine carboxyl methyltransferase gene causes mislocalization of K-Ras in mammalian cells. *J Biol Chem* 275:17605–17610.
21. Beck, L.A., Hosick, T.J., and Sinensky, M. (1988). Incorporation of a product of mevalonic acid metabolism into proteins of Chinese hamster ovary cell nuclei. *J Cell Biol* 107:1307–1316.
22. Wolda, S.L., and Glomset, J.A. (1988). Evidence for modification of lamin B by a product of mevalonic acid. *J Biol Chem* 263:5997–6000.
23. Broers, J.L., *et al.* (1997). A- and B-type lamins are differentially expressed in normal human tissues. *Histochem Cell Biol* 107:505–517.
24. Rober, R.A., Sauter, H., Weber, K., and Osborn, M. (1990). Cells of the cellular immune and hemopoietic system of the mouse lack lamins A/C: distinction versus other somatic cells. *J Cell Sci* 95(Pt 4):587–598.
25. Rober, R.A., Weber, K., and Osborn, M. (1989). Differential timing of nuclear lamin A/C expression in the various organs of the mouse embryo and the young animal: a developmental study. *Development* 105:365–378.
26. Stewart, C., and Burke, B. (1987). Teratocarcinoma stem cells and early mouse embryos contain only a single major lamin polypeptide closely resembling lamin B. *Cell* 51:383–392.
27. Padiath, Q.S., *et al.* (2006). Lamin B1 duplications cause autosomal dominant leukodystrophy. *Nat Genet* 38:1114–1123.
28. Vergnes, L., PÇterfy, M., Bergo, M.O., Young, S.G., and Reue, K. (2004). Lamin B1 is required for mouse development and nuclear integrity. *Proc Natl Acad Sci USA* 101:10428–10433.

29. Izumi, M., Vaughan, O.A., Hutchison, C.J., and Gilbert, D.M. (2000). Head and/or CaaX domain deletions of lamin proteins disrupt preformed lamin A and C but not lamin B structure in mammalian cells. *Mol Biol Cell* 11:4323–4337.

30. Krohne, G., Waizenegger, I., and Hoger, T.H. (1989). The conserved carboxy-terminal cysteine of nuclear lamins is essential for lamin association with the nuclear envelope. *J Cell Biol* 109:2003–2011.

31. Dalton, M.B., *et al.* (1995). The farnesyl protein transferase inhibitor BZA-5B blocks farnesylation of nuclear lamins and p21ras but does not affect their function or localization. *Cancer Res* 55:3295–3304.

32. Vorburger, K., Lehner, C.F., Kitten, G.T., Eppenberger, H.M., and Nigg, E.A. (1989). A second higher vertebrate B-type lamin. cDNA sequence determination and in vitro processing of chicken lamin B2. *J Mol Biol* 208:405–415.

33. Hoger, T.H., Zatloukal, K., Waizenegger, I., and Krohne, G. (1990). Characterization of a second highly conserved B-type lamin present in cells previously thought to contain only a single B-type lamin. *Chromosoma* 99:379–390.

34. Coffinier, C., *et al.* (2010). Abnormal development of the cerebral cortex and cerebellum in the setting of lamin B2 deficiency. *Proc Natl Acad Sci USA* 107:5076–5081.

35. Solecki, D.J., Govek, E.E., Tomoda, T., and Hatten, M.E. (2006). Neuronal polarity in CNS development. *Genes Dev* 20:2639–2647.

36. Vallee, R.B., and Tsai, J.W. (2006). The cellular roles of the lissencephaly gene LIS1, and what they tell us about brain development. *Genes Dev* 20:1384–1393.

37. Wynshaw-Boris, A., and Gambello, M.J. (2001). LIS1 and dynein motor function in neuronal migration and development. *Genes Dev* 15:639–651.

38. Ayala, R., Shu, T., and Tsai, L.H. (2007). Trekking across the brain: the journey of neuronal migration. *Cell* 128:29–43.

39. Gleeson, J.G., and Walsh, C.A. (2000). Neuronal migration disorders: from genetic diseases to developmental mechanisms. *Trends Neurosci* 23:352–359.

40. Wynshaw-Boris, A. (2007). Lissencephaly and LIS1: insights into the molecular mechanisms of neuronal migration and development. *Clin Genet* 72:296–304.

41. Kilic, F., Johnson, D.A., and Sinensky, M. (1999). Subcellular localization and partial purification of prelamin A endoprotease: an enzyme which catalyzes the conversion of farnesylated prelamin A to mature lamin A. *FEBS Lett* 450:61–65.

42. Fong, L.G., *et al.* (2004). Heterozygosity for Lmna deficiency eliminates the progeria-like phenotypes in Zmpste24-deficient mice. *Proc Natl Acad Sci USA* 101:18111–18116.

43. Pendas, A.M., *et al.* (2002). Defective prelamin A processing and muscular and adipocyte alterations in Zmpste24 metalloproteinase-deficient mice. *Nat Genet* 31:94–99.

44. Sinensky, M., Fantle, K., Trujillo, M., McLain, T., Kupfer, A., and Dalton, M. (1994). The processing pathway of prelamin A. *J Cell Sci* 107(Pt 1):61–67.

45. Vaughan, O.A., *et al.* (2001). Both emerin and lamin C depend on lamin A for localization at the nuclear envelope. *J Cell Sci* 114:2577–2590.

46. Fong, L.G., *et al.* (2006). Prelamin A and lamin A appear to be dispensable in the nuclear lamina. *J Clin Invest* 116:743–752.

47. Yang, S.H., *et al.* (2005). Blocking protein farnesyltransferase improves nuclear blebbing in mouse fibroblasts with a targeted Hutchinson-Gilford progeria syndrome mutation. *Proc Natl Acad Sci USA* 102:10291–10296.

48. Yang, S.H., *et al.* (2006). A farnesyltransferase inhibitor improves disease phenotypes in mice with a Hutchinson-Gilford progeria syndrome mutation. *J Clin Invest* 116:2115–2121.

49. Yang, S.H., Andres, D.A., Spielmann, H.P., Young, S.G., and Fong, L.G. (2008). Progerin elicits disease phenotypes of progeria in mice whether or not it is farnesylated. *J Clin Invest* 118:3291–3300.

50. Yang, S.H., *et al.* (2011). Absence of progeria-like disease phenotypes in knock-in mice expressing a nonfarnesylated version of progerin. *Hum Mol Genet* 20:436–444.

51. Davies, B.S., *et al.* (2010). An accumulation of non-farnesylated prelamin A causes cardiomyopathy but not progeria. *Hum Mol Genet* 19:2682–2694.

52. Coffinier, C., *et al.* (2010). Direct synthesis of lamin A, bypassing prelamin a processing, causes misshapen nuclei in fibroblasts but no detectable pathology in mice. *J Biol Chem* 285:20818–20826.

53. Davies, B.S., *et al.* (2011). Investigating the purpose of prelamin A processing. *Nucleus* 2:4–9.

54. Sullivan, T., *et al.* (1999). Loss of A-type lamin expression compromises nuclear envelope integrity leading to muscular dystrophy. *J Cell Biol* 147:913–920.

55. Lammerding, J., *et al.* (2004). Lamin A/C deficiency causes defective nuclear mechanics and mechanotransduction. *J Clin Invest* 113:370–378.

56. Wolf, C.M., *et al.* (2008). Lamin A/C haploinsufficiency causes dilated cardiomyopathy and apoptosis-triggered cardiac conduction system disease. *J Mol Cell Cardiol* 44:293–303.

57. Bonne, G., *et al.* (1999). Mutations in the gene encoding lamin A/C cause autosomal dominant Emery-Dreifuss muscular dystrophy. *Nat Genet* 21:285–288.

58. Sebillon, P., *et al.* (2003). Expanding the phenotype of LMNA mutations in dilated cardiomyopathy and functional consequences of these mutations. *J Med Genet* 40:560–567.

59. van Engelen, B.G., Muchir, A., Hutchison, C.J., van der Kooi, A.J., Bonne, G., and Lammens, M. (2005). The lethal phenotype of a homozygous nonsense mutation in the lamin A/C gene. *Neurology* 64:374–376.

60. Leung, G.K., *et al.* (2001). Biochemical studies of Zmpste24-deficient mice. *J Biol Chem* 276:29051–29058.

61. Toth, J.I., *et al.* (2005). Blocking protein farnesyltransferase improves nuclear shape in fibroblasts from humans with progeroid syndromes. *Proc Natl Acad Sci USA* 102:12873–12878.

62. Coffinier, C., *et al.* (2008). A potent HIV protease inhibitor, Darunavir, does not inhibit ZMPSTE24 or lead to an accumulation of farnesyl-prelamin A in cells. *J Biol Chem* 283:9797–9804.

63. Moulson, C.L., Go, G., van der Wal, A.C., Smitt, J.H., van Hagen, J.M., and Miner, J.H. (2005). Homozygous and compound heterozygous mutations in ZMPSTE24 cause the laminopathy restrictive dermopathy. *J Invest Dermatol* 125:913–919.

64. Navarro, C.L., *et al.* (2005). Loss of ZMPSTE24 (FACE-1) causes autosomal recessive restrictive dermopathy and accumulation of Lamin A precursors. *Hum Mol Genet* 14:1503–1513.

65. Batstone, M.D., and Macleod, A.W.G. (2002). Oral and maxillofacial surgical considerations for a case of Hutchinson-Gilford progeria. *Int J Paediatr Dent* 12:429–432.

66. DeBusk, F.L. (1972). The Hutchinson-Gilford progeria syndrome. Report of 4 cases and review of the literature. *J Pediatr* 80:697–724.

67. Eriksson, M., *et al.* (2003). Recurrent de novo point mutations in lamin A cause Hutchinson-Gilford progeria syndrome. *Nature* 423:293–298.

68. Faivre, L., Kien, P.K.V., Madinier-Chappat, N., Nivelon-Chevallier, A., Beer, F., and LeMerrer, M. (1999). Can Hutchinson-Gilford progeria syndrome be a neonatal condition? *Am J Hum Genet* 87:450–452.

69. Fernandez-Palazzi, F., McLaren, A.T., and Slowie, D.F. (1992). Report on a case of Hutchinson-Gilford progeria, with special reference to orthopedic problems. *Eur J Pediatr Surg* 2:378–382.

70. Khalifa, M.M. (1989). Hutchinson-Gilford progeria syndrome: report of a Libyan family and evidence of autosomal recessive inheritance. *Clin Genet* 35:125–132.

71. Rodríguez, J.I., Perez-Alonso, P., Funes, R., and Perez-Rodríguez, J. (1999). Lethal neonatal Hutchinson-Gilford progeria syndrome. *Am J Med Genet* 82:242–248.

72. Dechat, T., et al. (2007). Alterations in mitosis and cell cycle progression caused by a mutant lamin A known to accelerate human aging. *Proc Natl Acad Sci USA* 104:4955–4960.

73. Goldman, R.D., et al. (2004). Accumulation of mutant lamin A causes progressive changes in nuclear architecture in Hutchinson-Gilford progeria syndrome. *Proc Natl Acad Sci USA* 101:8963–8968.

74. Yang, S.H., Qiao, X., Farber, E., Chang, S.Y., Fong, L.G., and Young, S.G. (2008). Eliminating the synthesis of mature lamin A reduces disease phenotypes in mice carrying a Hutchinson-Gilford Progeria Syndrome allele. *J Biol Chem* 283:7094–7099.

75. Capell, B.C., et al. (2005). Inhibiting farnesylation of progerin prevents the characteristic nuclear blebbing of Hutchinson-Gilford progeria syndrome. *Proc Natl Acad Sci USA* 102:12879–12884.

76. Glynn, M.W., and Glover, T.W. (2005). Incomplete processing of mutant lamin A in Hutchinson-Gilford progeria leads to nuclear abnormalities, which are reversed by far-nesyltransferase inhibition. *Hum Mol Genet* 14:2959–2969.

77. Mallampalli, M.P., Huyer, G., Bendale, P., Gelb, M.H., and Michaelis, S. (2005). Inhibiting farnesylation reverses the nuclear morphology defect in a HeLa cell model for Hutchinson-Gilford progeria syndrome. *Proc Natl Acad Sci USA* 102:14416–14421.

78. Fong, L.G., et al. (2006). A protein farnesyltransferase inhibitor ameliorates disease in a mouse model of progeria. *Science* 311:1621–1623.

79. Yang, S.H., Chang, S.Y., Andres, D.A., Spielmann, H.P., Young, S.G., and Fong, L.G. (2010). Assessing the efficacy of protein farnesyltransferase inhibitors in mouse models of progeria. *J Lipid Res* 51:400–405.

80. Yang, S.H., Qiao, X., Fong, L.G., and Young, S.G. (2008). Treatment with a farnesyl-transferase inhibitor improves survival in mice with a Hutchinson-Gilford progeria syndrome mutation. *Biochim Biophys Acta* 1781:36–39.

81. Muchir, A., et al. (2000). Identification of mutations in the gene encoding lamins A/C in autosomal dominant limb girdle muscular dystrophy with atrioventricular conduction disturbances (LGMD1B). *Hum Mol Genet* 9:1453–1459.

82. Vytopil, M., et al. (2003). Mutation analysis of the lamin A/C gene (LMNA) among patients with different cardiomuscular phenotypes. *J Med Genet* 40:e132.

83. Arimura, T., et al. (2005). Mouse model carrying H222P-Lmna mutation develops muscular dystrophy and dilated cardiomyopathy similar to human striated muscle laminopathies. *Hum Mol Genet* 14:155–169.

84. Mounkes, L.C., Kozlov, S.V., Rottman, J.N., and Stewart, C.L. (2005). Expression of an LMNA-N195K variant of A-type lamins results in cardiac conduction defects and death in mice. *Hum Mol Genet* 14:2167–2180.

85. Whyte, D.B., et al. (1997). K- and N-Ras are geranylgeranylated in cells treated with farnesyl protein transferase inhibitors. *J Biol Chem* 272:14459–14464.

86. Rusinol, A.E., and Sinensky, M.S. (2006). Farnesylated lamins, progeroid syndromes and farnesyl transferase inhibitors. *J Cell Sci* 119:3265–3272.

87. Lee, R., et al. (2010). Genetic studies on the functional relevance of the protein prenyl-transferases in skin keratinocytes. *Hum Mol Genet* 19:1603–1617.

88. Varela, I., *et al.* (2008). Combined treatment with statins and aminobisphosphonates extends longevity in a mouse model of human premature aging. *Nat Med* 14:767–772.

89. Davies, B.S., *et al.* (2009). Increasing the length of progerin's isoprenyl anchor does not worsen bone disease or survival in mice with Hutchinson-Gilford progeria syndrome. *J Lipid Res* 50:126–134.

90. Liu, M., *et al.* (2010). Targeting the protein prenyltransferases efficiently reduces tumor development in mice with K-RAS-induced lung cancer. *Proc Natl Acad Sci USA* 107:6471–6476.

91. Beck, L.A., Hosick, T.J., and Sinensky, M. (1990). Isoprenylation is required for the processing of the lamin A precursor. *J Cell Biol* 110:1489–1499.

92. Lutz, R.J., Trujillo, M.A., Denham, K.S., Wenger, L., and Sinensky, M. (1992). Nucleoplasmic localization of prelamin A: implications for prenylation-dependent lamin A assembly into the nuclear lamina. *Proc Natl Acad Sci USA* 89:3000–3004.

93. Vorburger, K., Kitten, G.T., and Nigg, E.A. (1989). Modification of nuclear lamin proteins by a mevalonic acid derivative occurs in reticulocyte lysates and requires the cysteine residue of the C-terminal CXXM motif. *EMBO J* 8:4007–4013.

94. Weber, K., Plessmann, U., and Traub, P. (1989). Maturation of nuclear lamin A involves a specific carboxy-terminal trimming, which removes the polyisoprenylation site from the precursor; implications for the structure of the nuclear lamina. *FEBS Lett* 257:411–414.

95. Hennekes, H., and Nigg, E.A. (1994). The role of isoprenylation in membrane attachment of nuclear lamins. A single point mutation prevents proteolytic cleavage of the lamin A precursor and confers membrane binding properties. *J Cell Sci* 107(Pt 4):1019–1029.

4

Prenylated Proteins in Peroxisome Biogenesis

ROBERT RUCKTÄSCHEL • REZEDA MIRGALIEVA • RALF ERDMANN

Institut für Physiologische Chemie
Abteilung für Systembiochemie
Medizinische Fakultät, Ruhr-Universität Bochum
Bochum, Germany

I. Abstract

Two peroxisomal proteins are known to be prenylated, Pex19p and Rho1p. The farnesylated Pex19p is an essential component of the transport machinery for peroxisomal membrane proteins (PMPs) and the *de novo* formation of peroxisomes. Pex19p functions as PMP-specific import receptor and chaperon, which binds newly synthesized PMPs in the cytosol in a farnesylation-dependent manner and directs them to the peroxisomal membrane. Pex19p is also involved in the endoplasmic reticulum (ER)-to-peroxisome transport of vesicles which appears to rely on a novel budding mechanism. The role of farnesylation of Pex19p in this process still awaits investigation. Prenylated Rho GTPases are involved in actin assembly on peroxisomes in yeast and peroxisome-microtubule association in mammalian cells.

II. Peroxisomes

Peroxisomes or microbodies are a family of functionally and structurally related organelles of eukaryotic cells. They are surrounded by a single lipid bilayer membrane and have a diameter of 0.1–1 μm [1]. The electron-dense

DOI: 10.1016/B978-0-12-381339-8.00004-4

matrix is built up by several proteins and contains no DNA. The family of peroxisomes comprises a subset of morphologically similar but functionally diverse organelles including the "typical" peroxisomes of plants and animals, the glyoxysomes of plants and fungi, the glycosomes of trypanosomes, and the Woronin-bodies of filamentous fungi [2]. With the exception of Woronin-bodies, whose sole function is to plug septal pores in case of hyphal injury [3], peroxisomes fulfill a variety of biochemical functions [4]. One of these functions is the β-oxidation of fatty acids, which is a typical feature of peroxisomes. In fact, β-oxidation of fatty acids is exclusively localized to peroxisomes in fungi and plants. In mammals, long-chain fatty acids are oxidized in mitochondria but very long chain fatty acids and other uncommon fatty acids like branched-chain fatty acids are oxidized in peroxisomes [5]. Further, peroxisomes are involved in the synthesis of plasmalogens, cholesterol, and bile acids as well as the oxidation of alcohols, catabolism of purines and polyamines, metabolism of prostaglandins in man, photorespiration in plants, and penicillin synthesis in fungi [1,6–12]. The importance of peroxisomes for human life is also exemplified by severe inborn peroxisomal diseases, which are categorized by two groups depending on their underlying defect. One group comprises diseases which are caused by single enzyme defects, like X-linked adrenoleukodystrophy or rhizomelic chondrodysplasia punctata type 2. The other group contains diseases which are characterized by defects in the biogenesis of peroxisomes, so-called PBDs (peroxisomal biogenesis disorders). This group comprises fatal diseases like the Zellweger spectrum, neonatal adrenoleukodystrophy, or Infantile Refsum's disease. These disorders are caused by mutations in so-called *PEX* genes encoding proteins that are involved in the biogenesis of peroxisomes, the peroxins [13]. Up today, 33 different *PEX* genes have been discovered which are required for the biogenesis and maintenance of functional peroxisomes. Peroxins play roles in different processes in peroxisome biogenesis: (1) the targeting and import of peroxisomal matrix proteins, (2) the proliferation and inheritance of the organelles, and proteins like the prenylated Pex19p play a role in (3) the topogenesis of peroxisomal membrane proteins (PMPs).

III.   Biogenesis of Peroxisomes

As peroxisomes do not contain their own DNA, all peroxisomal proteins are encoded in the nucleus. Thus most if not all peroxiomal proteins are synthesized on free ribosomes in the cytosol and imported posttranslationally into the organelle [14]. Two different models describing the biogenesis of peroxisomes are currently discussed. In the growth and division model, peroxisomes are seen as autonomous organelles like mitochondria or chloroplasts. Thus, peroxisomal matrix and membrane proteins are imported into preexisting organelles, which then grow up to a critical size and divide into two daughter organelles. In this model, the existence of one peroxisome per cell is a prerequisite [14]. In the more recent *de novo* synthesis model, peroxisomes are formed *de novo* by budding from the ER. In this model, PMPs are imported post- or cotranslationally into the ER) and pre-peroxisomal structures are formed from specialized subdomains from the ER which then start to import membrane and matrix proteins [15]. Recently, the budding process was reconstituted *in vitro*, and the ER-peroxisome transport vesicles appear to rely on a novel budding mechanism requiring Pex19p and additional unknown factors [16]. In the cell, the growth and division and the *de novo* pathways seem exist side by side, with growth and division pathway being the main route for the proliferation of peroxisomes, at least in yeast [17].

A. Import of Peroxisomal Matrix Proteins

A striking feature of peroxisomes which distinguishes them from other organelles like mitochondria and the ER is that they are able to import matrix proteins in a folded and even oligomeric state [18,19]. In this process, import receptors pass through a cycle which starts with the recognition of the cargo molecule in the cytosol followed by a docking to the peroxisomal membrane. Proteins designated for the peroxisomal matrix harbor (one of the two peroxisomal targeting signals) located at the extreme C-terminus (PTS1) or the N-terminal region of the proteins. PTS1 and PTS2 are recognized by the distinct import receptors Pex5p and Pex7p, respectively [20–22]. After docking of the receptor–cargo complexes, cargo protein is translocated across the membrane and released into the peroxisomal lumen in an unknown fashion [23]. In a last step, the receptor molecules are ubiquitinated and exported back into the cytosol in an ATP-dependent manner by AAA-ATPases of the peroxisomal protein import machinery [24,25]. In the cytosol, the receptors are available for a new round of import [26,27]. Interestingly, there is a surprising similarity between the machinery that imports proteins into peroxisomes and the machinery that degrades ER-associated proteins. Recently, it was reported that this similarity lies in the fact that both machineries make use of the same basic mechanistic principle: the tagging of a substrate by mono- or polyubiquitylation and its subsequent recognition and ATP-dependent removal from a membrane by ATPases of the AAA-family of proteins. For peroxisomal protein import, it was proposed that the ERAD-like removal of the peroxisomal import receptor is mechanically coupled to protein translocation into the organelle giving rise to the export-driven import model [28].

B. Peroxisomal Membrane Biogenesis

In all models which are discussed for the formation of peroxisomes, the import of PMPs is a major aspect. The import of peroxisomal matrix proteins and the topogenesis of PMPs are facilitated by distinct import machineries [29,30]. This is supported by the fact that most *pex*-mutants are defective in the import of peroxisomal matrix proteins but the import of PMPs is still functional. These mutants contain peroxisomal membrane remnants, called ghosts, in which PMPs are imported [31–33]. Only a few mutants are characterized by the lack of any detectable peroxisomal structures. By functional complementation, three peroxins were identified involved in the biogenesis of the peroxisomal membrane, namely Pex3p, Pex19p, and in some organisms Pex16p [34–41].

1. The Membrane Biogenesis Factors Pex19p

Pex19p is the only known prenylated protein involved in the biogenesis of peroxisomes. Pex19p was first discovered in 1994 as PxP by a screen for farnesylated proteins and described as a protein associated to the outer surface of peroxisome [42]. Later on, Pex19p was identified by functional complementation of yeast mutants affected in peroxisome biogenesis [43]. The cells lacking Pex19p are characterized not only by a defect in the import of peroxisomal matrix proteins but also by the complete absence of peroxisomal membranes, indicating that Pex19p is essential for peroxisome membrane formation [37]. Up to now, Pex19p orthologues have been iden-tified in all species analyzed. Pex19p is a mostly soluble protein with a small portion associated with the peroxisomal membrane and has the ability to interact with most PMPs [37,44–50]. Structurally, Pex19p consists of an unstructured N-terminal- and a structured C-terminal domain [51]. The N-terminal domain is responsible for the membrane targeting of Pex19p as it mediates the interaction with Pex3p at the peroxisomal membrane. The C-terminal domain harbors the binding sites for most PMPs [51–54]. Crystal structure of the C-terminal part revealed that it consists of a bundle of three long helices in an antiparallel arrangement [55].

Sequence comparisons of Pex19p from different species showed that Pex19p exhibits a CaaX-box at its extreme C-terminus which is conserved from yeast to man with trypanosomes being the only exception so far (Figure 4.1) [56]. Farnesylation sites have originally been described as "CaaX" ("Ca1a2X") boxes with "aa" being small aliphatic amino acids

Saccharomyces cerevisiae	LGNLDKELTDG **C** KQQ
Ashbya gossypii	LGNIDKEIQET **C** KQQ
Pichia pastoris	PEDLSKELDET **C** KQQ
Pichia angusta	PEDINKELEDT **C** KQT
Yarrowia lipolytica	KMPEMPENMPE **C** NQQ
Schizosaccharomyces pombe	PLIDNSMETAG **C** PTQ
Candida albicans	PPEFGKDLQEG **C** KQQ
Neurospora crassa	SAQNLDLPDDQ **C** APQ
Arabidopsis thaliana	ISPEMLESSPN **C** CIM
Caenorhabditis elegans	KVADAAAATEA **C** SIM
Drosophila melanogaster	AGVAAGGPGPQ **C** PTM
Danio rerio	PGAQGLPGADQ **C** SVM
Xenopus laevis	LAGSGSANGEQ **C** LIM
Cricetulus griseus	LSGPPGANGEQ **C** LIM
Mus musculus	LSGPPGANGEQ **C** LIM
Rattus norvegicus	LSGPPGANGEQ **C** LIM
Homo sapiens	LSGPPGASGEQ **C** LIM

FIG. 4.1. Conservation of the CaaX-box. Sequence alignment of Pex19p-C-termini from different species. The CaaX-box of Pex19p, representing the farnesylation motif, is conserved from yeast to man. While in yeast, Pex19p exhibits an unusual CaaX-box (CKQQ), a more classical farnesylation motif is present in higher eukaryotes (CLIM).

and X being any amino acid [57]. Pex19p from *Saccharomyces cerevisiae* contains an unusual farnesylation motif (CKQQ) with lysine and glutamine in the a1a2 positions not being "small aliphatic amino acids" (Figure 4.1). Despite the unusual motif, it was shown that yeast Pex19p is modified by farnesylation [37,58]. In higher eukaryotes, Pex19p evolved a more classical CaaX-box with the sequence CLIM (Figure 4.1). The role of the farnesylation for the function of Pex19p for a long time was a matter of debate. Interestingly, an involvement of farnesylation of Pex19p in attachment of the protein to the peroxisomal membrane has not been discovered. However, different studies revealed an effect of the farnesylation on the ability of Pex19p to bind PMPs [58,59]. It was shown that a nonfarnesylated form of Pex19p binds to PMPs with a significant lower affinity than farnesylated Pex19p. Here, binding of PMPs to the C-terminal domain is equally affected by the loss of farne-sylation as the interaction with Pex3p, which is bound by the N-terminus of Pex19p. Surprisingly, farnesylation of Pex19p was described to be dispensable for the function of Pex19p in peroxisome biogenesis in some species [49,54,59–62]. However, in baker's yeast, it was shown that mutants affected in Pex19p-farnesylation exhibit severe defects in the peroxisomal biogenesis. Mutants defective in Pex19p farnesylation are characterized by a significantly reduced steady-state concentration not only of prominent PMPs but also of essential components of the peroxisomal import machinery [58]. Moreover, it was shown that in baker's yeast, the complete pool of Pex19p is processed by farnesyltrans-ferase, indicating that the farnesylation might contribute to the structural integrity of Pex19p, which still awaits approval [58].

Many different functions in the biogenesis of peroxisomes have been proposed for Pex19p. Due to the fact that Pex19p binds most PMPs and based on its bimodal localization at the peroxisomal membrane and in the cytosol, it is thought to function as a soluble import receptor for newly synthesized PMPs (Figure 4.2). Accordingly, Pex19p binds PMPs in the cytosol and directs them to the peroxisomal membrane by docking to its membrane anchored binding partner Pex3p [63,64]. Additionally, Pex19p possesses the ability to stabilize PMPs by forming soluble complexes and thus preventing PMPs from aggregation. Therefore, it is also considered to function as a chaperone for PMPs [51,65]. A third proposed function of Pex19p is that of an insertion factor or an assembly/disassembly factor for peroxisomal membrane complexes at the peroxisomal membrane [66].

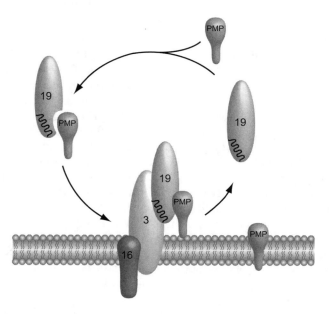

FIG. 4.2. Pex19p-dependent import of class I PMPs. Newly synthesized PMPs are recognized in the cytosol by the soluble receptor Pex19p. Accordingly, PMPs contain a Pex19p-binding site as an integral part of their mPTS. In a next step, the cargo-loaded Pex19p docks to the peroxisomal membrane by association with its membrane bound docking factor Pex3p. After docking, the PMP is integrated into the peroxisomal membrane in an unknown fashion, probably with the help of Pex3p and in some organisms, Pex16p. In a last step, Pex19p recycles back to the cytosol and can start a new round of import. The farnesylation contributes to the structural integrity of Pex19p and enhances the affinity of Pex19p to its binding sites in PMPs. (See color plate section in the back of the book.)

Finally, Pex19p is playing a role in the exit and transport of Pex3p and Pex15p containing vesicles from the ER and is in this way involved in the *de novo* biogenesis of peroxisomes from the ER [15,16].

Beside the well-established role of Pex19p in peroxisome biogenesis, additional functions for the protein are discussed. On one hand, Pex19p is interacting with renal type IIa Na/Pi cotransporter and thereby may be actively involved in controlling the internalization and trafficking of the Na/Pi IIa cotransporter. Interestingly, this interaction is also affected by the farnesylation of Pex19p [67]. On the other hand, an interaction between Pex19p and p19ARF was discovered, and an involvement of Pex19p in mechanisms for the downregulation of the p19ARF-p53 pathway was proposed [68].

2. Binding Partners of Pex19p

Pex19p is a mostly soluble protein, its binding partners Pex3p and Pex16p are membrane bound factors, and all three proteins function in concert in the biogenesis of the peroxisomal membrane.

Pex16p is a membrane protein which so far has only been identified in higher eukaryotes and the yeast *Yarrowia lipolytica*. Discovered in 1998 by complementation of Zellweger cell lines [39], its function is not clear and seems to differ between species. In mammals, Pex16p is described as an integral membrane protein with both termini facing the cytosolic site of the peroxisome [69]. Functionally, it is thought to be involved in the import of PMPs in the very early stages of peroxisome biogenesis [69,70]. In contrast, the yeast Pex16p is described as a membrane-associated protein that resides at the luminal side of the peroxisomal membrane and functions as a negative regulator of peroxisomal fission [35].

Pex3p is an integral PMP with a topology differing among species [36,71–73]. In baker's yeast, Pex3p exhibits one transmembrane span dividing the protein in a large cytosolic C-terminal domain and a small N-terminal domain projecting into the peroxisomal lumen [38]. The function of Pex3p is well conserved among species as the protein plays an important role in the import of PMPs and serves as a docking factor for Pex19p–PMP complexes at the peroxisomal membrane [52,53,74]. Additionally, Pex3p is involved in the regulation of peroxisome motility and inheritance as it acts as a peroxisomal receptor for class V myosin and the peroxisomal retention factor Inp1p [75,76]. Pex3p also plays an important role in the *de novo* formation of peroxisomes as it is thought to represent the starting point for this peroxisome forming process [15,77].

3. Role of Pex19p in the Topogenesis of PMPs

Based on different modes of topogenesis, two classes of PMPs are defined. Class I PMPs are transported to peroxisomes and inserted into the peroxisomal membrane in a Pex19p-dependent manner, whereas the topogenesis of class II PMPs occurs independent of Pex19p [63].

Most membrane proteins are class I PMPs. The first step in the topogenesis is the recognition of the PMP by Pex19p in the cytosol (Figure 4.2). Here, Pex19p is supposed to function as signal sequence receptor and chaperon which binds the PMP during or directly after translation on free ribosomes in the cytosol. PMPs contain Pex19p-binding sites which are an integral part of their targeting signal, the so-called mPTS [78]. This binding site is characterized by the presence of a cluster of basic and hydrophobic amino acids which are supposed to acquire an α-helical conformation. A reliable consensus sequence for the mPTS has not yet been defined. However, the mPTS from different PMPs and different species show similarities, which allowed to evolve an algorithm for the prediction of Pex19p-binding sites [78]. It turned out that the Pex19p-binding site alone is not sufficient to target a PMP to the peroxisome. In addition to the Pex19p-binding site, PMPs require a peroxisomal anchor sequence which is a transmembrane domain in case of an integral PMP or a protein interaction site, in case of peripheral membrane proteins [63,78–83]. Thus, the mPTS of class I PMPs consists of the Pex19p-binding site and a membrane anchor sequence. As the next step, the Pex19p–cargo complex docks to the peroxisomal membrane by binding to the docking protein Pex3p (Figure 4.2) [74,78,84]. This step is promoted by a higher affinity of cargo-loaded Pex19p to Pex3p compared to cargo-free Pex19p [85]. After this docking step, the PMP is inserted into the peroxisomal membrane with the basic mechanistic principle of this process still being unknown. Some concepts suggest that Pex19p might exhibit a similar behavior as the import receptor for peroxisomal matrix proteins Pex5p, which is supposed to shuttle between a soluble cytosolic and a membrane-integrated form. In analogy, Pex19p would integrate into the peroxisomal bilayer and release the PMP into the membrane. Subsequently, Pex19p is removed from the membrane and exported back into the cytosol for another round of import (Figure 4.2). Evidence has been provided for that the import step requires ATP, whereas the export step is independent of ATP [86]. However, other models suggest an ATP-independent import of PMPs [85,87]. The ATP-consuming factors in PMP-topogenesis are still unknown.

4. Role of Pex19p in ER-to-Peroxisome Transport of Vesicles

While most of the membrane proteins imported into peroxisomes are class I PMPs, some PMPs are targeted independent of Pex19p. Today, only a few of these class II PMPs are known, namely Pex3p, Pex16p (for review, see Ref. [88]), Pex22p [89] and Pex15p [16]. It is supposed that class II PMPs are not directly imported into peroxisomes but are directed to the ER membrane prior to their transport to the peroxisome. The mPTS of these proteins does not contain a Pex19p-binding site and is mostly located in the N-termini of the proteins [71,89]. Accordingly, Pex16p is cotranslationally imported into the ER and then transported to peroxisomes [70]. The mechanisms how these class II PMPs are imported into the ER are not clearly known. Earlier studies revealed that the ER translocon Sec61p is not required for the import step [90]. However, recent findings suggest an involvement of Sec61p as well as the Get-complex in the import of class II PMPs into the ER [91].

Our knowledge on the transport of class II PMPs from the ER to the peroxisomes is still scarce. A vesicular transport from the ER to peroxisomes has been described [92], and Pex19p has been reported to be required for the traffic of the PMPs Pex3p from the ER to the peroxisome [15]. However, inhibitors for COPI and COPII which block vesicle transport in the secretory pathway do not affect the transport of class II PMPs to peroxisomes [93,94]. However, essential components of the secretory pathway (Sec20p, Sec39p, and Dsl1p) have been identified as also being required for Pex3p-exit from the ER and thus being involved in the early stages of the *de novo* synthesis of peroxisomes [95]. Recent data showed the involvement of Pex19p in sorting of Pex3p and Pex15p, a tail-anchored PMP, to vesicles derived from the ER membrane [16]. The ER-to-peroxisome transport of vesicles appears to rely on a novel budding mechanism requiring Pex19p and additional unknown factors. However, the role of farnesylation of Pex19p in this process still awaits investigation.

C. Rho GTPases on Peroxisomes

Beside Pex19p, the prenylated Rho GTPases Rho1p in yeast and RhoA in mammals were shown to be involved in peroxisome functions. Rho GTPases (*R*as *ho*molog) are members of the Ras family of GTPases. The Rho GTPases act as GTP-binding and GTP-hydrolyzing proteins and function by cycling between GTP- and GDP-bound forms. Guanine nucleotide exchange factors (GEFs) catalyze the exchange of GDP by GTP which results in the activation of the GTPases. GTPase-activating proteins, or GAPs, stimulate GTP hydrolysis and thus switch the Rho GTPases to the inactive GDP-bound form [96,97]. Rho GTPases have a C-terminal CAAX-box—a signal peptide for prenylation. Rho GTPases can be modified by prenyltransferases by either a geranylgeranyl or a farnesyl isoprenoid group. These modifications serve as an anchor for the association of the proteins to specific membranes. Guanine nucleotide dissociation inhibitors contribute an additional level of GTPase regulation by extracting Rho GTPases from the membrane which removes them from their site of action [96–98]. Rho GTPases are involved in various cellular processes, such as actin reorganization, cell motility, cell proliferation, cell polarity, gene transcription, vesicular transport, and many others [96,99–101]. Most members of Rho GTPases are highly conserved in evolution, indicative for their fundamental role in the cell. For example, the mammalian RhoA and its orthologue in yeast Rho1p exhibit an identity of about 70% [102].

The Rho GTPases Rho1p is essential for baker's yeast, and it is involved in the actin reorganization, cell wall biosynthesis, and bud formation. Rho1p shows a bimodal localization in the cell as it is present in the cytosol and attached to specific membranes, for example, to the plasma membrane, endosomes, or golgi apparatus [100,101]. The attachment of Rho1p to membranes is facilitated by its geranylgeranyl moiety. Interestingly, yeast Rho1p and its mammalian orthologue RhoA were found to be localized to peroxisomes [103,104] as was the virulence factor SifA in *Salmonella* spp. which acts as GEF for Rho1p [105]. The association of Rho1p to peroxisomes is mediated by an interaction of Rho1p with Pex25p [103]. Pex25p is a member of the Pex11-protein family which is involved in the growth and division of peroxisomes [106]. Further investigations demonstrated that Rho1p regulates actin assembly on peroxisomes [103]. The GTP-bound form of RhoA is localized to mammalian peroxisomes, which show long-range saltatory movements along microtubules. It was shown that switching between the active and inactive state of RhoA leads to an attachment/detachment of peroxisomes from microtubules [104].

REFERENCES

1. van den Bosch, H., Schutgens, R.B., Wanders, R.J., and Tager, J.M. (1992). Biochemistry of peroxisomes. *Annu Rev Biochem* 61:157–197.
2. Baker, A., and Sparkes, I.A. (2005). Peroxisome protein import: some answers, more questions. *Curr Opin Plant Biol* 8:640–670.
3. Jedd, G., and Chua, N.H. (2000). A new self-assembled peroxisomal vesicle required for efficient resealing of the plasma membrane. *Nat Cell Biol* 2:226–231.
4. Islinger, M., Cardoso, M.J., and Schrader, M. (2010). Be different—the diversity of peroxisomes in the animal kingdom. *Biochim Biophys Acta* 1803:881–897.

5. Poirier, Y., Antonenkov, V.D., Glumoff, T., and Hiltunen, J.K. (2006). Peroxisomal beta-oxidation—a metabolic pathway with multiple functions. *Biochim Biophys Acta* 1763:1413–1426.
6. Biardi, L., and Krisans, S.K. (1996). Compartmentalization of cholesterol biosynthesis. Conversion of mevalonate to farnesyl diphosphate occurs in the peroxisomes. *J Biol Chem* 271:1784–1788.
7. Hajara, A.K., and Bishop, J.E. (1982). Glycerolipid biosynthesis in peroxisomes via the acyldihydroxacetone pathway. *Ann N Y Acad Sci* 386:170–182.
8. Krisans, S.K. (1992). The role of peroxisomes in cholesterol metabolism. *Am J Respir Cell Mol Biol* 7:358–364.
9. Krisans, S.K. (1996). Cell compartmentalization of cholesterol biosynthesis. *Ann N Y Acad Sci* 804:142–164.
10. Müller, W.H., van der Krift, T.P., Krouwer, A.J.J., Wosten, H.A.B., and van der Voort, L.H.M. (1991). Localisation of the pathway of the penicillin biosynthesis in *Penicillium chrysogenum*. *EMBO J* 10:489–496.
11. Tolbert, N.E., and Essner, E. (1981). Microbodies: peroxisomes and glyoxysomes. *J Cell Biol* 91:271–283.
12. Heupel, R., and Heldt, H.W. (1994). Protein organization in the matrix of leaf peroxisomes. A multi-enzyme complex involved in photorespiratory metabolism. *Eur J Biochem* 220:165–172.
13. Wanders, R.J., and Waterham, H.R. (2005). Peroxisomal disorders I: biochemistry and genetics of peroxisome biogenesis disorders. *Clin Genet* 67:107–133.
14. Lazarow, P.B., and Fujiki, Y. (1985). Biogenesis of peroxisomes. *Annu Rev Cell Biol* 1:489–530.
15. Hoepfner, D., Schildknegt, D., Braakman, I., Philippsen, P., and Tabak, H.F. (2005). Contribution of the endoplasmic reticulum to peroxisome formation. *Cell* 122:89–95.
16. Lam, S.K., Yoda, N., and Schekman, R. (2010). A vesicle carrier that mediates peroxisome protein traffic from the endoplasmic reticulum. *Proc Natl Acad Sci USA* 107:21523–21528.
17. Motley, A.M., and Hettema, E.H. (2007). Yeast peroxisomes multiply by growth and division. *J Cell Biol* 178:399–410.
18. McNew, J.A., and Goodman, J.M. (1994). An oligomeric protein is imported into peroxisomes *in vivo*. *J Cell Biol* 127:1245–1257.
19. Glover, J.R., Andrews, D.W., and Rachubinski, R.A. (1994). Saccharomyces cerevisiae peroxisomal thiolase is imported as a dimer. *Proc Natl Acad Sci USA* 91:10541–10545.
20. Brocard, C., Kragler, F., Simon, M.M., Schuster, T., and Hartig, A. (1994). The tetratricopeptide repeat-domain of the Pas10 protein of *Saccharomyces cerevisiae* is essential for binding the peroxisomal targeting signal-SKL. *Biochem Biophys Res Commun* 204:1016–1022.
21. Terlecky, S.R., Nuttley, W.M., McCollum, D., Sock, E., and Subramani, S. (1995). The *Pichia pastoris* peroxisomal protein Pas8p is the receptor for the C-terminal tripeptide peroxisomal targeting signal. *EMBO J* 14:3627–3634.
22. Rehling, P., Marzioch, M., Niesen, F., Wittke, E., Veenhuis, M., and Kunau, W.-H. (1996). The import receptor for the peroxisomal targeting signal 2 (PTS2) in *Saccharomyces cerevisiae* is encoded by the *PAS7* gene. *EMBO J* 15:2901–2913.
23. Meinecke, M., *et al.* (2010). The peroxisomal importomer constitutes a large and highly dynamic pore. *Nat Cell Biol* 12:273–277.
24. Platta, H.W., El Magraoui, F., Schlee, D., Grunau, S., Girzalsky, W., and Erdmann, R. (2007). Ubiquitination of the peroxisomal import receptor Pex5p is required for its recycling. *J Cell Biol* 177:197–204.

25. Platta, H.W., Debelyy, M.O., El Magraoui, F., and Erdmann, R. (2008). The AAA peroxins Pex1p and Pex6p function as dislocases for the ubiquitinated peroxisomal import receptor Pex5p. *Biochem Soc Trans* 36:99–104.
26. Dodt, G., and Gould, S.J. (1996). Multiple *PEX* genes are required for proper subcellular distribution and stability of Pex5p, the PTS1 receptor: evidence that PTS1 protein import is mediated by a cycling receptor. *J Cell Biol* 135:1763–1774.
27. Marzioch, M., Erdmann, R., Veenhuis, M., and Kunau, W.-H. (1994). *PAS7* encodes a novel yeast member of the WD-40 protein family essential for import of 3-oxoacyl-CoA thiolase, a PTS2-containing protein, into peroxisomes. *EMBO J* 13:4908–4918.
28. Schliebs, W., Girzalsky, W., and Erdmann, R. (2010). Peroxisomal protein import and ERAD: variations on a common theme. *Nat Rev* 11:885–890.
29. Erdmann, R., and Blobel, G. (1996). Identification of Pex13p a peroxisomal membrane receptor for the PTS1 recognition factor. *J Cell Biol* 135:111–121.
30. Gould, S.J., Kalish, J.E., Morrell, J.C., Bjorkman, J., Urquhart, A.J., and Crane, D.I. (1996). Pex13p is an SH3 protein of the peroxisome membrane and a docking factor for the predominantly cytoplasmic PTS1 receptor. *J Cell Biol* 135:85–95.
31. Santos, M.J., Imanaka, T., Shio, H., Small, G.M., and Lazarow, P.B. (1988). Peroxisomal membrane ghosts in Zellweger syndrome-aberrant organelle assembly. *Science* 239:1536–1538.
32. Brown, L.A., and Baker, A. (2003). Peroxisome biogenesis and the role of protein import. *J Cell Mol Med* 7:388–400.
33. Schrader, M., and Fahimi, H.D. (2008). The peroxisome: still a mysterious organelle. *Histochem Cell Biol* 129:421–440.
34. Baerends, R.J.S., *et al.* (1996). The *Hansenula polymorpha PER9* gene encodes a peroxisomal membrane protein essential for peroxisome assembly and integrity. *J Biol Chem* 271:8887–8894.
35. Eitzen, G.A., Szilard, R.K., and Rachubinski, R.A. (1997). Enlarged peroxisomes are present in oleic acid-grown *Yarrowia lipolytica* overexpressing the *PEX16* gene encoding an intraperoxisomal peripheral membrane peroxin. *J Cell Biol* 137:1265–1278.
36. Ghaedi, K., Tamura, S., Okumoto, K., Matsuzono, Y., and Fujiki, Y. (2000). The peroxin Pex3p initiates membrane assembly in peroxisome biogenesis. *Mol Biol Cell* 11:2085–2103.
37. Götte, K., *et al.* (1998). Pex19p, a farnesylated protein essential for peroxisome biogenesis. *Mol Cell Biol* 18:616–628.
38. Höhfeld, J., Veenhuis, M., and Kunau, W.-H. (1991). *PAS3*, a *Saccharomyces cerevisiae* gene encoding a peroxisomal integral membrane protein essential for peroxisome biogenesis. *J Cell Biol* 114:1167–1178.
39. Honsho, M., Tamura, S., Shimozawa, N., Suzuki, Y., Kondo, N., and Fujiki, Y. (1998). Mutation in *PEX16* is causal in the peroxisome-deficient zellweger syndrome of complementation group D. *Am J Hum Genet* 63:1622–1630.
40. Matsuzono, Y., *et al.* (1999). Human PEX19: cDNA cloning by functional complementation, mutation analysis in a patient with Zellweger syndrome, and potential role in peroxisomal membrane assembly. *Proc Natl Acad Sci USA* 96:2116–2121.
41. South, S.T., and Gould, S.J. (1999). Peroxisome synthesis in the absence of preexisting peroxisomes. *J Cell Biol* 144:255–266.
42. James, G.L., Goldstein, J.L., Pathak, R.K., Anderson, R.G.W., and Brown, M.S. (1994). PxF, a prenylated protein of peroxisomes. *J Biol Chem* 269:14182–14190.
43. Erdmann, R., Veenhuis, M., Mertens, D., and Kunau, W.-H. (1989). Isolation of peroxisome-deficient mutants of *Saccharomyces cerevisiae*. *Proc Natl Acad Sci USA* 86:5419–5423.

44. Kammerer, S., *et al.* (1997). Genomic organization and molecular characterization of a gene encoding HsPXF, a human peroxisomal farnesylated protein. *Genomics* 45:200–210.
45. Fransen, M., Wylin, T., Brees, C., Mannaerts, G.P., and Van Veldhoven, P.P. (2001). Human Pex19p binds peroxisomal integral membrane proteins at regions distinct from their sorting sequences. *Mol Cell Biol* 21:4413–4424.
46. Hadden, D.A., *et al.* (2006). Arabidopsis PEX19 is a dimeric protein that binds the peroxin PEX10. *Mol Membr Biol* 23:325–336.
47. Halbach, A., *et al.* (2006). Targeting of the tail-anchored peroxisomal membrane proteins PEX26 and PEX15 occurs through C-terminal PEX19-binding sites. *J Cell Sci* 119:2508–2517.
48. Halbach, A., Lorenzen, S., Landgraf, C., Volkmer-Engert, R., Erdmann, R., and Rottensteiner, H. (2005). Function of the PEX19-binding site of human ALDP as targeting motif in man and yeast: PMP targeting is evolutionarily conserved. *J Biol Chem* 280:21176–21182.
49. Sacksteder, K.A., Jones, J.M., South, S.T., Li, X., Liu, Y., and Gould, S.J. (2000). PEX19 binds multiple peroxisomal membrane proteins, is predominantly cytoplasmic, and is required for peroxisome membrane synthesis. *J Cell Biol* 148:931–944.
50. Snyder, W.B., Koller, A., Choy, A.J., and Subramani, S. (2000). The peroxin Pex19p interacts with multiple, integral membrane proteins at the peroxisomal membrane. *J Cell Biol* 149:1171–1178.
51. Shibata, H., Kashiwayama, Y., Imanaka, T., and Kato, H. (2004). Domain architecture and activity of human Pex19p, a chaperone-like protein for intracellular trafficking of peroxisomal membrane proteins. *J Biol Chem* 279:38486–38494.
52. Fransen, M., Vastiau, I., Brees, C., Brys, V., Mannaerts, G.P., and Van Veldhoven, P.P. (2005). Analysis of human Pex19p's domain structure by pentapeptide scanning mutagenesis. *J Mol Biol* 346:1275–1286.
53. Matsuzono, Y., Matsuzaki, T., and Fujiki, Y. (2006). Functional domain mapping of peroxin Pex19p: interaction with Pex3p is essential for function and translocation. *J Cell Sci* 119:3539–3550.
54. Mayerhofer, P.U., Kattenfeld, T., Roscher, A.A., and Muntau, A.C. (2002). Two splice variants of human PEX19 exhibit distinct functions in peroxisomal assembly. *Biochem Biophys Res Commun* 291:1180–1186.
55. Schueller, N., *et al.* (2010). The peroxisomal receptor Pex19p forms a helical mPTS recognition domain. *EMBO J* 29:2491–2500.
56. Banerjee, S.K., Kessler, P.S., Saveria, T., and Parsons, M. (2005). Identification of trypanosomatid PEX19: functional characterization reveals impact on cell growth and glycosome size and number. *Mol Biochem Parasitol* 142:47–55.
57. Zhang, F.L., and Casey, P.J. (1996). Protein prenylation: molecular mechanisms and functional consequences. *Annu Rev Biochem* 65:241–269.
58. Rucktäschel, R., *et al.* (2009). Farnesylation of pex19p is required for its structural integrity and function in peroxisome biogenesis. *J Biol Chem* 284:20885–20896.
59. Vastiau, I.M., *et al.* (2006). Farnesylation of Pex19p is not essential for peroxisome biogenesis in yeast and mammalian cells. *Cell Mol Life Sci* 63:1686–1699.
60. Otzen, M., *et al.* (2004). *Hansenula polymorpha* Pex19p is essential for the formation of functional peroxisomal membranes. *J Biol Chem* 279:19181–19190.
61. Lambkin, G.R., and Rachubinski, R.A. (2001). *Yarrowia lipolytica* cells mutant for the peroxisomal peroxin Pex19p contain structures resembling wild-type peroxisomes. *Mol Biol Cell* 12:3353–3364.
62. Snyder, W.B., *et al.* (1999). Pex19p interacts with Pex3p and Pex10p and is essential for peroxisome biogenesis in *Pichia pastoris*. *Mol Biol Cell* 10:1745–1761.

63. Jones, J.M., Morrell, J.C., and Gould, S.J. (2004). PEX19 is a predominantly cytosolic chaperone and import receptor for class 1 peroxisomal membrane proteins. *J Cell Biol* 164:57–67.

64. Schliebs, W., and Kunau, W.H. (2004). Peroxisome membrane biogenesis: the stage is set. *Curr Biol* 14:R397–R399.

65. Kashiwayama, Y., *et al.* (2005). Role of Pex19p in the targeting of PMP70 to peroxisome. *Biochim Biophys Acta* 1746:116–128.

66. Fransen, M., Vastiau, I., Brees, C., Brys, V., Mannaerts, G.P., and Van Veldhoven, P.P. (2004). Potential role for Pex19p in assembly of PTS-receptor docking complexes. *J Biol Chem* 279:12615–12624.

67. Ito, M., *et al.* (2004). Interaction of a farnesylated protein with renal type IIa Na/Pi co-transporter in response to parathyroid hormone and dietary phosphate. *Biochem J* 377:607–616.

68. Sugihara, T., Kaul, S.C., Kato, J., Reddel, R.R., Nomura, H., and Wadhwa, R. (2001). Pex19p dampens the p19ARF-p53-p21WAF1 tumor suppressor pathway. *J Biol Chem* 276:18649–18652.

69. Honsho, M., Hiroshige, T., and Fujiki, Y. (2002). The membrane biogenesis peroxin Pex16p. Topogenesis and functional roles in peroxisomal membrane assembly. *J Biol Chem* 277:44513–44524.

70. Kim, P.K., Mullen, R.T., Schumann, U., and Lippincott-Schwartz, J. (2006). The origin and maintenance of mammalian peroxisomes involves a de novo PEX16-dependent pathway from the ER. *J Cell Biol* 173:521–532.

71. Soukupova, M., Sprenger, C., Gorgas, K., Kunau, W.-H., and Dodt, G. (1999). Identification and characterization of the human peroxin PEX3. *Eur J Cell Biol* 78:357–374.

72. Hunt, J.E., and Trelease, R.N. (2004). Sorting pathway and molecular targeting signals for the Arabidopsis peroxin 3. *Biochem Biophys Res Commun* 314:586–596.

73. Haan, G.J., *et al.* (2002). *Hansenula polymorpha* Pex3p is a peripheral component of the peroxisomal membrane. *J Biol Chem* 277:26609–26617.

74. Fang, Y., Morrell, J.C., Jones, J.M., and Gould, S.J. (2004). PEX3 functions as a PEX19 docking factor in the import of class I peroxisomal membrane proteins. *J Cell Biol* 164:863–875.

75. Chang, J., *et al.* (2009). Pex3 peroxisome biogenesis proteins function in peroxisome inheritance as class V myosin receptors. *J Cell Biol* 187:233–246.

76. Munck, J.M., Motley, A.M., Nuttall, J.M., and Hettema, E.H. (2009). A dual function for Pex3p in peroxisome formation and inheritance. *J Cell Biol* 187:463–471.

77. Rucktäschel, R., Halbach, A., Girzalsky, W., Rottensteiner, H., and Erdmann, R. (2010). De novo synthesis of peroxisomes upon mitochondrial targeting of Pex3p. *Eur J Cell Biol* 89:947–954.

78. Rottensteiner, H., *et al.* (2004). Peroxisomal membrane proteins contain common Pex19p-binding sites that are an integral part of their targeting signals (mPTS). *Mol Biol Cell* 7:3406–3417.

79. Baerends, R.J., Faber, K.N., Kram, A.M., Kiel, J.A., van Der Klei, I.J., and Veenhuis, M. (2000). A stretch of positively charged amino acids at the N terminus of *hansenula polymorpha* pex3p is involved in incorporation of the protein into the peroxisomal membrane. *J Biol Chem* 275:9986–9995.

80. Jones, J.M., Morrell, J.C., and Gould, S.J. (2001). Multiple distinct targeting signals in integral peroxisomal membrane proteins. *J Cell Biol* 153:1141–1150.

81. Wang, X., Unruh, M.J., and Goodman, J.M. (2001). Discrete targeting signals direct Pmp47 to oleate-induced peroxisomes in *Saccharomyces cerevisiae*. *J Biol Chem* 276:10897–10905.

82. Honsho, M., and Fujiki, Y. (2000). Topogenesis of peroxisomal membrane protein requires a short, positively charged intervening-loop sequence and flanking hydrophobic segments: study using human membrane protein PMP34. *J Biol Chem* 276:9375–9382.
83. Girzalsky, W., Hoffmann, L.S., Schemenewitz, A., Nolte, A., Kunau, W.H., and Erdmann, R. (2006). Pex19p-dependent targeting of Pex17p, a peripheral component of the peroxisomal protein import machinery. *J Biol Chem* 281:19417–19425.
84. Van Ael, E., and Fransen, M. (2006). Targeting signals in peroxisomal membrane proteins. *Biochim Biophys Acta* 1763:1629–1638.
85. Pinto, M.P., et al. (2006). The import competence of a peroxisomal membrane protein is determined by Pex19p before the docking step. *J Biol Chem* 281:34492–344502.
86. Matsuzono, Y., and Fujiki, Y. (2006). In vitro transport of membrane proteins to peroxisomes by shuttling receptor Pex19p. *J Biol Chem* 281:36–42.
87. Diestelkötter, P., and Just, W.W. (1993). *In vitro* insertion of the 22-kD peroxisomal membrane protein into isolated rat liver peroxisomes. *J Cell Biol* 123:1717–1725.
88. Fujiki, Y., Matsuzono, Y., Matsuzaki, T., and Fransen, M. (2006). Import of peroxisomal membrane proteins: the interplay of Pex3p- and Pex19p-mediated interactions. *Biochim Biophys Acta* 1763:1639–1646.
89. Halbach, A., Rucktaschel, R., Rottensteiner, H., and Erdmann, R. (2009). The N-domain of Pex22p can functionally replace the Pex3p N-domain in targeting and peroxisome formation. *J Biol Chem* 284:3906–3916.
90. South, S.T., Baumgart, E., and Gould, S.J. (2001). Inactivation of the endoplasmic reticulum protein translocation factor, Sec61p, or its homolog, Ssh1p, does not affect peroxisome biogenesis. *Proc Natl Acad Sci USA* 98:12027–12031.
91. van der Zand, A., Braakman, I., and Tabak, H.F. (2010). Peroxisomal membrane proteins insert into the endoplasmic reticulum. *Mol Biol Cell* 21:2057–2065.
92. Titorenko, V.I., and Rachubinski, R.A. (1998). Mutants of the yeast *Yarrowia lipolytica* defective in protein exit from the endoplasmic reticulum are also defective in peroxisome biogenesis. *Mol Cell Biol* 18:2789–2803.
93. Salomons, F.A., van der Klei, I.J., Kram, A.M., Harder, W., Veenhuis, M., and Brefeldin, A. (1997). interferes with peroxisomal protein sorting in the yeast *Hansenula polymorpha*. *FEBS Lett* 411:133–139.
94. Voorn-Brouwer, T., Kragt, A., Tabak, H.F., and Distel, B. (2001). Peroxisomal membrane proteins are properly targeted to peroxisomes in the absence of COPI- and COPII-mediated vesicular transport. *J Cell Sci* 114:2199–2204.
95. Perry, R.J., Mast, F.D., and Rachubinski, R.A. (2009). Endoplasmic reticulum-associated secretory proteins Sec20p, Sec39p, and Dsl1p are involved in peroxisome biogenesis. *Eukaryot Cell* 8:830–843.
96. Ridley, A.J. (1997). The GTP-binding protein Rho. *Int J Biochem Cell Biol* 29:1225–1229.
97. Wennerberg, K., and Der, C.J. (2004). Rho-family GTPases: it's not only Rac and Rho (and I like it). *J Cell Sci* 117:1301–1312.
98. Michaelson, D., Silletti, J., Murphy, G., D'Eustachio, P., Rush, M., and Philips, M.R. (2001). Differential localization of Rho GTPases in live cells: regulation by hypervariable regions and RhoGDI binding. *J Cell Biol* 152:111–126.
99. Symons, M., and Settleman, J. (2000). Rho family GTPases: more than simple switches. *Trends Cell Biol* 10:415–419.
100. Yamochi, W., Tanaka, K., Nonaka, H., Maeda, A., Musha, T., and Takai, Y. (1994). Growth site localization of Rho1 small GTP-binding protein and its involvement in bud formation in Saccharomyces cerevisiae. *J Cell Biol* 125:1077–1093.

101. McCaffrey, M., *et al.* (1991). The small GTP-binding protein Rho1p is localized on the Golgi apparatus and post-Golgi vesicles in Saccharomyces cerevisiae. *J Cell Biol* 115:309–319.

102. Qadota, H., Anraku, Y., Botstein, D., and Ohya, Y. (1994). Conditional lethality of a yeast strain expressing human RHOA in place of RHO1. *Proc Natl Acad Sci USA* 91:9317–9321.

103. Marelli, M., *et al.* (2004). Quantitative mass spectrometry reveals a role for the GTPase Rho1p in actin organization on the peroxisome membrane. *J Cell Biol* 167:1099–1112.

104. Schollenberger, L., *et al.* (2010). RhoA regulates peroxisome association to microtubules and the actin cytoskeleton. *PLoS ONE* 5:e13886.

105. Vinh, D.B., Ko, D.C., Rachubinski, R.A., Aitchison, J.D., and Miller, S.I. (2010). Expression of the Salmonella spp. virulence factor SifA in yeast alters Rho1 activity on peroxisomes. *Mol Biol Cell* 21:3567–3577.

106. Rottensteiner, H., Stein, K., Sonnenhol, E., and Erdmann, R. (2003). Conserved function of pex11p and the novel pex25p and pex27p in peroxisome biogenesis. *Mol Biol Cell* 14:4316–4328.

5

Lipid Modification of Ras Superfamily GTPases: Not Just Membrane Glue

EMILY J. CHENETTE[a] • CHANNING J. DER[b]

[a]*Nature Publishing Group*
Cambridge, Massachusetts, USA

[b]*Department of Pharmacology, Lineberger Comprehensive Cancer Center*
University of North Carolina at Chapel Hill, Chapel Hill
North Carolina, USA

I.   Abstract

The activity of Ras-related GTPases supports a myriad of physiological and pathophysiological cellular events. Ras-mediated signaling is regulated by guanine nucleotide binding and hydrolysis, which governs binding to and activation of downstream effector proteins. However, the subcellular localization of active Ras proteins—and hence the identity of nearby effector and regulatory proteins—can determine the biological outcome of these signaling events. Most Ras superfamily members associate with cellular membranes, and precise subcellular localization is determined by residues at the C-terminus of the protein (and modification of these residues) that increase the affinity for membranes; these can include palmitoylation of cysteines, stretches of basic residues, and prenylation of the extreme C-terminal cysteine residue. In this chapter, we discuss the process, regulation, and biological consequence of lipid modification of Ras superfamily small GTPases, with a focus on the role of isoprenoid modification of Rho and Rab family

DOI: 10.1016/B978-0-12-381339-8.00005-6

members. For each subfamily, we describe the responsible prenyltransferases and the effects of prenylation on protein localization and activity.
Finally, we chronicle the development of prenylation inhibitors as potential
anticancer therapeutics. From the literature reviewed in this chapter, prenylation emerges as one mechanism that imparts unique localization and
function to these otherwise highly conserved small GTPases.

II.   Introduction

The Ras superfamily of small GTPases includes more than 150 human
members that can be subdivided into several structurally or functionally
related families conserved in evolution: Ras, Rho, Rab, Arf, and Ran
(Figure 5.1A) [1,2]. Ras superfamily proteins are highly conserved within
the core G domain and share a common topology that governs their activity
and interaction with regulatory and effector proteins (Figure 5.1B). The
activity of all Ras superfamily proteins is regulated by binding to and
hydrolysis of guanine nucleotides. In the GTP-bound state, Ras proteins
adopt a conformation that permits binding to downstream effectors, stimulating cytoplasmic signaling cascades to affect a range of developmental and
biological outcomes. Hydrolysis of the bound GTP molecule stimulates a
conformational change that precludes effector access and attenuates Ras-
mediated signaling. The nucleotide status of Ras proteins is regulated by
both their own slow intrinsic GTP exchange and hydrolysis activities, which
are greatly accelerated by guanine nucleotide exchange factors (GEFs) and
GTPase-activating proteins (GAPs), respectively. GEFs activate Ras proteins by stimulating the release of bound GDP, permitting reassociation
with GTP. GAPs stimulate the intrinsic hydrolysis activity of Ras proteins to
inhibit Ras-mediated signaling [3–5]. GAPs and GEFs have not been identified for all Ras family proteins and might not be essential for activity in
every case. Structurally distinct but biochemically similar GEFs and GAPs
regulate the GDP–GTP cycle of each Ras superfamily branch (Figure 5.1C).

In sharp contrast to the sequence and structural similarity found
throughout the core G domain, the C-terminal hypervariable region of
Ras, Rho, and Rab family members exhibits significant sequence differences (Figure 5.2A). With the exception of the Arf family members, it is this
region that governs membrane association and subcellular localization of
Ras proteins. Membrane targeting of canonical Ras, Rho, and Rab proteins
is mediated by C-terminal cysteine-containing motifs that signal for prenyltransferase posttranslational modification by either C15 farnesyl or C20
geranylgeranyl lipid addition (see Section III).

While the isoprenoid modification is necessary for membrane association, it is not sufficient for proper subcellular localization required for

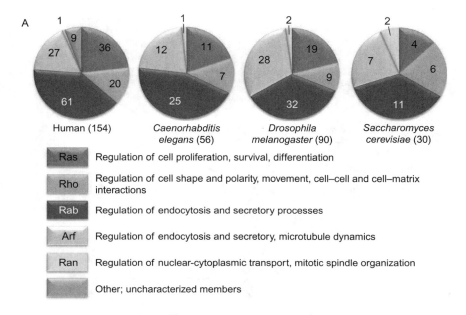

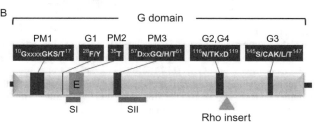

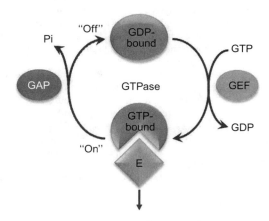

FIG. 5.1. (Continued)

C

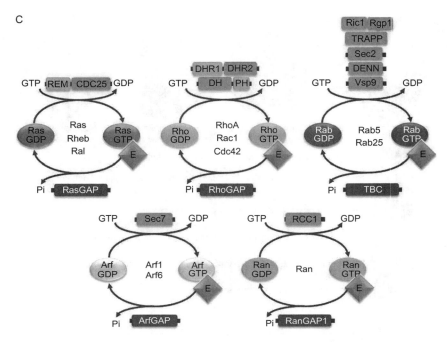

FIG. 5.1. Ras superfamily small GTPases. (A) The human Ras superfamily comprised over 150 members which is divided into five major branches on the basis of sequence and functional similarities [1], with functional orthologs found in *Caenorhabditis elegans*, *Drosophila melanogaster*, and *Saccharomyces cerevisiae* [179,180]. The three Ras proteins (H-Ras, N-Ras, K-Ras) are the founding members of the superfamily and the best studied of the Ras family (36 human members). RhoA, Rac1, and Cdc42 are the best studied of the Rho family (20 members). Rab proteins comprise the largest family (61 human members) and together with the Arf family (27 human members) are regulators of vesicular trafficking. The Ran family is the smallest, with only one human member. (B) The ∼20-kDa core G domain (corresponding to Ras residues 4–166) is conserved among all Ras superfamily proteins and is involved in GTP binding and hydrolysis [181]. This domain comprised five conserved guanine nucleotide consensus sequence elements (Ras residue numbering) involved in binding phosphate/Mg^{2+} (PM) or the guanine base (G). The switch I (SI; Ras residues 30–38) and II (SII; 59–76) regions change in conformation during GDP–GTP cycling and contribute to preferential effector (E) binding to the GTP-bound state and the core effector domain (Ras residues 32–40). Ras superfamily proteins possess intrinsic guanine nucleotide exchange and GTP hydrolysis activities. However, these activities are too weak to allow efficient and rapid cycling between their active GTP-bound and inactive GDP-bound states. Guanine nucleotide exchange factors (GEFs) and GTPase-activating proteins (GAPs) catalyze and regulate these intrinsic activities. SI and SII sequences are also involved in GEF and GAP interaction. Rho proteins are characterized by an up to 13 amino acid "Rho insert" sequence corresponding to Ras residues 122/123 and not found in any other superfamily branches. (C) Members of the different branches of the Ras superfamily are regulated by GEFs and GAPs with structurally distinct catalytic domains [3,4,182–185]. RasGEFs are characterized by CDC25 homology

GTPase biological activity. Additionally, a second membrane-targeting signal, comprising basic residues and/or the modification of cysteine residues—either through palmitoylation or prenylation—present in the C-terminal hypervariable region, is needed (e.g., Ras; Figure 5.2C). Positively charged lysine amino acids increase Ras membrane association by interacting with the negatively charged membrane phospholipids. Palmitoylation of cysteine residues within the Ras and Rho hypervariable regions also increases membrane association, as the palmitate moieties integrate into target membranes (Figure 5.2A). However, palmitoylation is reversible, and the electrostatic interactions between the basic residues and acidic phospholipids are not sufficient to anchor Ras to cellular membranes. Therefore, most Ras proteins require prenylation of the C-terminal cysteine residue for stable association with cellular membranes [6–9]. Indeed, preventing prenylation of Ras, either by mutation of the key cysteine residue or by pharmacologic inhibition of prenyl biosynthesis, causes its cytoplasmic mislocalization and suppresses Ras-mediated signaling [10]. Though prenylation is necessary and sufficient for membrane targeting, proper subcellular localization and signaling activity depend on the secondary signal (palmitoylation of cysteines or the presence of basic residues) that together with flanking sequences target the GTPase to its specific membrane subdomain [11–15]. Thus, the signaling and biological diversity achieved by otherwise highly similar Ras family members is potentially mediated by their different subcellular distributions, which in turn influences their access to GEFs, GAPs, and effector proteins [16–18]. Here, we focus primarily on the prenylation of Ras and Rho family members.

III. Ras Superfamily Prenyltransferases

The addition of farnesyl or geranylgeranyl moieties is catalyzed by enzymes known as farnesyltransferase (FTase) and geranylgeranyltransferase (GGTase), respectively (Figure 5.2D). There are two GGTase enzymes:

---

domains (named after the *S. cerevisiae* CDC25 gene product). RhoGEFs are subdivided into two major structurally distinct families. Dbl family RhoGEFs possess a tandem Dbl homology (DH) catalytic and pleckstrin homology (PH) regulatory domain topology. DOCK family RhoGEFs are characterized by two regions of high-sequence conservation that are designated Dock-homology region regulatory DHR-1 and catalytic DHR-2 domains. RabGEFs are more heterogeneous and are characterized by several structurally distinct GEFs that include Vps9, DENN (differentially expressed in normal and neoplastic cells) [186] and Sec2 homology domain-containing proteins, and the yeast TRAPP and Ric1/Rgp1 complexes. ArfGEFs possess a Sec7 domain (named after the *S. cerevisiae* SEC7 gene product) [187]. Each Ras superfamily branch is also regulated by structurally distinct GAPs, including TBC (Tre-2/Bub2/Cdc16) domain-containing proteins that act as RabGEFs [188].

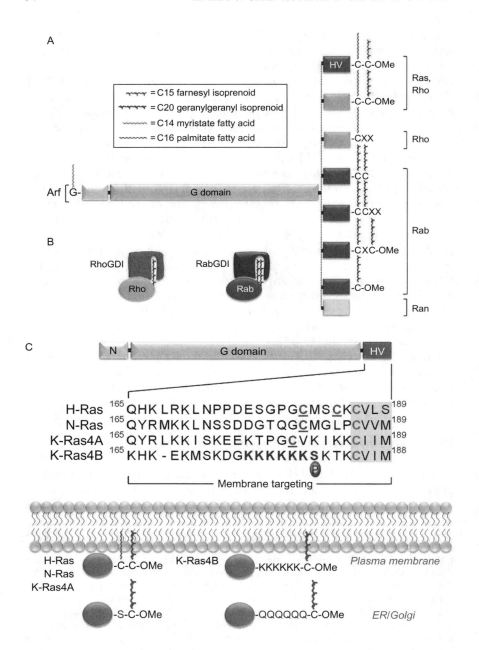

D

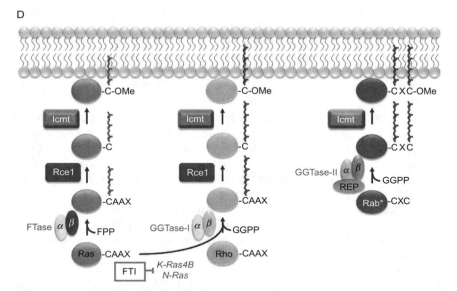

Fig. 5.2. Posttranslational lipid modification of Ras superfamily proteins. (A) Lipid modification of Ras superfamily proteins. Ras and Rho family proteins have additional C-terminal sequences that commonly terminate with a CAAX tetrapeptide sequence that signals for a series of posttranslational modifications: isoprenoid addition to the cysteine residue, proteolytic removal of the AAX residues, and carboxylmethylation of the prenylated cysteine. Some members are further modified posttranslationally by palmitoyl acyltransferase-catalyzed addition of a palmitate fatty acid in the hypervariable (HV) sequence that contributes to membrane association. Rab proteins also terminate with cysteine-containing motifs (-CXC, -CC, -CCXX, and -CXXX) that are modified by addition of one or two geranylgeranyl lipids, with some undergoing carboxylmethylation. Arf family proteins are characterized by an N-terminal extension involved in membrane interaction, with some modified by addition of a myristate fatty acid. After the N-terminal methionine has been removed by the Met-amino peptidase, N-myristoyltransferase catalyzes cotranslational addition of a myristoyl moiety to the now N-terminal glycine residue. The N-terminus changes conformation during GDP–GTP cycling, with GTP binding exposing the myristate to facilitate membrane binding. Ran is not lipid modified but contains a C-terminal extension that changes conformation during GDP–GTP cycling and is essential for its function in nuclear transport. (B) Guanine nucleotide dissociation inhibitors (GDI). Rho and Rab protein membrane association is regulated by Rho- or Rab-specific GDIs that recognize SI/SII GTPase sequences and additionally mask the isoprenoid modification(s) to prevent membrane association. (C) Ras protein C-terminal sequence divergence. The Ras C-terminal 24–25 residues, with the exception of the cysteine residue (highlighted in bolded red text) of the CAAX motif (yellow shaded box), are completely divergent and comprise the hypervariable (HV) membrane-binding domain. Additional membrane-targeting information includes the palmitoylated cysteine residue (underlined in bolded green text), a polybasic stretch and a protein kinase C phosphorylation site in K-Ras4B. Mutational disruption of these second targeting signals results in mislocalization of Ras proteins to endoplasmic reticulum (ER)/Golgi membranes. (D) Ras superfamily C-terminal

GGTase-I and GGTase-II (also called RabGGTase). The former catalyzes geranylgeranylation of Ras and Rho family proteins. Together with an accessory protein known as Rab escort protein (REP), GGTase-II geranyl-geranylates Rab proteins. FTase and GGTase enzymes are composed of one catalytic β subunit and one regulatory α subunit, which is common between FTase and GGTase-I (but not GGTase-II) [19]. Chapter 13 describes these prenyltransferases and their structure in detail.

Prenylated Ras and Rho proteins terminate in a common sequence motif called the CAAX box, where C is cysteine, A is any aliphatic amino acid, and X is the terminal residue that dictates prenyltransferase specificity (Figure 5.2C and D; Table 5.1). The CAAX tetrapeptide sequence is necessary and sufficient for recognition by the two CAAX prenyltrans-ferases. In contrast, most Rab family proteins instead terminate in tandem (-CC, -CXC, or -CCXX) cysteine-containing motifs that are doubly mod-ified by geranylgeranyl group addition. A few Rab proteins have single cysteine-containing CAAX motifs (e.g., Rab8 and Rab13 instead possess a canonical CAAX motif [20,21]).

Prenyltransferase enzymes catalyze the addition of FPP or GGPP prenyl groups to the cysteine residue(s) (Figure 5.2D). CAAX-containing proteins undergo further C-terminal processing at the endoplasmic reticulum (ER) following prenylation: the Rce1 (Ras converting enzyme 1) endoprotease cleaves the -AAX residues [22], and the now-terminal prenylated cysteine residue is methylated by Icmt (isoprenylcysteine carboxyl methyltransfer-ase) [23]. CXC- but not CC-terminating Rab proteins are also Icmt sub-strates. These series of modifications increase the overall hydrophobicity of the Ras and Rho C-terminus and facilitate membrane association.

Given that all prenylated Ras and Rho proteins terminate in a conserved CAAX motif, what determines the identity of prenyl modification? Bio-chemical and structural studies have suggested that the terminal X amino acid in the CAAX box imparts specificity [24–29]. Bioinformatics

---

prenylation. The two CAAX prenyltransferases, FTase and GGTase-I, utilize farnesyl pyro-phosphate (FPP) and geranylgeranyl pyrophosphate (GGPP) and catalyze addition of a farnesyl or geranylgeranyl group, respectively, to CAAX-terminating Ras or Rho family proteins. They then undergo modification by Golgi-associated Rce1 and Icmt. The Rab pre-nyltransferase catalyzes addition of one (CXXX) or two (CC, CXC, CCXX) geranylgeranyl groups to C-terminal cysteine-containing sequences, with a subset additionally modified by Rce1 (CXXX) and/or Icmt (CXXX, CXC). In contrast to the heterodimeric CAAX prenyl-transferases which directly interact with their GTPase substrates, the heterodimeric GGTase-II requires interaction with REP, with then facilitates Rab GTPase recognition. Hence, unlike the sufficiency of the Ras and Rho CAAX sequences to act as a substrate, the Rab C-terminal peptides alone are not sufficient for GGTase-II recognition. (See color plate section in the back of the book.)

TABLE 5.1

The C-Termini of Ras and Rho GTPases are Substrates for FTase and GGTase-I

GTPase	C-terminal 25 amino acids[a]	FTase	GGTase-I	Reference
H-Ras	QHKLRKLNPPDESGPGCMSCKCVLS	+	–	[36]
K-Ras4A	QYRLKKISKEEKTPGCVKIKKCIIM	+	–	[36]
K-Ras4B	RKHKEKMSKDGKKKKKKSKTKCVIM	+	+	[36–39]
N-Ras	QYRMKKLNSSDDGTQGCMGLPCVVM	+	+	[36–39]
Rap1a	DLVRQINRKTPVEKKKPKKKSCLLL	–	+	[62,63]
Rap1b	DLVRQINRKTPVPGKARKKSSCQLL	–	+	[62,63]
Rap2a	EIVRQMNYAAQPDKDDPCCSACNIQ	+	+	[62,63]
Rap2b	EIVRQMNYAAQPNGDEGCCSACVII	+	+	[62,63]
Rap2c	EIVRQMNYSSLPEKQDQCCTTCVVQ	–	+	[62,63]
RalA	DSKEKNGKKKRKSLAKRIRERCCIL	–	+	[67]
RalB	SENKDKNGKKSSKNKKSFKERCCLL	–	+	[68]
Rheb1	FRRIILEAEKMDGAASOGKSSCSVM	+	–	[76–79]
Rheb2	IFTKVIQEIARVENSYGOERRCHLM	+	–	[78]
R-Ras	QEQELPPSPSAPRKKGGGCPCVLL	+	+	[82]
TC21/R-Ras2	EQECPPSPEPTRKEKDKKGCHCVIF	+	+	[89]
M-Ras/R-Ras3	QKKKKTKWRGDRATGTHKLQCVIL	–	+	[86]
Di-Ras1/Rig	SLNIDGKRSGKQKRTDRVKGKCTLM	+	+	[92]
Di-Ras2	SLQIDGKKSKQQKREKLKGKCVIM	Unknown	Unknown	
E-Ras	EPMARSCREKTRHOKATCHCGCSVA	Unknown	Unknown	
Noey2/ARHI/Di-Ras3	GLQEPEKKSOMPNTTEKLLLDKCIIM	+	–	E.J. Chenette and C.J. Der (unpublished observations)
RasD1/DexRas/AGS1	SDLMYIREKASAGSQAKDKERCVIS	Unknown	Unknown	
RasD2/Rhes	SDLKYIKAKVLREGQARERDKCTIQ	+	–	[96,97]
Rerg	GKTRRSSTTHVKQAINKMLTKISS	–	–	[98]
Rad	RIVARNSRKMAFRAKSKSCHDLSVL	–	–	[99–102]
Gem/Kir	KIVAKNNKNMAFKLKSKSCHDLSVL	–	–	[99–102]

(Continued)

TABLE 5.1 (*CONTINUED*)

GTPase	C-terminal 25 amino acids[a]	FTase	GGTase-I	Reference
Rem1/Ges	RLTARSARRRALKARSKSCHNLAVL	–	–	[99–102]
Rem2	LANLVPRNAKFFKQRSRSCHDLSVL	–	–	[99–102]
NKIRas1/κB-Ras1	LSQPQSKSSFPLPGRKNKGNSNSEN	–	–	[103]
NKIRas2/κB-Ras2	KMTQPQSKSAFPLSRKNKGSGSLDG	–	–	[103]
Rit1	KKSKPKNSVWKRLKSPFRKKKDSVT	–	–	[15,104–106]
Rin/Rit2	KKLKRKDSLWKKLKGSLKKKRENMT	–	–	[15,104–106]
RasL10A/RRP22	LVRARPAHPALRLQGALHPARCSLM	+	–	[95]
RasL10B	CARCKHVHAALRFQGALRRNRCAIM	Unknown	Unknown	
RasL11A	PNMQDLKRRFKQALSPKVKAPSALG	–	–	
RasL11B	SPNMQDLKRRFKQALSAKVRTVTSV	–	–	
Ris/RasL12	TVKSSRAQSKRKAPTLTLLKGFKIF	–	–	
Rergl/FLJ22655	KRRPSGSKSMAKLINNVFGKRRKSV	–	–	

[a]Human sequences (see Ref. [1]): CAAX prenylation motifs or palmitoylated cysteines are underlined.

approaches have also found that a short linker sequence upstream of the CAAX motif regulates prenylation [24]. From these reports, it is clear that GGTase-I almost exclusively modifies CAAX sequences that terminate in Phe or Leu; FTase enzymes tolerate a much broader selection of terminal amino acids. From the available structural data on FTase and GGTase-I, the specificity appears to be attributed to the interaction of the terminal X amino acid of Ras or Rho with the binding pocket of the prenyltransferase. Notably, some Ras superfamily proteins can be modified normally by either FTase or GGTase-I [30].

Unlike Ras and Rho family members, where the C-terminal X residue is critically important in determining association with the prenyltransferase, Rab GTPases are all geranylgeranylated by GGTase-II. Newly synthesized Rab proteins associate with REP, which interacts with GGTase-II. As such, it is REP, rather than GGTase-II, that recognizes the Rab protein [31]. Thus the C-terminal cysteine motifs are not sufficient for enzyme recognition. Structural analyses have shown that REP guides the C-terminus of Rab into the GGTase-II active site, which geranylgeranylates both cysteine residues [32]. As with prenylated Rho (but not Ras) GTPases, the membrane localization of geranylgeranylated Rab proteins is further regulated by their association with guanine nucleotide dissociation inhibitors (RabG-DIs), which are distinct from the RhoGDIs but similarly sequester Rab proteins in the cytosol by masking the geranylgeranyl groups (Figure 5.2D) [33].

IV. Ras Family Members

H-, N-, and K-Ras are the canonical members of the 36-member Ras subfamily [1]. A majority of Ras family proteins terminate with a tetrapeptide CAAX motif (23 of 36) and are known or likely prenylated (Table 5.1). In the following sections, we discuss prenylation of Ras (Section IV.A) and its closely related family members, including Rap proteins (Section IV.B), RalA and RalB (Section IV.C), and R-Ras proteins (Section IV.D).

A. Farnesylation of H-Ras, N-Ras, K-Ras4A, and K-Ras4B

Ras GTPases are key modulators of cellular survival proliferation. Indeed, activating mutations in Ras proteins have been detected in a significant fraction of human cancers [34]. Posttranslational prenylation, palmitoylation, and carboxymethylation of Ras was initially described 20 years ago [9,35,36], and it was immediately obvious that prenylation was essential for membrane localization and transforming activity of Ras proteins [9,10]. As such, compounds that block Ras prenylation have been under investigation as potential anticancer therapeutics (discussed in Section VIII). H-Ras, N-Ras, and both isoforms of K-Ras (4A and 4B) terminate in a CAAX motif and are normally farnesylated (Table 5.1) [37]. In addition, N-Ras and K-Ras4B can also be geranylgeranylated by GGTase-I. Such alternative prenylation of N- and K-Ras4B is observed only in cells treated with FTase inhibitors [38–40] and was the unexpected finding that led to the demise of the clinical development of FTase inhibitors.

Prenylated Ras proteins are then cleaved and carboxylmethylated by the ER-resident Rce1 and Icmt proteins and are dispatched to various subcellular locales. Work from the Philips lab and others over the past two decades has shown that H- and N-Ras associate with the plasma membrane, perinuclear Golgi, and endosomes, whereas K-Ras localizes exclusively to the plasma membrane. Although prenylation was sufficient to stimulate Ras association with the ER and Golgi, proper subcellular localization required the presence of basic residues (in the case of K-Ras) or palmitoylated cysteines (in the case of N-Ras and H-Ras) [8,41,42]. Palmitoylated and prenylated H-Ras and N-Ras transit to the plasma membrane via exocytic Golgi-derived vesicles, whereas K-Ras traffics to the plasma membrane through an alternative, nonexocytic pathway [41,43]. Recent work has firmly established the importance of palmitoylation in N- and H-Ras association with endomembrane structures and intracellular trafficking [44,45]. Intriguingly, a mutant H-Ras protein that is prenylated but not palmitoylated accumulated at the Golgi but was nonetheless able to elicit cellular transformation. Further, an ER-targeted H-Ras mutant could also transform cells; however, this effect was dependent on an intact CAAX motif, suggesting a potential role for prenylation in Ras function beyond its membrane-targeting abilities [42].

CAAX-directed prenylation is not reversible, and yet it is well established that H-Ras, N-Ras, and K-Ras can transit between the plasma membrane and endomembranes. Reversible palmitoylation of H-Ras and N-Ras has been shown to permit relocalization of these GTPases; depalmitoylation of these proteins allows their reassociation with endomembranes via a nonvesicular pathway [46–48]. K-Ras, which is not palmitoylated, can be induced to dissociate from membranes by binding to calmodulin [49,50]. Similarly, cytoplasmic redistribution of K-Ras is induced by PKC-directed phosphorylation of a key serine residue in the hypervariable domain [51]; this modification presumably interferes with the charge–charge interaction between the basic residues in K-Ras and the acidic membrane phospholipids. These findings raise the question of whether there are specific GDIs for Ras proteins that might bind to and shield the prenyl group, permitting extraction of Ras from membranes in a manner analogous to Rho- and RabGDI. Several proteins have been identified that interact specifically with prenylated Ras proteins, including phosphodiesterase-δ (PDE-δ) [52,53], prenylated Rab acceptor protein (PRA1) [54], and galectin-1 [55]. PDE-δ, in particular, exhibits sequence similarities with RhoGDI1, can extract H-, N-, and K-Ras from membranes, and can interfere with Ras membrane localization when overexpressed, strengthening a potential role for PDE-δ as a bona fide Ras GDI [52].

Thus, prenylation of Ras is essential for membrane association and signal transduction, but a secondary membrane-targeting signal is required for proper subcellular localization. Understanding the precise subcellular localization of Ras proteins, and the formation of Ras "nanoclusters" containing upstream activating proteins and downstream effectors, has been an area of active research, and it is clear that C-terminal modifications, including prenylation, contribute to membrane specificity [56,57].

B.   PRENYLATION OF RAP PROTEINS

Rap family GTPases have important roles in cell–cell adhesion and junction formation, and cell polarity [5,58,59]. The Rap small GTPases are highly similar to the canonical Ras proteins; indeed, Rap1 was first isolated as a suppressor of Ras-mediated transformation by virtue of its ability to bind and supposedly sequester key downstream Ras effector proteins [60,61]. Subsequent studies have shown that overexpression of Rap can transform cultured cells and dysregulation of Rap signaling has been observed in some cancers [62]. Five Rap proteins have been described: Rap1A, Rap1B, Rab2B, and Rap2C, which are geranylgeranylated, and Rap2A, which is farnesylated [63,64] (Table 5.1). Similar to H- and N-Ras, Rap proteins have been identified at the plasma membrane, Golgi, and endosomes, and palmitoylation is essential for targeting to precise subcellular locations [65,66].

Intriguingly, SmgGDS (small GTPase dissociation stimulator; an atypical GEF) binds nonprenylated, GDP-bound Rap1A (and RhoA and Rac1 as well) and facilitates its geranylgeranylation and membrane association [67]. Prenylation of small GTPases was previously thought to be a constitutive process. This report suggests that prenylation might instead be dynamically regulated and uncovers additional avenues for therapeutic intervention.

C.   GERANYLGERANYLATION OF RALA AND RALB

RalA and RalB share 85% amino acid identity but diverge in the hypervariable region at the C-terminus. This sequence divergence supports discrete biological roles for RalA and RalB. Both proteins are geranylgeranylated [68] (Table 5.1), and both reside at the plasma membrane but are most prominently associated with intracellular vesicles. As such, Ral proteins have key functions in polarized trafficking and vesicle sorting, secretion, endocytosis, and cancer [69–71]. A clever assay that takes advantage of the fact that RalA and RalB are exclusively geranylgeranylated has helped to tease apart the contribution of these proteins to tumorigenesis. Sebti and colleagues noted that a RalA mutant that is insensitive to GGTI inhibitors (achieving membrane localization instead via farnesylation) rescued anchorage-independent cell proliferation but could not prevent apoptosis, in GGTI-treated pancreatic cancer cells. Conversely, a farnesylated RalB mutant protected against GGTI-induced apoptosis but could not promote anchorage-independent proliferation [72]. These data show that RalA and RalB have important and nonredundant roles in tumorigenesis and suggest that these proteins are key targets of GGTIs in cancer cells.

D. FARNESYLATION OF RHEB1 AND RHEB2

The small GTPase Rheb is well known for its ability to activate mTOR and as such has an important role in the physiological response to nutrient and energy availability [73–75]. Rheb activity is regulated by the TSC1–TSC2 Rheb-specific GAP complex (a GEF has been identified, although with conflicting observations); in its GTP-bound state, Rheb stimulates the kinase activity of the mTORC1 complex to promote cell growth and proliferation. Nutrient withdrawal activates TSC1–TSC2 GAP activity, inactivating Rheb and attenuating mTORC1 signaling. Before the link to mTOR signaling was made, Rheb was shown to be farnesylated [76] (Table 5.1). Indeed, farnesyltransferase inhibitors (FTIs) were subsequently found to inhibit Rheb-dependent mTOR signaling [77–79].

Constitutively, active lab-generated Rheb mutants can elicit oncogenic transformation of cultured cells in a farnesylation-sensitive manner [80]. However, Rheb is not mutated in cancer, and instead, is activated by germline or somatic mutation and inactivation of TSC1 or TSC2. Rheb is localized to the Golgi and ER, and proper subcellular localization depends on Rce1-mediated proteolysis and Icmt-dependent carboxymethylation. However, the ability of Rheb to stimulate mTORC1 signaling was not affected by Rce1 or Icmt deficiency. This suggests that, unlike other Ras-related GTPases, proper membrane localization of Rheb is not necessarily critical for its signaling activity [81]. In light of its central importance to energy homeostasis, it will be important to uncover other factors that regulate Rheb localization and interaction with mTORC1.

E.   PRENYLATION OF R-RAS PROTEINS

The R-Ras subfamily of Ras-related GTPases includes R-Ras, TC21 (R-Ras2), and M-Ras (R-Ras3) (Table 5.1). R-Ras is geranylgeranylated [82] and localizes to the plasma membrane. Functional studies have implicated R-Ras activity in membrane ruffle formation, and cell migration and adhesion [83–85]. By virtue of its CAAX sequence, M-Ras is thought to be geranylgeranylated [86]. M-Ras appears to regulate actin dynamics (particularly in neurons) and might also have a role in differentiation [87,88]. In contrast to the other R-Ras family members, TC21 is alternatively prenylated [89]. Like Ras, analogous tumor-associated mutants of TC21 exhibit potent transforming activity in cell culture assays, yet surprisingly, have not been found to be mutationally activated to any significant frequency [90,91].

F.   OTHER PRENYLATED RAS FAMILY PROTEINS

There are several other Ras subfamily members that are less well characterized functionally. These include Di-Ras1 (also called Rig), which is alternatively prenylated [92]. The prenylation status of the related Di-Ras2 protein is not known [93]. Noey2/ARHI was shown to be prenylated and an intact CAAX motif was required for its tumor suppressor function [94]. Experiments with FTIs suggest that Noey2 requires farnesylation for membrane association (E.J. Chenette and C.J. Der, unpublished observations). RRP22 [95] and RasD2 (also called Rhes) [96,97] are farnesylated, while the identity of the prenyl residue that modifies RasD1 and E-Ras is not known.

G.   NONPRENYLATED RAS FAMILY PROTEINS

As described in Section III above, most Ras subfamily members terminate in a CAAX motif and are prenylated. The subset of Ras proteins (13 human members) that lack CAAX sequences is, in general, cytosolic. Indeed, Rerg terminates in a -KISS motif and is not lipidated, and is present in the cytosol [98]. Members of the RGK (Rad/Gem/Kir) subfamily—Rad, Gem/Kir, Rem1/Ges, and Rem2—also lack a CAAX motif and accumulate in the cytoplasm [99–102]. NKIRas1 (also called κB-Ras1) and NKIRas2 (also called κB-Ras2) both lack CAAX motifs and are not prenylated [103] (Table 5.1). Finally, the uncharacterized RasL11A-B, RasL12, and Rergl proteins lack CAAX motifs and are therefore not prenylated.

However, the small GTPases Rit1 and Rin (also called Rit2) are plasma membrane localized, despite lacking a CAAX box. C-terminal basic residues are important for Rit1 and Rin membrane localization [104–106], and Tobias Meyer's group has shown that depletion of the PtdIns(4,5)P$_2$ and PtdIns(3,4,5)P$_3$ causes dissociation of Rit and Rin from the plasma membrane [15]. A hydrophobic tryptophan residue present in the hypervariable domain of both proteins was also shown to be important for their plasma membrane localization. Intriguingly, a Trp residue is also present in RhoD, Wrch1, and Chp/Wrch-2 (and, in the case of Chp, is essential for membrane localization [107]), providing support for the idea that hydrophobic residues in the C-terminal tail might also mediate membrane association.

V. Rho Family Members

RhoA, Rac1, and Cdc42 are canonical members of the 20-member Rho subfamily [1] (Table 5.2). Including the two Cdc42 isoforms, 17 of 21 terminate with CAAX motifs and are known or predicted to be prenylated and two terminate with CXX motifs that are palmitoylated. As with Ras family proteins, polybasic sequences or palmitoylated cysteines comprise second membrane-targeting elements that complement CAAX-signaled modifications. In the following sections, we discuss prenylation of Rho, Rac, and Cdc42 (Section IV.A); Rnd1, Rnd2, and Rnd3 (Section IV.B); and other Rho-related proteins (Section IV.C).

A. Geranylgeranylation of Rho, Rac, and Cdc42 Proteins

The Rho, Rac, and Cdc42 GTPases transmit upstream signals through their effector proteins to affect changes in the actin cytoskeleton. As such, they regulate cell shape, polarity, adhesion, and migration and can support tumorigenic activities including invasion and metastasis, angiogenesis, and proliferation. CAAX-containing Rho proteins are not only targeted to cellular membranes by virtue of the prenyl modification but also require a secondary signal in the hypervariable domain for proper subcellular localization [108].

Work from our laboratory has defined the contribution of prenylation, and Rce1- or Icmt-directed processing, in Rho protein membrane localization [108]. Membrane association of RhoA, RhoC, Cdc42 (brain and placental isoforms), and all three Rac isoforms was sensitive to GGTI treatment. In contrast, RhoB was shown to dissociate from membranes only following combined FTI and GGTI treatment, suggesting that it is alternatively prenylated (Table 5.2). Further biochemical analyses confirmed that RhoB could incorporate either FPP or GGPP moieties. Though

TABLE 5.2

The C-Termini of Rho Family GTPases are Substrates for FTase and GGTase-I

GTPase	C-terminal 25 amino acids[a]	FTase	GGTase-I	Reference
RhoA	EVFEMATRAALQARRGKKKSGCLVL	−	+	[108]
RhoB	ETATRAALQKRYGSQNGCINCCKVL	+	+	[108]
RhoC	EVFEMATRAGLQVRKNKRRRGCPIL	−	+	[108]
Rac1 (Rac1b)[b]	VFDEAIRAVLCPPPVKKRKRKCLLL	−	+	[108]
Rac2[b]	VFDEAIRAVLCPQPTRQQKRACSLL	−	+	[108]
Rac3[b]	VFDEAIRAVLCPPPVKPGKKCTVF	−	+	[108]
Cdc42 (brain)	NVFDEAILAALEPPETQPKRKCCIF	−	+	[108]
Cdc42 (placental)	NVFDEAILAALEPPEPKKSRRCVLL	−	+	[108]
Rnd1	LHLPSRSELISSTFKKEKAKSCSIM	+	−	[108]
Rnd2	QLSGRPDRGNEGEIHKDRAKSCNLM	+	−	[108]
Rnd3/RhoE	MPSRPELSAVATDLRKDKAKSCTVM	+	−	[108]
RhoD	AAEVALSSRGRNFWRRITOGFCVVT	+	−	[108]
RhoG	EVFAEAVRAVLNPTPIKRGRSCILL	Unknown	Unknown	[108]
RhoH/TTF	VRTAVNQARRRNRRRLFSINECKIF	+	+	[108]
Rif/RhoF	EAAKVALSALKKAQROKKRRLCLLL	±	+	[108]
TC10	AILTPKKHTVKKRIGSRCINCCLIT	+	−	[108]
TCL[b]	AILTIFHPKKKKKRCSEGHSCCSII	+	−	[108]
Wrch-1/RhoU	KSKSRTPDKMKNLSKSWVKKYCCFV	−	−	[128]
Wrch-2/Chp	LEKKLNAKGVRTLSRCRWKKFFCFV	−	−	[129]
RhoBTB1	KEDIALNKIHRSRRKWCFWNSSPAVA	−	−	[108]
RhoBTB2/DBC2	FWNSPSPSSSAASSSSPSSSAVV	−	−	[108]
RhoBTB3	SNMYLKQLAEYRKYIHSRKCRCLVM	+	−	J. Cha and C.J. Der (unpublished observations)

[a]Human sequences (see Ref. [1]); CAAX prenylation motifs or palmitoylated cysteines are underlined.
[b]Although upstream cysteine residues are present, there is no evidence for palmitate modification and/or involvement in subcellular localization and membrane association [108].

prenylation is sufficient for membrane targeting, secondary membrane-targeting signals are again important for dictating precise subcellular localization. In addition to geranylgeranylation, RhoA and RhoC associate with cellular membranes via basic residues in the C-terminus; both localize to the plasma membrane and are also found in the cytosol. Cdc42 also possesses basic residues in its C-terminus but accumulates at the plasma membrane, Golgi, ER, and nuclear envelope. RhoB, Rac2, and Rac3 localize to the plasma membrane, Golgi, and endosomes. Finally, Rac1 possesses a polybasic stretch in its hypervariable region and is targeted to the plasma membrane [109]. Despite the presence of a conserved cysteine residue upstream of the CAAX motif for Rac1-3 proteins, there is no evidence for palmitate modification or in regulation of subcellular localization and membrane association [108].

The membrane localization of Rho GTPases is also regulated by RhoGDI proteins [110]. As their name suggests, the RhoGDI family of proteins, which includes RhoGDI1 (also called GDIα), GDI2 (also called GDIβ), and GDI3 (also called GDIγ), can bind to Rho-GDP and prevent guanine nucleotide dissociation, thus locking the GTPase in an inactive state. Conversely, GDIs have also been shown to prevent GTP hydrolysis and block effector binding to the active GTPase. Finally, GDI proteins can extract prenylated Rho proteins from lipid bilayers, sequestering them in the cytosol [111–114]. The activity of RhoGDI thus affords another level of regulation, as Rho GTPases can be actively shuttled between the cytoplasm and cellular membranes. Structural studies of the Cdc42–GDI1 complex have shown that the Cdc42 prenyl group becomes embedded within a hydrophobic pocket in the GDI, which stimulates release of Cdc42 from the membrane [115]; this mechanism has been extrapolated to other Rho-related proteins as well. Rho GTPases are uniquely sensitive to the different GDI proteins, though the hydrophobic binding pocket within the GDI is able to accommodate both farnesylated and geranylgeranylated proteins [116]. Of the canonical Rho proteins, RhoB was not found to be regulated by RhoGDI1 but is instead sensitive to RhoGDI3 [110,117]. RhoA, Cdc42, Rac1, and Rac2 can all be regulated by GDI1 and GDI2; intriguingly, the Rac1b splice variant found in breast and colon tumor cells does not interact with RhoGDI1 [116].

A recent study by Burridge and colleagues provides an additional function for RhoGDIs in protecting cytoplasmic Rho GTPases from degradation [118]. Depletion of RhoGDI1 induced misfolding and subsequent degradation of endogenous, geranylgeranylated Rho. Intriguingly, overexpression of exogenous Rho proteins appeared to displace endogenous Rho from GDI1, again leading to misfolding and degradation of Rho. Thus, RhoGDIs are multifunctional regulators of Rho localization, activity, and stability.

B. Farnesylation of Rnd1, Rnd2, and Rnd3/RhoE

Unlike the canonical Rho family members, the Rnd proteins (Rnd1, Rnd2, and Rnd3/RhoE) are constitutively GTP bound as they do not hydrolyze GTP. No GEFs or GAPs have been identified for Rnd proteins, with transcriptional and posttranslational mechanisms of regulation described. Rnd GTPases regulate actin cytoskeletal dynamics and have roles in axon extension and neurite outgrowth via interactions with plexins. In addition, upregulation of Rnd1 and Rnd2 has been detected during inflammation and in gastric cancer, whereas Rnd3 expression is decreased in some tumors. This antiproliferative effect has been attributed to the ability of Rnd3 to antagonize RhoA by increasing p190 RhoGAP activity and to induce cell-cycle arrest at the G1 phase [119,120]. Each Rnd GTPase possesses a CAAX box that terminates with Met; as such, all Rnd proteins are farnesylated [108] (Table 5.1). Rnd proteins also possess basic residues, but not palmitoylated cysteines, in their C-termini. Rnd1 localizes to the plasma membrane; Rnd2 is targeted to endosomes and is also cytosolic; and Rnd3 has been detected at the plasma membrane, endosomes, and cytosol [109].

C. Prenylation of Other Rho Subfamily Members

The use of FTIs and GGTIs has elucidated the contribution of prenylation to membrane localization of other Rho-related proteins [108]. RhoD membrane localization was weakly perturbed by FTI treatment, and unaffected by GGTI, suggesting that it undergoes farnesylation, but that prenylation might not be essential for membrane localization. RhoD also possesses C-terminal basic amino acids; however, given that basic residues are generally not sufficient for membrane association, it is possible that an as-of-yet undefined motif in RhoD might facilitate membrane targeting. RhoH dissociates from its endomembrane localization only following combined FTI/GGTI treatment, suggesting that it is alternatively prenylated. Rif localization is weakly sensitive to FTIs and is more strongly perturbed by GGTIs, suggesting that geranylgeranylation is critical for its membrane association. Membrane localization of both TC10 and the related GTPase TCL is sensitive to FTI treatment [108] (Table 5.1). TC10 is not extracted from membranes by RhoGDI1 [110]. Interestingly, despite both possessing upstream cysteine residues, TC10 but not TCL is modified by palmitoylation [108]. Thus, the presence of a cysteine residue upstream of CAAX prenylation motifs alone is not an absolute indication of palmitate modification and reflects the absence of well-defined palmitoylation signal sequences.

The activity of the RhoG small GTPase has been implicated in proliferation, migration, and membrane remodeling [121–124]. Though the prenylation status of RhoG has not been determined experimentally, it terminates in a -CILL tetrapeptide sequence and is likely geranylgeranylated. Importantly, RhoG has been shown to associate with both RhoGDI1 and RhoGDI3. Binding to either RhoGDI1 or RhoGDI3 inhibits RhoG activation [125,126], whereas RhoGDI3 could additionally target RhoG to the Golgi [127].

D. NONPRENYLATED RHO-RELATED SMALL GTPASES

Wrch and Chp (also called Wrch-2) are atypical Rho-related GTPases that lack functional CAAX motifs and are not prenylated (Table 5.2) but are nonetheless targeted to the plasma membrane and endosomes. Both proteins depend on a combination of palmitoylation of a CXX motif and upstream basic residues for membrane localization and signaling activity [107,128,129].

RhoBTB1 and RhoBTB2 are atypical Rho GTPases that lack C-terminal CAAX motifs (Table 5.2), and instead, possess tandem BTB (Bric-a-brac, Tramtrack, Broad-complex) domains downstream of the GTPase domain [130,131]. The GTPase domains of RhoBTB1/2 are thought to be inactive. Although RhoBTB3 is not considered a Rho family protein, it shares overall domain topology with RhoBTB1 and RhoBTB2. However, its putative GTPase domain has been described to bind ATP instead and, additionally, terminates in a CAAX motif. Surprisingly, Pfeffer and colleagues found that the CAAX box was dispensable for RhoBTB3 localization to the Golgi, suggesting that RhoBTB3 might not be prenylated [132].

VI.   Rab Family Members

The Rab subfamily is the largest of the Ras superfamily, containing more than 60 human members (Figure 5.1A). Rabs are best known for their roles in regulating membrane trafficking pathways, and as such their recruitment to membranes is tightly regulated. As mentioned above, most Rab proteins terminate in a -CC, -CXC, -CCXX, or -CXXX motif and as such are substrates for GGTase-II. Newly synthesized Rab proteins interact with the REP • GGTase-II complex [133]. Structural studies have confirmed that it is REP, and not the Rab protein *per se*, that is recognized by the GGTase enzyme [32]. Both cysteines in the Rab C terminus are geranylgeranylated [134,135]. Indeed, Rabs that contain dicysteine motifs are mislocalized to the ER or Golgi when the membrane-targeting sequence is replaced by a traditional CAAX sequence [136]. In contrast, Rab proteins that naturally terminate in a CAAX motif are obviously monogeranylgeranylated and can in theory be recognized by either FTase/GGTase-I or GGTase-II. However, there is evidence that some CAAX-containing Rab GTPases, like Rab8, are preferentially modified by GGTase-II [20,21,137]. Prenylation of Rab proteins is discussed in detail in Chapter 8.

Rab proteins that possess -CC or -CCXX motifs are not processed further following geranylgeranylation, and REP escorts the Rab protein to the target membrane. Intriguingly, Rab proteins terminating in a -CXC motif are carboxymethylated by Icmt on the terminal cysteine [138]. Those Rab proteins that instead possess tetrapeptide CXXX motifs are delivered to the Golgi following prenylation by REP and then processed by both Rce1 and Icmt [139].

Membrane association of Rab proteins is highly dynamic and choreographed by RabGDIs, GAPs, GEFs, and GDI displacement factors (GDFs). The traditional model proposed that REP escorted newly prenylated Rabs to target membranes. Once there, Rabs were targets for GAPs, and GDP-bound Rab proteins would then be extracted by RabGDIs in a manner analogous to Rho proteins. Cytosolic GDI–Rab–GDP complexes could then interact with GDFs present in the membrane, which would displace the GDI and permit reassociation of Rab-GDP with the target membrane. GEFs would then stimulate nucleotide exchange, forming signaling-competent GTP-bound Rab [140]. A recent study calls this model into question, however, by showing that GEFs are able to dissociate the GDI–Rab–GDP complexes, which could negate the need for GDFs in promoting Rab membrane association [141]. Rab membrane localization is a complex, multistep process that is regulated at many levels, and it will be important to elucidate the relative contributions of Rab regulatory proteins to this process.

VII. The Arf Family of Small GTPases

Arf and Arl (Arf-like) small GTPases regulate vesicle biogenesis at the Golgi and, to some extent, at the plasma membrane [142,143]. Despite their membrane localization, Arf proteins are not prenylated. Instead, they are tethered to membranes by the addition of a 14-carbon myristate fatty acid to the N-terminus. Nucleotide status also regulates Arf membrane localization; in the GDP-bound conformation, the N-terminus is tucked within the protein and is not accessible to the solvent. Nucleotide exchange frees the myristoylated N-terminus for insertion into the lipid bilayer [144].

VIII.   The Ran Small GTPase

There is only a single human Ran GTPase. Ran is a predominantly nuclear protein that lacks a CAAX motif. It regulates nucleocytoplasmic trafficking, and its precise subcellular localization is determined by nucleotide status. Ran-GTP is primarily nuclear, whereas Ran-GDP is thought to be predominately cytoplasmic. However, Ran GEFs have also been identified in the cytoplasm, suggesting that Ran-GTP might also function in extranuclear locales [145,146].

IX.   Prenyltransferases as Therapeutic Targets

A.   RATIONALE

The discovery that Ras proteins were modified by farnesylation, and that Ras membrane association and transforming activity were critically dependent on this modification, prompted considerable excitement that inhibitors of this modification may be effective anti-Ras inhibitors and useful for cancer treatment. The isolation and characterization of the enzyme by Brown, Goldstein, and colleagues in 1990 stimulated an intensive effort by the pharmaceutical industry to develop FTIs (Figure 5.3A) [147].

B.   CLINICAL EVALUATION OF FTIs

A wide spectrum of different structural classes of FTIs was developed, including CAAX peptidomimetics, nopeptide peptidomimetics, farnesyl diphosphate analogs, and bisubstrate inhibitors [148], with several advancing into clinical testing for oncology, either alone or in combination with conventional cytotoxic drugs [149] (Figure 5.3A). Of these, two nonpeptide peptidomimetics, tipifarnib (R115777) and lonafarnib (SCH66336) (Figure 5.3B), underwent the most significant clinical evaluation. The first cell culture analyses showed potent growth inhibitory activity for H-Ras-transformed cells [150,151]. Subsequent studies showed inhibitory activity for human tumor cells and for mutant *RAS*-driven mouse tumor models. The impressive success of these preclinical studies supported the start of the intensive clinical evaluation of multiple FTIs. However, the general lack of efficacy seen, in particular, for pancreatic [152–154] and colorectal [155] cancers, where *KRAS* mutation frequency is highest, led to the disappointing conclusion that FTIs were not effective therapies for *RAS*-mutant cancers.

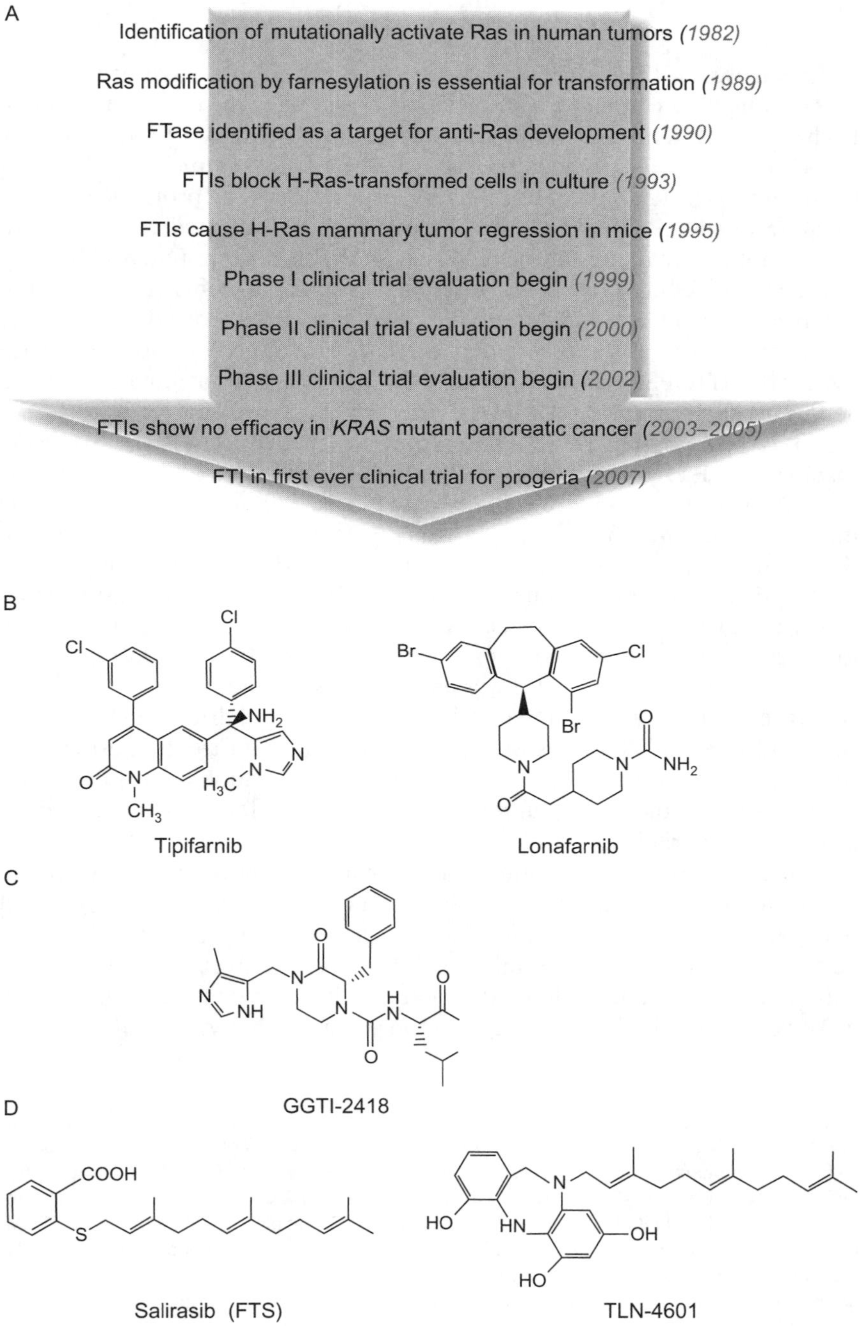

FIG. 5.3. Inhibition of Ras membrane association. (A) Farnesyltransferase inhibitors (FTIs). Key steps in the chronology of FTI development for cancer and progeria treatment are indicated. (B) Farnesyltransferase inhibitors. Lonafarnib (FTase $IC_{50} = 1.9$ nM, GGTase-1

In retrospect, the failure of FTIs in the clinic should have been anticipated in light of cell culture studies that found that, in contrast to H-Ras, both K-Ras4B and N-Ras proteins undergo alternative prenylation by GGTase-I when FTase activity is blocked by FTI treatment [38,156,157] (Figure 5.2D). That the alternatively prenylated Ras proteins retained membrane association and transforming activity indicated that FTIs are simply not effective inhibitors of K-Ras or N-Ras protein function [6,158].

Since FTase modifies >50 additional human proteins, including the Rheb small GTPase and lamin A, FTIs have been considered for the treatment of other diseases. In particular, Hutchinson-Gilford Progeria Syndrome (HGPS; also called progeria) is caused by a mutation in the gene encoding lamin A (LMNA), resulting in expression of a defective lamin A protein that retains the farnesyl modification [159]. Promising results with FTI treatment in cell culture [160–163] and mouse models [164] support their clinical value for this disease. Since progeria patients number fewer than 50 worldwide, that FTIs may fortuitously serve as a therapeutic approach for this disorder prompted the first ever clinical trial for this disease in 2007 (Figure 5.3A). Completed in 2009, the results from this trial have not been reported. A second clinical trial is ongoing, where lonafarnib will be used in combination with pravastatin (an inhibitor of the mevalonate biosynthetic pathway and hence all protein prenylation) and zoledronic acid, a biphosphonate which is an approved drug used to prevent skeletal fractures in patients with cancers, as well as for treating osteoporosis. The therapeutic value of the combination treatment of statins and aminobisphosphonates is supported by increased longevity in a mouse model of progeria [165].

The alternative prenylation of Ras proteins, when coupled with the implicated involvement of other Ras superfamily proteins in Ras-mediated oncogenesis that are substrates for GGTase-I, prompted interest in the development of GGTase-I inhibitors (GGTI). Cell culture and tumor xenograft studies, and additional genetic studies in mouse models of *RAS*-driven oncogenesis [166,167], support the antitumor activity of GGTIs, with

---

$IC_{50} > 50,000$ nM) and tipifarnib (FTase $IC_{50} = 7.9$ nM, GGTase-1 = 40% inhibition at 50,000 nM) are highly selective CAAX competitive inhibitors [149]. (C) GGTase-I inhibitor. GGTI-2418 showed 5600-fold selectivity toward inhibition of GGTase-I versus FTase *in vitro*. GGTI-2418 inhibited GGTase-I and FTase activities with IC_{50}s of 9.5 ± 2.0 nM and 53 ± 11 μM, respectively. (D) Farnesyl-containing inhibitors of Ras. Two farnesyl-containing small molecules have been evaluated in Phase I clinical trials. One mechanism of their action is believed to be through competition with Ras for farnesyl docking proteins that facilitated Ras membrane association.

one highly selective GGTI (GGTI-2418; Figure 5.3C) currently in Phase I clinical evaluation. Early Phase I results found GGTI-2418 well tolerated with minimal toxicity, supporting expansion of the trial. One FTI evaluated in clinical trials, L-778,123, also possessed dual inhibitory activity for GGTase-I (FTase $IC_{50} = 2$ nM, GGTase-I $IC_{50} = 98$ nM) [149] and inhibited GGTase-I activity in the patient, but nevertheless still failed to block K-Ras prenylation [168].

There is increasing evidence for the involvement of Rab proteins, in particular, Rab25, in cancer [169,170]. A few GGTase-II inhibitors have been described [134,135]. Fortuitously, analogs of one clinically evaluated selective FTI (BMS-214662; FTase $IC_{50} = 1.35$ nM, GGTase-I $IC_{50} > 1000$ nM) [149] was determined to have GGTase-II inhibitory activity as an off-target activity responsible for the unique apoptotic activity of this unusual FTI [171].

Another related approach for blocking farnesylation-mediated Ras membrane association involves small molecules with farnesyl moieties (Figure 5.3D). Salirasib (farnesylthiosalicylic acid/FTS) and TLN-4601 are proposed to function by competing with Ras proteins for membrane binding, and additionally mTOR inhibition [81,172,173], with both evaluated in Phase I clinical trials.

C. INHIBITORS OF OTHER CAAX-DEPENDENT POST-TRANSLATIONAL MODIFICATIONS

Supported by genetic knockout studies in mice [174–176], inhibitors of Rce1 and Icmt to block Ras and Rho GTPase function have also been considered [177] (Figure 5.2D), but potent and selective inhibitors remain to be identified. Similarly, inhibitors of palmitoylation have also been considered for inhibition of Ras and Rho family GTPase function. However, the palmitoyl acyltransferases responsible for protein palmitoylation are complex and their activities and specificities continue to be elucidated [178].

X. Conclusion

The existence of >150 human small GTPases emphasizes the amazing versatility of a "simple" GDP–GTP-regulated binary switch in controlling virtually all key aspects of normal cellular physiology. In addition to the diversification of different Ras superfamily GTPase branches by their control by unique regulators and their activation of distinct effectors, their biological roles are further diversified by their posttranslational modification by isoprenoid and fatty acid lipids. These lipids do not simply increase hydrophobicity and promote a static association with a single membrane compartment. Instead, together with other CAAX-signaled and additional posttranslational modifications and (phosphorylation, ubiquitination, SUMOylation), further diversification of the subcellular localization and biological roles of otherwise highly structurally and biochemically related small GTPases can be achieved. These modifications dictate distinct biological functions by impacting the spatiotemporal localization and activity of small GTPases. In summary, the field has advanced considerably from the once simple notion that Ras was positioned persistently and exclusively at the inner face of the plasma membrane, with more surprises yet to come.

ACKNOWLEDGMENTS

Due to space limitations, we apologize to colleagues whose work we could not include. Our research was supported in part by grants (CA042978, CA129610, CA127152, CA67771, and CA106991) from the US National Institutes of Health.

REFERENCES

1. Wennerberg, K., Rossman, K.L., and Der, C.J. (2005). The Ras superfamily at a glance. *J Cell Sci* 118:843–846.
2. Colicelli, J. (2004). Human RAS superfamily proteins and related GTPases. *Sci STKE* 2004:RE13.
3. Bos, J.L., Rehmann, H., and Wittinghofer, A. (2007). GEFs and GAPs: critical elements in the control of small G proteins. *Cell* 129:865–877.
4. Vigil, D., Cherfils, J., Rossman, K.L., and Der, C.J. (2010). Ras superfamily GEFs and GAPs: validated and tractable targets for cancer therapy? *Nat Rev Cancer* 10:842–857.
5. Raaijmakers, J.H., and Bos, J.L. (2009). Specificity in Ras and Rap signaling. *J Biol Chem* 284:10995–10999.
6. Hancock, J.F., Cadwallader, K., Paterson, H., and Marshall, C.J. (1991). A CAAX or a CAAL motif and a second signal are sufficient for plasma membrane targeting of ras proteins. *EMBO J* 10:4033–4039.
7. Ghomashchi, F., Zhang, X., Liu, L., and Gelb, M.H. (1995). Binding of prenylated and polybasic peptides to membranes: affinities and intervesicle exchange. *Biochemistry* 34:11910–11918.
8. Hancock, J.F., Paterson, H., and Marshall, C.J. (1990). A polybasic domain or palmitoylation is required in addition to the CAAX motif to localize p21ras to the plasma membrane. *Cell* 63:133–139.
9. Hancock, J.F., Magee, A.I., Childs, J.E., and Marshall, C.J. (1989). All ras proteins are polyisoprenylated but only some are palmitoylated. *Cell* 57:1167–1177.
10. Jackson, J.H., Cochrane, C.G., Bourne, J.R., Solski, P.A., Buss, J.E., and Der, C.J. (1990). Farnesol modification of Kirsten-ras exon 4B protein is essential for transformation. *Proc Natl Acad Sci USA* 87:3042–3046.

11. Laude, A.J., and Prior, I.A. (2008). Palmitoylation and localisation of RAS isoforms are modulated by the hypervariable linker domain. *J Cell Sci* 121:421–427.
12. Dudler, T., and Gelb, M.H. (1996). Palmitoylation of Ha-Ras facilitates membrane binding, activation of downstream effectors, and meiotic maturation in Xenopus oocytes. *J Biol Chem* 271:11541–11547.
13. Coats, S.G., Booden, M.A., and Buss, J.E. (1999). Transient palmitoylation supports H-Ras membrane binding but only partial biological activity. *Biochemistry* 38:12926–12934.
14. Jackson, J.H., Li, J.W., Buss, J.E., Der, C.J., and Cochrane, C.G. (1994). Polylysine domain of K-ras 4B protein is crucial for malignant transformation. *Proc Natl Acad Sci USA* 91:12730–12734.
15. Heo, W.D., Inoue, T., Park, W.S., Kim, M.L., Park, B.O., Wandless, T.J., and Meyer, T. (2006). PI(3,4,5)P3 and PI(4,5)P2 lipids target proteins with polybasic clusters to the plasma membrane. *Science* 314:1458–1461.
16. Philips, M.R. (2005). Compartmentalized signalling of Ras. *Biochem Soc Trans* 33:657–661.
17. ten Klooster, J.P., and Hordijk, P.L. (2007). Targeting and localized signalling by small GTPases. *Biol Cell* 99:1–12.
18. Omerovic, J., and Prior, I.A. (2009). Compartmentalized signalling: Ras proteins and signalling nanoclusters. *FEBS J* 276:1817–1825.
19. Seabra, M.C., Reiss, Y., Casey, P.J., Brown, M.S., and Goldstein, J.L. (1991). Protein farnesyltransferase and geranylgeranyltransferase share a common alpha subunit. *Cell* 65:429–434.
20. Wilson, A.L., Erdman, R.A., Castellano, F., and Maltese, W.A. (1998). Prenylation of Rab8 GTPase by type I and type II geranylgeranyl transferases. *Biochem J* 333(Pt 3): 497–504.
21. Joberty, G., Tavitian, A., and Zahraoui, A. (1993). Isoprenylation of Rab proteins possessing a C-terminal CaaX motif. *FEBS Lett* 330:323–328.
22. Otto, J.C., Kim, E., Young, S.G., and Casey, P.J. (1999). Cloning and characterization of a mammalian prenyl protein-specific protease. *J Biol Chem* 274:8379–8382.
23. Ashby, M.N., King, D.S., and Rine, J. (1992). Endoproteolytic processing of a farnesylated peptide in vitro. *Proc Natl Acad Sci USA* 89:4613–4617.
24. Maurer-Stroh, S., and Eisenhaber, F. (2005). Refinement and prediction of protein prenylation motifs. *Genome Biol* 6:R55.
25. Caplin, B.E., Hettich, L.A., and Marshall, M.S. (1994). Substrate characterization of the Saccharomyces cerevisiae protein farnesyltransferase and type-I protein geranylgeranyltransferase. *Biochim Biophys Acta* 1205:39–48.
26. Boutin, J.A., Marande, W., Petit, L., Loynel, A., Desmet, C., Canet, E., and Fauchere, J.L. (1999). Investigation of S-farnesyl transferase substrate specificity with combinatorial tetrapeptide libraries. *Cell Signal* 11:59–69.
27. Reid, T.S., Terry, K.L., Casey, P.J., and Beese, L.S. (2004). Crystallographic analysis of CaaX prenyltransferases complexed with substrates defines rules of protein substrate selectivity. *J Mol Biol* 343:417–433.
28. Moores, S.L., Schaber, M.D., Mosser, S.D., Rands, E., O'Hara, M.B., Garsky, V.M., Marshall, M.S., Pompliano, D.L., and Gibbs, J.B. (1991). Sequence dependence of protein isoprenylation. *J Biol Chem* 266:14603–14610.
29. Roskoski, R., Jr., and Ritchie, P. (1998). Role of the carboxyterminal residue in peptide binding to protein farnesyltransferase and protein geranylgeranyltransferase. *Arch Biochem Biophys* 356:167–176.
30. Fiordalisi, J.J., Johnson, R.L., 2nd, Weinbaum, C.A., Sakabe, K., Chen, Z., Casey, P.J., and Cox, A.D. (2003). High affinity for farnesyltransferase and alternative prenylation

contribute individually to K-Ras4B resistance to farnesyltransferase inhibitors. *J Biol Chem* 278:41718–41727.

31. Andres, D.A., Seabra, M.C., Brown, M.S., Armstrong, S.A., Smeland, T.E., Cremers, F.P., and Goldstein, J.L. (1993). cDNA cloning of component A of Rab geranylgeranyl transferase and demonstration of its role as a Rab escort protein. *Cell* 73:1091–1099.

32. Guo, Z., Wu, Y.W., Das, D., Delon, C., Cramer, J., Yu, S., Thuns, S., Lupilova, N., Waldmann, H., Brunsveld, L., Goody, R.S., Alexandrov, K., and Blankenfeldt, W. (2008). Structures of RabGGTase-substrate/product complexes provide insights into the evolution of protein prenylation. *EMBO J* 27:2444–2456.

33. Pylypenko, O., Rak, A., Durek, T., Kushnir, S., Dursina, B.E., Thomae, N.H., Constantinescu, A.T., Brunsveld, L., Watzke, A., Waldmann, H., Goody, R.S., and Alexandrov, K. (2006). Structure of doubly prenylated Ypt1:GDI complex and the mechanism of GDI-mediated Rab recycling. *EMBO J* 25:13–23.

34. Karnoub, A.E., and Weinberg, R.A. (2008). Ras oncogenes: split personalities. *Nat Rev Mol Cell Biol* 9:517–531.

35. Stimmel, J.B., Deschenes, R.J., Volker, C., Stock, J., and Clarke, S. (1990). Evidence for an S-farnesylcysteine methyl ester at the carboxyl terminus of the Saccharomyces cerevisiae RAS2 protein. *Biochemistry* 29:9651–9659.

36. Clarke, S., Vogel, J.P., Deschenes, R.J., and Stock, J. (1988). Posttranslational modification of the Ha-ras oncogene protein: evidence for a third class of protein carboxyl methyltransferases. *Proc Natl Acad Sci USA* 85:4643–4647.

37. Wright, L.P., and Philips, M.R. (2006). Thematic review series: lipid posttranslational modifications. CAAX modification and membrane targeting of Ras. *J Lipid Res* 47:883–891.

38. Whyte, D.B., Kirschmeier, P., Hockenberry, T.N., Nunez-Oliva, I., James, L., Catino, J.J., Bishop, W.R., and Pai, J.K. (1997). K- and N-Ras are geranylgeranylated in cells treated with farnesyl protein transferase inhibitors. *J Biol Chem* 272:14459–14464.

39. Zhang, F.L., Kirschmeier, P., Carr, D., James, L., Bond, R.W., Wang, L., Patton, R., Windsor, W.T., Syto, R., Zhang, R., and Bishop, W.R. (1997). Characterization of Ha-ras, N-ras, Ki-Ras4A, and Ki-Ras4B as in vitro substrates for farnesyl protein transferase and geranylgeranyl protein transferase type I. *J Biol Chem* 272:10232–10239.

40. Lerner, E.C., Zhang, T.T., Knowles, D.B., Qian, Y., Hamilton, A.D., and Sebti, S.M. (1997). Inhibition of the prenylation of K-Ras, but not H- or N-Ras, is highly resistant to CAAX peptidomimetics and requires both a farnesyltransferase and a geranylgeranyltransferase I inhibitor in human tumor cell lines. *Oncogene* 15:1283–1288.

41. Choy, E., Chiu, V.K., Silletti, J., Feoktistov, M., Morimoto, T., Michaelson, D., Ivanov, I.E., and Philips, M.R. (1999). Endomembrane trafficking of ras: the CAAX motif targets proteins to the ER and Golgi. *Cell* 98:69–80.

42. Chiu, V.K., Bivona, T., Hach, A., Sajous, J.B., Silletti, J., Wiener, H., Johnson, R.L., 2nd, Cox, A.D., and Philips, M.R. (2002). Ras signalling on the endoplasmic reticulum and the Golgi. *Nat Cell Biol* 4:343–350.

43. Apolloni, A., Prior, I.A., Lindsay, M., Parton, R.G., and Hancock, J.F. (2000). H-ras but not K-ras traffics to the plasma membrane through the exocytic pathway. *Mol Cell Biol* 20:2475–2487.

44. Misaki, R., Morimatsu, M., Uemura, T., Waguri, S., Miyoshi, E., Taniguchi, N., Matsuda, M., and Taguchi, T. (2010). Palmitoylated Ras proteins traffic through recycling endosomes to the plasma membrane during exocytosis. *J Cell Biol* 191:23–29.

45. Valero, R.A., Oeste, C.L., Stamatakis, K., Ramos, I., Herrera, M., Boya, P., and Perez-Sala, D. (2010). Structural determinants allowing endolysosomal sorting and degradation of endosomal GTPases. *Traffic* 11:1221–1233.

46. Rocks, O., Peyker, A., Kahms, M., Verveer, P.J., Koerner, C., Lumbierres, M., Kuhlmann, J., Waldmann, H., Wittinghofer, A., and Bastiaens, P.I. (2005). An acylation cycle regulates localization and activity of palmitoylated Ras isoforms. *Science* 307:1746–1752.

47. Magee, A.I., Gutierrez, L., McKay, I.A., Marshall, C.J., and Hall, A. (1987). Dynamic fatty acylation of p21N-ras. *EMBO J* 6:3353–3357.

48. Goodwin, J.S., Drake, K.R., Rogers, C., Wright, L., Lippincott-Schwartz, J., Philips, M.R., and Kenworthy, A.K. (2005). Depalmitoylated Ras traffics to and from the Golgi complex via a nonvesicular pathway. *J Cell Biol* 170:261–272.

49. Villalonga, P., Lopez-Alcala, C., Bosch, M., Chiloeches, A., Rocamora, N., Gil, J., Marais, R., Marshall, C.J., Bachs, O., and Agell, N. (2001). Calmodulin binds to K-Ras, but not to H- or N-Ras, and modulates its downstream signaling. *Mol Cell Biol* 21:7345–7354.

50. Sidhu, R.S., Clough, R.R., and Bhullar, R.P. (2003). Ca2+/calmodulin binds and dissociates K-RasB from membrane. *Biochem Biophys Res Commun* 304:655–660.

51. Bivona, T.G., Quatela, S.E., Bodemann, B.O., Ahearn, I.M., Soskis, M.J., Mor, A., Miura, J., Wiener, H.H., Wright, L., Saba, S.G., Yim, D., Fein, A., Perez de Castro, I., Li, C., Thompson, C.B., Cox, A.D., and Philips, M.R. (2006). PKC regulates a farnesyl-electrostatic switch on K-Ras that promotes its association with Bcl-XL on mitochondria and induces apoptosis. *Mol Cell* 21:481–493.

52. Nancy, V., Callebaut, I., El Marjou, A., and de Gunzburg, J. (2002). The delta subunit of retinal rod cGMP phosphodiesterase regulates the membrane association of Ras and Rap GTPases. *J Biol Chem* 277:15076–15084.

53. Hanzal-Bayer, M., Renault, L., Roversi, P., Wittinghofer, A., and Hillig, R.C. (2002). The complex of Arl2-GTP and PDE delta: from structure to function. *EMBO J* 21:2095–2106.

54. Figueroa, C., Taylor, J., and Vojtek, A.B. (2001). Prenylated Rab acceptor protein is a receptor for prenylated small GTPases. *J Biol Chem* 276:28219–28225.

55. Paz, A., Haklai, R., Elad-Sfadia, G., Ballan, E., and Kloog, Y. (2001). Galectin-1 binds oncogenic H-Ras to mediate Ras membrane anchorage and cell transformation. *Oncogene* 20:7486–7493.

56. Kholodenko, B.N., Hancock, J.F., and Kolch, W. (2010). Signalling ballet in space and time. *Nat Rev Mol Cell Biol* 11:414–426.

57. Harding, A., and Hancock, J.F. (2008). Ras nanoclusters: combining digital and analog signaling. *Cell Cycle* 7:127–134.

58. Boettner, B., and Van Aelst, L. (2009). Control of cell adhesion dynamics by Rap1 signaling. *Curr Opin Cell Biol* 21:684–693.

59. Pannekoek, W.J., Kooistra, M.R., Zwartkruis, F.J., and Bos, J.L. (2009). Cell-cell junction formation: the role of Rap1 and Rap1 guanine nucleotide exchange factors. *Biochim Biophys Acta* 1788:790–796.

60. Kitayama, H., Sugimoto, Y., Matsuzaki, T., Ikawa, Y., and Noda, M. (1989). A ras-related gene with transformation suppressor activity. *Cell* 56:77–84.

61. Cook, S.J., Rubinfeld, B., Albert, I., and McCormick, F. (1993). RapV12 antagonizes Ras-dependent activation of ERK1 and ERK2 by LPA and EGF in Rat-1 fibroblasts. *EMBO J* 12:3475–3485.

62. Minato, N., and Hattori, M. (2009). Spa-1 (Sipa1) and Rap signaling in leukemia and cancer metastasis. *Cancer Sci* 100:17–23.

63. Buss, J.E., Quilliam, L.A., Kato, K., Casey, P.J., Solski, P.A., Wong, G., Clark, R., McCormick, F., Bokoch, G.M., and Der, C.J. (1991). The COOH-terminal domain of the Rap1A (Krev-1) protein is isoprenylated and supports transformation by an H-Ras: Rap1A chimeric protein. *Mol Cell Biol* 11:1523–1530.

64. Farrell, F.X., Yamamoto, K., and Lapetina, E.G. (1993). Prenyl group identification of rap2 proteins: a ras superfamily member other than ras that is farnesylated. *Biochem J* 289:349–355.

65. Beranger, F., Tavitian, A., and de Gunzburg, J. (1991). Post-translational processing and subcellular localization of the Ras-related Rap2 protein. *Oncogene* 6:1835–1842.

66. Pizon, V., Desjardins, M., Bucci, C., Parton, R.G., and Zerial, M. (1994). Association of Rap1a and Rap1b proteins with late endocytic/phagocytic compartments and Rap2a with the Golgi complex. *J Cell Sci* 107(Pt 6):1661–1670.

67. Berg, T.J., Gastonguay, A.J., Lorimer, E.L., Kuhnmuench, J.R., Li, R., Fields, A.P., and Williams, C.L. (2010). Splice variants of SmgGDS control small GTPase prenylation and membrane localization. *J Biol Chem* 285:35255–35266.

68. Kinsella, B.T., Erdman, R.A., and Maltese, W.A. (1991). Carboxyl-terminal isoprenylation of ras-related GTP-binding proteins encoded by rac1, rac2, and ralA. *J Biol Chem* 266:9786–9794.

69. Feig, L.A. (2003). Ral-GTPases: approaching their 15 minutes of fame. *Trends Cell Biol* 13:419–425.

70. Bodemann, B.O., and White, M.A. (2008). Ral GTPases and cancer: linchpin support of the tumorigenic platform. *Nat Rev Cancer* 8:133–140.

71. Wu, H., Rossi, G., and Brennwald, P. (2008). The ghost in the machine: small GTPases as spatial regulators of exocytosis. *Trends Cell Biol* 18:397–404.

72. Falsetti, S.C., Wang, D.A., Peng, H., Carrico, D., Cox, A.D., Der, C.J., Hamilton, A.D., and Sebti, S.M. (2007). Geranylgeranyltransferase I inhibitors target RalB to inhibit anchorage-dependent growth and induce apoptosis and RalA to inhibit anchorage-independent growth. *Mol Cell Biol* 27:8003–8014.

73. Zoncu, R., Efeyan, A., and Sabatini, D.M. (2011). mTOR: from growth signal integration to cancer, diabetes and ageing. *Nat Rev Mol Cell Biol* 12:21–35.

74. Avruch, J., Long, X., Lin, Y., Ortiz-Vega, S., Rapley, J., Papageorgiou, A., Oshiro, N., and Kikkawa, U. (2009). Activation of mTORC1 in two steps: Rheb-GTP activation of catalytic function and increased binding of substrates to raptor. *Biochem Soc Trans* 37:223–226.

75. Huang, J., and Manning, B.D. (2009). A complex interplay between Akt, TSC2 and the two mTOR complexes. *Biochem Soc Trans* 37:217–222.

76. Clark, G.J., Kinch, M.S., Rogers-Graham, K., Sebti, S.M., Hamilton, A.D., and Der, C.J. (1997). The Ras-related protein Rheb is farnesylated and antagonizes Ras signaling and transformation. *J Biol Chem* 272:10608–10615.

77. Castro, A.F., Rebhun, J.F., Clark, G.J., and Quilliam, L.A. (2003). Rheb binds tuberous sclerosis complex 2 (TSC2) and promotes S6 kinase activation in a rapamycin- and farnesylation-dependent manner. *J Biol Chem* 278:32493–32496.

78. Basso, A.D., Mirza, A., Liu, G., Long, B.J., Bishop, W.R., and Kirschmeier, P. (2005). The farnesyl transferase inhibitor (FTI) SCH66336 (lonafarnib) inhibits Rheb farnesylation and mTOR signaling. Role in FTI enhancement of taxane and tamoxifen anti-tumor activity. *J Biol Chem* 280:31101–31108.

79. Finlay, G.A., Malhowski, A.J., Liu, Y., Fanburg, B.L., Kwiatkowski, D.J., and Toksoz, D. (2007). Selective inhibition of growth of tuberous sclerosis complex 2 null cells by atorvastatin is associated with impaired Rheb and Rho GTPase function and reduced mTOR/S6 kinase activity. *Cancer Res* 67:9878–9886.

80. Jiang, H., and Vogt, P.K. (2008). Constitutively active Rheb induces oncogenic transformation. *Oncogene* 27:5729–5740.

81. Hanker, A.B., Mitin, N., Wilder, R.S., Henske, E.P., Tamanoi, F., Cox, A.D., and Der, C.J. (2010). Differential requirement of CAAX-mediated posttranslational processing for Rheb localization and signaling. *Oncogene* 29:380–391.

82. Berzat, A.C., Brady, D.C., Fiordalisi, J.J., and Cox, A.D. (2006). Using inhibitors of prenylation to block localization and transforming activity. *Methods Enzymol* 407:575–597.

83. Conklin, M.W., Ada-Nguema, A., Parsons, M., Riching, K.M., and Keely, P.J. (2010). R-Ras regulates beta1-integrin trafficking via effects on membrane ruffling and endocytosis. *BMC Cell Biol* 11:14.

84. Lehto, M., Mayranpaa, M.I., Pellinen, T., Ihalmo, P., Lehtonen, S., Kovanen, P.T., Groop, P.H., Ivaska, J., and Olkkonen, V.M. (2008). The R-Ras interaction partner ORP3 regulates cell adhesion. *J Cell Sci* 121:695–705.

85. Gawecka, J.E., Griffiths, G.S., Ek-Rylander, B., Ramos, J.W., and Matter, M.L. (2010). R-Ras regulates migration through an interaction with filamin A in melanoma cells. *PLoS ONE* 5:e11269.

86. Matsumoto, K., Asano, T., and Endo, T. (1997). Novel small GTPase M-Ras participates in reorganization of actin cytoskeleton. *Oncogene* 15:2409–2417.

87. Watanabe-Takano, H., Takano, K., Keduka, E., and Endo, T. (2010). M-Ras is activated by bone morphogenetic protein-2 and participates in osteoblastic determination, differentiation, and transdifferentiation. *Exp Cell Res* 316:477–490.

88. Saito, Y., Oinuma, I., Fujimoto, S., and Negishi, M. (2009). Plexin-B1 is a GTPase activating protein for M-Ras, remodelling dendrite morphology. *EMBO Rep* 10:614–621.

89. Carboni, J.M., Yan, N., Cox, A.D., Bustelo, X., Graham, S.M., Lynch, M.J., Weinmann, R., Seizinger, B.R., Der, C.J., Barbacid, M., *et al.* (1995). Farnesyltransferase inhibitors are inhibitors of Ras but not R-Ras2/TC21, transformation. *Oncogene* 10:1905–1913.

90. Graham, S.M., Cox, A.D., Drivas, G., Rush, M.G., D'Eustachio, P., and Der, C.J. (1994). Aberrant function of the Ras-related protein TC21/R-Ras2 triggers malignant transformation. *Mol Cell Biol* 14:4108–4115.

91. Erdogan, M., Pozzi, A., Bhowmick, N., Moses, H.L., and Zent, R. (2007). Signaling pathways regulating TC21-induced tumorigenesis. *J Biol Chem* 282:27713–27720.

92. Ellis, C.A., Vos, M.D., Howell, H., Vallecorsa, T., Fults, D.W., and Clark, G.J. (2002). Rig is a novel Ras-related protein and potential neural tumor suppressor. *Proc Natl Acad Sci USA* 99:9876–9881.

93. Kontani, K., Tada, M., Ogawa, T., Okai, T., Saito, K., Araki, Y., and Katada, T. (2002). Di-Ras, a distinct subgroup of ras family GTPases with unique biochemical properties. *J Biol Chem* 277:41070–41078.

94. Luo, R.Z., Fang, X., Marquez, R., Liu, S.Y., Mills, G.B., Liao, W.S., Yu, Y., and Bast, R.C. (2003). ARHI is a Ras-related small G-protein with a novel N-terminal extension that inhibits growth of ovarian and breast cancers. *Oncogene* 22:2897–2909.

95. Elam, C., Hesson, L., Vos, M.D., Eckfeld, K., Ellis, C.A., Bell, A., Krex, D., Birrer, M.J., Latif, F., and Clark, G.J. (2005). RRP22 is a farnesylated, nucleolar, Ras-related protein with tumor suppressor potential. *Cancer Res* 65:3117–3125.

96. Benetka, W., Koranda, M., Maurer-Stroh, S., Pittner, F., and Eisenhaber, F. (2006). Farnesylation or geranylgeranylation? Efficient assays for testing protein prenylation in vitro and in vivo *BMC Biochem* 7:6.

97. Vargiu, P., De Abajo, R., Garcia-Ranea, J.A., Valencia, A., Santisteban, P., Crespo, P., and Bernal, J. (2004). The small GTP-binding protein, Rhes, regulates signal transduction from G protein-coupled receptors. *Oncogene* 23:559–568.

98. Finlin, B.S., Gau, C.L., Murphy, G.A., Shao, H., Kimel, T., Seitz, R.S., Chiu, Y.F., Botstein, D., Brown, P.O., Der, C.J., Tamanoi, F., Andres, D.A., and Perou, C.M. (2001). RERG is a novel ras-related, estrogen-regulated and growth-inhibitory gene in breast cancer. *J Biol Chem* 276:42259–42267.

99. Reynet, C., and Kahn, C.R. (1993). Rad: a member of the Ras family overexpressed in muscle of type II diabetic humans. *Science* 262:1441–1444.

100. Cohen, L., Mohr, R., Chen, Y.Y., Huang, M., Kato, R., Dorin, D., Tamanoi, F., Goga, A., Afar, D., Rosenberg, N., *et al.* (1994). Transcriptional activation of a ras-like gene (kir) by oncogenic tyrosine kinases. *Proc Natl Acad Sci USA* 91:12448–12452.

101. Maguire, J., Santoro, T., Jensen, P., Siebenlist, U., Yewdell, J., and Kelly, K. (1994). Gem: an induced, immediate early protein belonging to the Ras family. *Science* 265:241–244.

102. Bilan, P.J., Moyers, J.S., and Kahn, C.R. (1998). The ras-related protein rad associates with the cytoskeleton in a non-lipid-dependent manner. *Exp Cell Res* 242:391–400.

103. Fenwick, C., Na, S.Y., Voll, R.E., Zhong, H., Im, S.Y., Lee, J.W., and Ghosh, S. (2000). A subclass of Ras proteins that regulate the degradation of IkappaB. *Science* 287:869–873.

104. Lee, C.H., Della, N.G., Chew, C.E., and Zack, D.J. (1996). Rin, a neuron-specific and calmodulin-binding small G-protein, and Rit define a novel subfamily of ras proteins. *J Neurosci* 16:6784–6794.

105. Shao, H., Kadono-Okuda, K., Finlin, B.S., and Andres, D.A. (1999). Biochemical characterization of the Ras-related GTPases Rit and Rin. *Arch Biochem Biophys* 371:207–219.

106. Calissano, M., and Latchman, D.S. (2003). Functional interaction between the small GTP-binding protein Rin and the N-terminal of Brn-3a transcription factor. *Oncogene* 22:5408–5414.

107. Chenette, E.J., Mitin, N.Y., and Der, C.J. (2006). Multiple sequence elements facilitate Chp Rho GTPase subcellular location, membrane association, and transforming activity. *Mol Biol Cell* 17:3108–3121.

108. Roberts, P.J., Mitin, N., Keller, P.J., Chenette, E.J., Madigan, J.P., Currin, R.O., Cox, A.D., Wilson, O., Kirschmeier, P., and Der, C.J. (2008). Rho Family GTPase modification and dependence on CAAX motif-signaled posttranslational modification. *J Biol Chem* 283:25150–25163.

109. Ridley, A.J. (2006). Rho GTPases and actin dynamics in membrane protrusions and vesicle trafficking. *Trends Cell Biol* 16:522–529.

110. Michaelson, D., Silletti, J., Murphy, G., D'Eustachio, P., Rush, M., and Philips, M.R. (2001). Differential localization of Rho GTPases in live cells: regulation by hypervariable regions and RhoGDI binding. *J Cell Biol* 152:111–126.

111. Hart, M.J., Maru, Y., Leonard, D., Witte, O.N., Evans, T., and Cerione, R.A. (1992). A GDP dissociation inhibitor that serves as a GTPase inhibitor for the Ras-like protein CDC42Hs. *Science* 258:812–815.

112. Nomanbhoy, T.K., and Cerione, R. (1996). Characterization of the interaction between RhoGDI and Cdc42Hs using fluorescence spectroscopy. *J Biol Chem* 271:10004–10009.

113. Fukumoto, Y., Kaibuchi, K., Hori, Y., Fujioka, H., Araki, S., Ueda, T., Kikuchi, A., and Takai, Y. (1990). Molecular cloning and characterization of a novel type of regulatory protein (GDI) for the rho proteins, ras p21-like small GTP-binding proteins. *Oncogene* 5:1321–1328.

114. Leonard, D., Hart, M.J., Platko, J.V., Eva, A., Henzel, W., Evans, T., and Cerione, R.A. (1992). The identification and characterization of a GDP-dissociation inhibitor (GDI) for the CDC42Hs protein. *J Biol Chem* 267:22860–22868.

115. Hoffman, G.R., Nassar, N., and Cerione, R.A. (2000). Structure of the Rho family GTP-binding protein Cdc42 in complex with the multifunctional regulator RhoGDI. *Cell* 100:345–356.

116. DerMardirossian, C., and Bokoch, G.M. (2005). GDIs: central regulatory molecules in Rho GTPase activation. *Trends Cell Biol* 15:356–363.

117. Bilodeau, D., Lamy, S., Desrosiers, R.R., Gingras, D., and Beliveau, R. (1999). Regulation of Rho protein binding to membranes by rhoGDI: inhibition of releasing activity by physiological ionic conditions. *Biochem Cell Biol* 77:59–69.

118. Boulter, E., Garcia-Mata, R., Guilluy, C., Dubash, A., Rossi, G., Brennwald, P.J., and Burridge, K. (2010). Regulation of Rho GTPase crosstalk, degradation and activity by RhoGDI1. *Nat Cell Biol* 12:477–483.

119. Chardin, P. (2006). Function and regulation of Rnd proteins. *Nat Rev Mol Cell Biol* 7:54–62.

120. Riou, P., Villalonga, P., and Ridley, A.J. (2010). Rnd proteins: multifunctional regulators of the cytoskeleton and cell cycle progression. *Bioessays* 32:986–992.

121. Meller, J., Vidali, L., and Schwartz, M.A. (2008). Endogenous RhoG is dispensable for integrin-mediated cell spreading but contributes to Rac-independent migration. *J Cell Sci* 121:1981–1989.

122. Prosser, D.C., Tran, D., Schooley, A., Wendland, B., and Ngsee, J.K. (2010). A novel, retromer-independent role for sorting nexins 1 and 2 in RhoG-dependent membrane remodeling. *Traffic* 11:1347–1362.

123. Hiramoto-Yamaki, N., Takeuchi, S., Ueda, S., Harada, K., Fujimoto, S., Negishi, M., and Katoh, H. (2010). Ephexin4 and EphA2 mediate cell migration through a RhoG-dependent mechanism. *J Cell Biol* 190:461–477.

124. Prieto-Sanchez, R.M., Berenjeno, I.M., and Bustelo, X.R. (2006). Involvement of the Rho/Rac family member RhoG in caveolar endocytosis. *Oncogene* 25:2961–2973.

125. Zalcman, G., Closson, V., Camonis, J., Honore, N., Rousseau-Merck, M.F., Tavitian, A., and Olofsson, B. (1996). RhoGDI-3 is a new GDP dissociation inhibitor (GDI). Identification of a non-cytosolic GDI protein interacting with the small GTP-binding proteins RhoB and RhoG. *J Biol Chem* 271:30366–30374.

126. Elfenbein, A., Rhodes, J.M., Meller, J., Schwartz, M.A., Matsuda, M., and Simons, M. (2009). Suppression of RhoG activity is mediated by a syndecan 4-synectin-RhoGDI1 complex and is reversed by PKCalpha in a Rac1 activation pathway. *J Cell Biol* 186:75–83.

127. Brunet, N., Morin, A., and Olofsson, B. (2002). RhoGDI-3 regulates RhoG and targets this protein to the Golgi complex through its unique N-terminal domain. *Traffic* 3:342–357.

128. Berzat, A.C., Buss, J.E., Chenette, E.J., Weinbaum, C.A., Shutes, A., Der, C.J., Minden, A., and Cox, A.D. (2005). Transforming activity of the Rho family GTPase, Wrch-1, a Wnt-regulated Cdc42 homolog, is dependent on a novel carboxyl-terminal palmitoylation motif. *J Biol Chem* 280:33055–33065.

129. Chenette, E.J., Abo, A., and Der, C.J. (2005). Critical and distinct roles of amino- and carboxyl-terminal sequences in regulation of the biological activity of the Chp atypical Rho GTPase. *J Biol Chem* 280:13784–13792.

130. Ramos, S., Khademi, F., Somesh, B.P., and Rivero, F. (2002). Genomic organization and expression profile of the small GTPases of the RhoBTB family in human and mouse. *Gene* 298:147–157.

131. Berthold, J., Schenkova, K., and Rivero, F. (2008). Rho GTPases of the RhoBTB subfamily and tumorigenesis. *Acta Pharmacol Sin* 29:285–295.

132. Espinosa, E.J., Calero, M., Sridevi, K., and Pfeffer, S.R. (2009). RhoBTB3: a Rho GTPase-family ATPase required for endosome to Golgi transport. *Cell* 137:938–948.

133. Baron, R.A., and Seabra, M.C. (2008). Rab geranylgeranylation occurs preferentially via the pre-formed REP-RGGT complex and is regulated by geranylgeranyl pyrophosphate. *Biochem J* 415:67–75.

134. Leung, K.F., Baron, R., and Seabra, M.C. (2006). Thematic review series: lipid posttranslational modifications. Geranylgeranylation of Rab GTPases. *J Lipid Res* 47:467–475.

135. Shen, F., and Seabra, M.C. (1996). Mechanism of digeranylgeranylation of Rab proteins. Formation of a complex between monogeranylgeranyl-Rab and Rab escort protein. *J Biol Chem* 271:3692–3698.

136. Gomes, A.Q., Ali, B.R., Ramalho, J.S., Godfrey, R.F., Barral, D.C., Hume, A.N., and Seabra, M.C. (2003). Membrane targeting of Rab GTPases is influenced by the prenylation motif. *Mol Biol Cell* 14:1882–1899.

137. Soldati, T., Riederer, M.A., and Pfeffer, S.R. (1993). Rab GDI: a solubilizing and recycling factor for rab9 protein. *Mol Biol Cell* 4:425–434.

138. Smeland, T.E., Seabra, M.C., Goldstein, J.L., and Brown, M.S. (1994). Geranylgeranylated Rab proteins terminating in Cys-Ala-Cys, but not Cys-Cys, are carboxyl-methylated by bovine brain membranes in vitro. *Proc Natl Acad Sci USA* 91:10712–10716.

139. Leung, K.F., Baron, R., Ali, B.R., Magee, A.I., and Seabra, M.C. (2007). Rab GTPases containing a CAAX motif are processed post-geranylgeranylation by proteolysis and methylation. *J Biol Chem* 282:1487–1497.

140. Pfeffer, S. (2005). A model for Rab GTPase localization. *Biochem Soc Trans* 33:627–630.

141. Wu, Y.W., Oesterlin, L.K., Tan, K.T., Waldmann, H., Alexandrov, K., and Goody, R.S. (2010). Membrane targeting mechanism of Rab GTPases elucidated by semisynthetic protein probes. *Nat Chem Biol* 6:534–540.

142. Kahn, R.A. (2009). Toward a model for Arf GTPases as regulators of traffic at the Golgi. *FEBS Lett* 583:3872–3879.

143. Pucadyil, T.J., and Schmid, S.L. (2009). Conserved functions of membrane active GTPases in coated vesicle formation. *Science* 325:1217–1220.

144. Liu, Y., Kahn, R.A., and Prestegard, J.H. (2009). Structure and membrane interaction of myristoylated ARF1. *Structure* 17:79–87.

145. Pemberton, L.F., and Paschal, B.M. (2005). Mechanisms of receptor-mediated nuclear import and nuclear export. *Traffic* 6:187–198.

146. Yudin, D., and Fainzilber, M. (2009). Ran on tracks—cytoplasmic roles for a nuclear regulator. *J Cell Sci* 122:587–593.

147. Reiss, Y., Goldstein, J.L., Seabra, M.C., Casey, P.J., and Brown, M.S. (1990). Inhibition of purified p21ras farnesyl:protein transferase by Cys-AAX tetrapeptides. *Cell* 62:81–88.

148. Rowinsky, E.K., Windle, J.J., and Von Hoff, D.D. (1999). Ras protein farnesyltransferase: a strategic target for anticancer therapeutic development. *J Clin Oncol* 17:3631–3652.

149. Basso, A.D., Kirschmeier, P., and Bishop, W.R. (2006). Lipid posttranslational modifications. Farnesyl transferase inhibitors. *J Lipid Res* 47:15–31.

150. James, G.L., Goldstein, J.L., Brown, M.S., Rawson, T.E., Somers, T.C., McDowell, R.S., Crowley, C.W., Lucas, B.K., Levinson, A.D., and Marsters, J.C., Jr. (1993). Benzodiazepine peptidomimetics: potent inhibitors of Ras farnesylation in animal cells. *Science* 260:1937–1942.

151. Kohl, N.E., Mosser, S.D., deSolms, S.J., Giuliani, E.A., Pompliano, D.L., Graham, S.L., Smith, R.L., Scolnick, E.M., Oliff, A., and Gibbs, J.B. (1993). Selective inhibition of ras-dependent transformation by a farnesyltransferase inhibitor. *Science* 260:1934–1937.

152. Cohen, S.J., Ho, L., Ranganathan, S., Abbruzzese, J.L., Alpaugh, R.K., Beard, M., Lewis, N.L., McLaughlin, S., Rogatko, A., Perez-Ruixo, J.J., Thistle, A.M., Verhaeghe, T., Wang, H., Weiner, L.M., Wright, J.J., Hudes, G.R., and Meropol, N.J. (2003). Phase II and pharmacodynamic study of the farnesyltransferase inhibitor R115777

as initial therapy in patients with metastatic pancreatic adenocarcinoma. *J Clin Oncol* 21:1301–1306.

153. Macdonald, J.S., McCoy, S., Whitehead, R.P., Iqbal, S., Wade, J.L., 3rd, Giguere, J.K., and Abbruzzese, J.L. (2005). A phase II study of farnesyl transferase inhibitor R115777 in pancreatic cancer: a Southwest oncology group (SWOG 9924) study. *Invest New Drugs* 23:485–487.

154. Van Cutsem, E., van de Velde, H., Karasek, P., Oettle, H., Vervenne, W.L., Szawlowski, A., Schoffski, P., Post, S., Verslype, C., Neumann, H., Safran, H., Humblet, Y., Perez Ruixo, J., Ma, Y., and Von Hoff, D. (2004). Phase III trial of gemcitabine plus tipifarnib compared with gemcitabine plus placebo in advanced pancreatic cancer. *J Clin Oncol* 22:1430–1438.

155. Rao, S., Cunningham, D., de Gramont, A., Scheithauer, W., Smakal, M., Humblet, Y., Kourteva, G., Iveson, T., Andre, T., Dostalova, J., Illes, A., Belly, R., Perez-Ruixo, J.J., Park, Y.C., and Palmer, P.A. (2004). Phase III double-blind placebo-controlled study of farnesyl transferase inhibitor R115777 in patients with refractory advanced colorectal cancer. *J Clin Oncol* 22:3950–3957.

156. James, G., Goldstein, J.L., and Brown, M.S. (1996). Resistance of K-RasBV12 proteins to farnesyltransferase inhibitors in Rat1 cells. *Proc Natl Acad Sci USA* 93:4454–4458.

157. Rowell, C.A., Kowalczyk, J.J., Lewis, M.D., and Garcia, A.M. (1997). Direct demonstration of geranylgeranylation and farnesylation of Ki-Ras in vivo. *J Biol Chem* 272:14093–14097.

158. Cox, A.D., Hisaka, M.M., Buss, J.E., and Der, C.J. (1992). Specific isoprenoid modification is required for function of normal, but not oncogenic, Ras protein. *Mol Cell Biol* 12:2606–2615.

159. Worman, H.J., Fong, L.G., Muchir, A., and Young, S.G. (2009). Laminopathies and the long strange trip from basic cell biology to therapy. *J Clin Invest* 119:1825–1836.

160. Yang, S.H., Bergo, M.O., Toth, J.I., Qiao, X., Hu, Y., Sandoval, S., Meta, M., Bendale, P., Gelb, M.H., Young, S.G., and Fong, L.G. (2005). Blocking protein farnesyltransferase improves nuclear blebbing in mouse fibroblasts with a targeted Hutchinson-Gilford progeria syndrome mutation. *Proc Natl Acad Sci USA* 102:10291–10296.

161. Mallampalli, M.P., Huyer, G., Bendale, P., Gelb, M.H., and Michaelis, S. (2005). Inhibiting farnesylation reverses the nuclear morphology defect in a HeLa cell model for Hutchinson-Gilford progeria syndrome. *Proc Natl Acad Sci USA* 102:14416–14421.

162. Capell, B.C., Erdos, M.R., Madigan, J.P., Fiordalisi, J.J., Varga, R., Conneely, K.N., Gordon, L.B., Der, C.J., Cox, A.D., and Collins, F.S. (2005). Inhibiting farnesylation of progerin prevents the characteristic nuclear blebbing of Hutchinson-Gilford progeria syndrome. *Proc Natl Acad Sci USA* 102:12879–12884.

163. Glynn, M.W., and Glover, T.W. (2005). Incomplete processing of mutant lamin A in Hutchinson-Gilford progeria leads to nuclear abnormalities, which are reversed by farnesyltransferase inhibition. *Hum Mol Genet* 14:2959–2969.

164. Fong, L.G., Frost, D., Meta, M., Qiao, X., Yang, S.H., Coffinier, C., and Young, S.G. (2006). A protein farnesyltransferase inhibitor ameliorates disease in a mouse model of progeria. *Science* 311:1621–1623.

165. Varela, I., Pereira, S., Ugalde, A.P., Navarro, C.L., Suárez, M.F., Cau, P., Cadiñanos, J., Osorio, F.G., Foray, N., Cobo, J., de Carlos, F., Lévy, N., Freije, J.M., and López-Otín, C. (2008). Combined treatment with statins and aminobisphosphonates extends longevity in a mouse model of human premature aging. *Nat Med* 14:767–772.

166. Liu, M., Sjogren, A.K., Karlsson, C., Ibrahim, M.X., Andersson, K.M., Olofsson, F.J., Wahlstrom, A.M., Dalin, M., Yu, H., Chen, Z., Yang, S.H., Young, S.G., and Bergo, M.O.

(2010). Targeting the protein prenyltransferases efficiently reduces tumor development in mice with K-RAS-induced lung cancer. *Proc Natl Acad Sci USA* 107:6471–6476.

167. Sjogren, A.K., Andersson, K.M., Khan, O., Olofsson, F.J., Karlsson, C., and Bergo, M.O. (2011). Inactivating GGTase-I reduces disease phenotypes in a mouse model of K-RAS-induced myeloproliferative disease. *Leukemia* 25:186–189

168. Lobell, R.B., Liu, D., Buser, C.A., Davide, J.P., DePuy, E., Hamilton, K., Koblan, K.S., Lee, Y., Mosser, S., Motzel, S.L., Abbruzzese, J.L., Fuchs, C.S., Rowinsky, E.K., Rubin, E.H., Sharma, S., Deutsch, P.J., Mazina, K.E., Morrison, B.W., Wildonger, L., Yao, S.L., and Kohl, N.E. (2002). Preclinical and clinical pharmacodynamic assessment of L-778,123, a dual inhibitor of farnesyl:protein transferase and geranylgeranyl:protein transferase type-I. *Mol Cancer Ther* 1:747–758.

169. Chia, W.J., and Tang, B.L. (2009). Emerging roles for Rab family GTPases in human cancer. *Biochim Biophys Acta* 1795:110–116.

170. Agarwal, R., Jurisica, I., Mills, G.B., and Cheng, K.W. (2009). The emerging role of the RAB25 small GTPase in cancer. *Traffic* 10:1561–1568.

171. Lackner, M.R., Kindt, R.M., Carroll, P.M., Brown, K., Cancilla, M.R., Chen, C., de Silva, H., Franke, Y., Guan, B., Heuer, T., Hung, T., Keegan, K., Lee, J.M., Manne, V., O'Brien, C., Parry, D., Perez-Villar, J.J., Reddy, R.K., Xiao, H., Zhan, H., Cockett, M., Plowman, G., Fitzgerald, K., Costa, M., and Ross-Macdonald, P. (2005). Chemical genetics identifies Rab geranylgeranyl transferase as an apoptotic target of farnesyl transferase inhibitors. *Cancer Cell* 7:325–336.

172. McMahon, L.P., Yue, W., Santen, R.J., and Lawrence, J.C.J. (2005). Farnesylthiosalicylic acid inhibits mammalian target of rapamycin (mTOR) activity both in cells and in vitro by promoting dissociation of the mTOR-raptor complex. *Mol Endocrinol* 19:175–183.

173. Yue, W., Wang, J., Li, Y., Fan, P., and Santen, R.J. (2005). Farnesylthiosalicylic acid blocks mammalian target of rapamycin signaling in breast cancer cells. *Int J Cancer* 117:746–754.

174. Bergo, M.O., Ambroziak, P., Gregory, C., George, A., Otto, J.C., Kim, E., Nagase, H., Casey, P.J., Balmain, A., and Young, S.G. (2002). Absence of the CAAX endoprotease Rce1: effects on cell growth and transformation. *Mol Cell Biol* 22:171–181.

175. Bergo, M.O., Gavino, B.J., Hong, C., Beigneux, A.P., McMahon, M., Casey, P.J., and Young, S.G. (2004). Inactivation of Icmt inhibits transformation by oncogenic K-Ras and B-Raf. *J Clin Invest* 113:539–550.

176. Wahlstrom, A.M., Cutts, B.A., Liu, M., Lindskog, A., Karlsson, C., Sjogren, A.K., Andersson, K.M., Young, S.G., and Bergo, M.O. (2008). Inactivating Icmt ameliorates K-RAS-induced myeloproliferative disease. *Blood* 112:1357–1365.

177. Winter-Vann, A.M., and Casey, P.J. (2005). Post-prenylation-processing enzymes as new targets in oncogenesis. *Nat Rev Cancer* 5:405–412.

178. Draper, J.M., and Smith, C.D. (2009). Palmitoyl acyltransferase assays and inhibitors. *Mol Membr Biol* 26:5–13.

179. Jiang, S.Y., and Ramachandran, S. (2006). Comparative and evolutionary analysis of genes encoding small GTPases and their activating proteins in eukaryotic genomes. *Physiol Genomics* 24:235–251.

180. Lundquist, E.A. (2006). Small GTPases. In WormBook (Ed.), The *C. elegans* Research Community. WormBook. doi:10.1895/wormbook.1.67.1. http://www.wormbook.org.

181. Vetter, I.R., and Wittinghofer, A. (2001). The guanine nucleotide-binding switch in three dimensions. *Science* 294:1299–1304.

182. Kahn, R.A., Bruford, E., Inoue, H., Logsdon, J.M., Jr., Nie, Z., Premont, R.T., Randazzo, P.A., Satake, M., Theibert, A.B., Zapp, M.L., and Cassel, D. (2008). Consensus nomenclature for the human ArfGAP domain-containing proteins. *J Cell Biol* 182:1039–1044.

183. Buday, L., and Downward, J. (2008). Many faces of Ras activation. *Biochim Biophys Acta* 1786:178–187.
184. Tcherkezian, J., and Lamarche-Vane, N. (2007). Current knowledge of the large RhoGAP family of proteins. *Biol Cell* 99:67–86.
185. Rossman, K.L., Der, C.J., and Sondek, J. (2005). GEF means go: turning on RHO GTPases with guanine nucleotide-exchange factors. *Nat Rev Mol Cell Biol* 6:167–180.
186. Marat, A.L., Dokainish, H., and McPherson, P.S. (2011). DENN domain proteins: regulators of Rab GTPases. *J Biol Chem* 286:13791–13800.
187. Cox, R., Mason-Gamer, R.J., Jackson, C.L., and Segev, N. (2004). Phylogenetic analysis of Sec7-domain-containing Arf nucleotide exchangers. *Mol Biol Cell* 15:1487–1505.
188. Fukuda, M. (2011). TBC proteins: GAPs for mammalian small GTPase Rab? *Biosci Rep* 31:159–168.

6

Heterogeneous Prenyl Processing of the Heterotrimeric G protein Gamma Subunits

JOHN D. HILDEBRANDT

Department of Cell Molecular Pharmacology
Medical University of South Carolina
Charleston, South Carolina, USA

I. Abstract

The heterotrimeric G proteins contain one member each from three small gene families with 16 Gα, 5 Gβ, and 12 Gγ genes coded for in the human genome. They mediate the cellular effects of many hormones, neurotransmitters, sensory stimuli, and other extracellular regulators, including the effects of many drugs. The small (7–9 kDa) Gγ subunits are all prenylated proteins. Prenylation is necessary for or a component of most membrane and protein–protein interactions required for signaling by these proteins. Characterization of native G proteins purified from bovine brain indicates significant variation in the expression of each of the three constituent reactions characteristic of prenyl modifications: attachment of the prenyl group, proteolytic removal of the last three amino acids of the CaaX prenylation signal sequence, and carboxymethylation of the new C-terminus. At least some of these variations, such as resistance to proteolytic processing, are determined by the sequence of the CaaX motif, which is evolutionarily conserved for a number of Gγ proteins at all four positions in the signal sequence. Heterogeneity in prenyl processing associated with these proteins generates variation at a site that is instrumental in many

ISSN NO: 1874-6047
DOI: 10.1016/B978-0-12-381339-8.00006-8

interactions of the G protein with both membranes and other proteins. The functional significance of this variation is discussed in terms of the past biochemical characterization of the constituent reactions of the protein prenylation modifications.

II. Heterotrimeric G Proteins and Prenylation of the Gγ Subunit

A. G PROTEIN STRUCTURE AND SIGNALING DYNAMICS

G proteins are named for their regulation by guanine nucleotides. Most basically, they have two states: an "inactive" state with GDP bound, which does not generate the cellular response normally associated with the protein, and an "active" state with GTP bound, which does generate a downstream signaling response [1,2]. Guanine nucleotide exchange factors (GEFs) activate G proteins by catalyzing the exchange of GDP for GTP. The functional lifetime of the G protein is determined by its inherent GTPase activity that hydrolyzes GTP to GDP, converting it from the "active" to the "inactive" state. Two broad classes of G proteins are sometimes referred to as the small (or monomeric) G proteins, such as Ras and its many homologs (discussed in several other chapters of this volume), and the heterotrimeric G proteins [3,4] that generate responses to hormones, neurotransmitters, and other extracellular signals through their activation by G protein-coupled receptors (GPCRs) (Figure 6.1). GPCRs are the primary GEFs for the heterotrimeric G proteins. They are called heterotrimeric because they contain three nonidentical subunits [8–11]: Gα (37–46 kDa) with the guanine nucleotide binding site; Gβ (~37 kDa), which is a seven-bladed beta propeller protein; and a small (7–9 kDa) Gγ subunit. The classical view is that heterotrimeric G proteins are activated by a dual reaction involving nucleotide exchange (GTP for GDP on Gα) and subunit dissociation, whereby Gα-GTP separates from a stable Gβγ dimer [3]. There has long been debate over the obligatory role of subunit dissociation in G protein activation [12–14]. Current evidence supports the idea that, whereas some G protein dissociate upon activation, others do not [6,15]. Regardless of the mechanism of activation, Gα and Gβγ often behave as if they independently regulate intracellular enzymes and/or ion channels to generate intracellular responses [4,16,17]. Throughout the remainder of this discussion, unless specified otherwise, G protein will refer to the heterotrimeric G proteins.

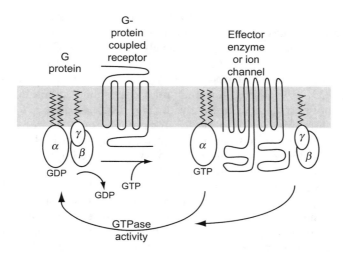

FIG. 6.1. Components and "classical" mechanism of action of the G protein signaling system. The shaded area represents the plane of the membrane. Most Gα subunits have palmitoyl and/or myristoyl residues at or near the N-terminus that could serve as membrane anchors. Membrane localization of Gβγ requires a C-terminal prenyl group on Gγ that may directly anchor the dimer to membranes [5] or may do so indirectly by (prenyl)protein–protein interactions [6,7].

B.   COMBINATORIAL ASSEMBLY OF HETEROTRIMERIC G PROTEINS

There are about 400 nonolfactory GPCR genes in the human genome [18]. The products of a subset of about 100 of these are directly or indirectly the targets of 30–50% of clinically useful drugs [19]. G protein signaling is widespread throughout most eukaryotic organisms and is particularly prominent within metazoans (animals). To generate responses to a wide range of receptors, the heterotrimeric G proteins within vertebrates are also diverse, based upon their combinatorial assembly into heterotrimers from (within humans) 16 Gα, 5 Gβ, and 12 Gγ genes [20]. G proteins require lipid modifications for membrane targeting [21–25]: the Gα subunits through isoform-specific patterns of myristoyl and/or palmitoyl residues at or near their N-termini, and the Gβγ dimer through a prenyl group attached to the C-terminus of the Gγ subunit.

C.   Gγ SUBUNIT DIVERSITY AND PRENYLATION

Prenylation is a complex, sequence-dependent C-terminal modification of proteins most commonly coded for by the last four (but sometimes alternatively three) amino acids, most frequently as a CaaX motif [26–32].

The prenyl group is attached through a thioether linkage to a Cys (C of CaaX) four residues from the C-terminus. The prenyl group attached is determined primarily, but not exclusively [33], by the C-terminal amino acid (X); where a 20-carbon geranylgeranyl moiety is added if X is Leu, or (sometimes) Phe, or a 15-carbon farnesyl moiety if X is Ser, Met, Gln, Cys, or Ala [25,34]. For other C-terminal amino acids, and sometimes Phe, the protein is not prenylated. Classically, the two residues between C (Cys) and X are aliphatic amino acids (a). Following prenylation, the three C-terminal amino acids are removed by a specific protease called Rce1 (Ras converting enzyme 1) [35,36] and the protein is carboxymethylated by a prenyl protein-specific carboxymethylase (Icmt) [37–39].

Early recognition of the similarity of C-terminal sequences of Ras and Gγ proteins [40], and the realization that these sequences included protein prenylation signals [41–43], led to the rapid characterization of prenylation of the Gγ subunits [44–51]. All 12 (human) Gγ isoforms are modified by prenylation, but some (Gγ 1, Gγ 9, and Gγ 11) contain the 15-carbon farnesyl moiety, while others (Gγ2, Gγ3, Gγ4, Gγ5, Gγ7, Gγ8, Gγ10, Gγ12, and Gγ13) contain the 20-carbon geranylgeranyl moiety. Several of the Gγ CaaX motifs are evolutionarily well conserved at all four positions (Figure 6.2), suggesting that all four of these residues are functionally significant.

III.   Functional Role of G Protein γ Subunit Prenylation

Prenyl modifications have multiple and complex functions [25,28,52]. At least four functional roles can be ascribed to protein prenylation in the context of the heterotrimeric G proteins. These different functions are not mutually exclusive and are interrelated, making their distinction somewhat arbitrary. Nevertheless, this is a convenient framework for discussing the roles of prenylation in the function of heterotrimeric G proteins.

A.   MEMBRANE ATTACHMENT OF PRENYLATED PROTEINS

An early idea about the role of prenylation was that it targeted proteins to membranes where they are active [33,53–57]. One perceived difference between geranylgeranylation and farnesylation is the strength of their membrane association when not coupled with other membrane attachment mechanisms [57,58]. Many studies indicate that prenylation is required for membrane localization and targeting of Gβγ [59–66] and that carboxy-methylation increases membrane association, particularly of G proteins with a farnesylated Gγ [67]. Membrane association is a basic property of

Geranylgeranylated proteins / Farnesylated proteins — CaaX motif table

Common name	Species name	Gγ1	Gγ9	Gγ11	Gγ2	Gγ3	Gγ4	Gγ5	Gγ7	Gγ8	Gγ10	Gγ12	Gγ13
Mammal													
Human	*Homo sapiens*	CVIS	CLIS	CVIS	CAIL	CALL	CTIL	CSFL	CTIL	CVLL	CALL	CIIL	CTIL
Chimpanzee	*Pan troglodytes*	CVIS	CLIS	CVIS	CAIL	CALL	CTIL	CSFL	CIIL	xxxx	CALL	– – –	CTIL
Rhesus monkey	*Macaca mulatta*	CVIS	CLIS	CVVS	CAIL	– – –	CTIL	CSFL	CIIL	CVLL	CALL	CTIL	CTIL
Cow	*Bos taurus*	CVIS	CIIS	CIIS	CAIL	CALL	CTIL	CSFL	CIIL	CVLL	CALL	CTIL	CAIL
Dog	*Canis lupus familiaris*	CVIS	CMIS	CIIS	CAIL	– – –	CTIL	xxxx	CIII	CVLL	– – –	CAIL	CAIL
Horse	*Equus caballus*	CVIS	CVLS	CVIS	CAIL	CALL	CTIL	CSFL	CIIL	CVLL	CALL	CIIL	CTIL
Mouse	*Mus musculus*	CVIS	CVIS	CVIS	CAIL	CALL	CTIL	CSFL	CIIL	CTLL	CALL	CIIL	CTIL
Rat	*Rattus norvegicus*	CVIS	CVIS	CVIS	CAIL	CALL	CTIL	CSFL	CIIL	CTIL	CALL	CIIL	CTIL
Pig	*Sus scrofa*	CVIS	– – –	CIIS	CAIL	CALL	CTIL	– – –	– – –	– – –	CALL	CIIL	CAIL
Marsupial													
Opossum	*Monodelphis domestica*	– – –	CMIS	– – –	– – –	CALL	CTIL	CSFL	CIIL	CVLL	CVLL	CIIL	CTIL
Monotreme													
Duck-billed platypus	*Ornithorhynchus anatinus*	CAIS	– – –	– – –	CAIL	CALL	– – –	CSFL	CIIL	– – –	CALL	CIIL	CAIL
Bird													
Chicken	*Gallus gallus*	CVIA	CIIT	– – –	CVIL	– – –	CTIL	CSFL	– – –	– – –	CTIL	CVLL	CSIL
Zebra finch	*Taeniopygia guttata*	CVIA	CTIT	– – –	CVIL	– – –	CTIL	CSFL	CIIL	– – –	CALL	CIIL	– – –
Amphibian													
Western clawed frog	*Xenopus tropicalis*	CVIA	CLIS	– – –	CAIL	CALL	CTIL	CSFL	CSIL	– – –	CALL	CIIL	CSIL
Fish													
Zebra fish	*Danio rerio*	CVIC	CIIT	– – –	CAIL	CALL	CTIL	CSFL	CTIL	CTVL	CTLL	CTIL	CVLL
Puffer fish	*Takifugu Fugu rubripes*	CVLS	CTIT	– – –	CAIL	– – –	– – –	CSFL	CIIL	CTIL	CTVV	CALL	CVIL
Freshwater puffer fish	*Tetraodon nigroviridis*	CVLS	CIIT	– – –	CAIL	CALL	CTIL	CSFL	CIIL	CIIL	CSLF	CTIL	CVIL

Fᴵɢ. 6.2. Evolutionary conservation within vertebrates of the CaaX motif of Gγ subunits. Shown is the CaaX motif of the closest homolog of the human Gγ in vertebrate genome databases. Sequences noted as "– – –" indicate that a close homolog was not found in the indicated genome database either because the gene was lost, as is likely the case for some bird Gγ, because the gene was generated latter due to a gene duplication, as in absence of Gγ11 in fish, or because of an incomplete state of annotation of a genome, as for the pig. Sequences noted as "xxxx" indicate that a close (essentially identical) homolog for the Gγ protein exists based upon sequence corresponding to one exon of the gene, but that the exon containing the prenylation signal has not yet been annotated, as for the Gγ5 gene in the dog genome.

the heterotrimeric G proteins from which is generated the other functions ascribed to Gγ prenylation that give these proteins their role in cell signaling.

B. INTRACELLULAR MEMBRANE TRAFFICKING OF PRENYLATED PROTEINS

Prenylation is often one step in an integrated mechanism for targeting modified proteins to specific membranes. For example, for the Ras proteins, prenylation appears to constitute one of two signals required for (plasma) membrane targeting [58,68]. The other signal is provided either by basic residues in a hypervariable region near the C-terminus, or by palmitoylation of this region, a modification that requires previous prenylation [42]. For the heterotrimeric G proteins, prenylation is a component of a sequential intracellular trafficking mechanism. Prenylation of Gγ subunits of Gβγ dimers targets them to the ER [69] where they form heterotrimers with Gα subunits, a process that is required for transit of the complex to the plasma membrane [59]. For heterotrimeric G proteins, the dual signals for transit to the plasma membrane appears to be intimately tied to hetero-trimer formation where one signal, prenylation, is on the Gβγ dimer, while the other, myristoylation or palmitoylation, is on the Gα subunit.

C. REVERSIBLE TRAFFICKING BETWEEN PLASMA MEMBRANES AND ENDOMEMBRANES

Classically, the heterotrimeric G proteins were considered to be functionally active from the inner leaflet of the plasma membrane [3,4]. A more recent idea is that G protein function is modulated both by constitutive and by regulated trafficking of the G protein subunits between the plasma membrane and internal membranes [70]. Based upon studies using fluorescence recovery after photobleaching (FRAP), intact G protein heterotrimers constitutively shuttle between the plasma membrane and internal membranes [71]. This constitutive trafficking appears to be related to a cycle of palmitoylation–depalmitoylation (presumably of Gα) similar to what has been shown for Ras proteins [72,73], but with faster kinetics ($t_{1/2} < 1$ min vs. $t_{1/2} > 10$ min) [71]. The role of prenylation in this process may be only indirect in that it is required for membrane targeting, but prenylation may also be a more direct contributor depending upon the role of the prenyl group in protein–protein interactions associated with this trafficking process (see below).

Superimposed upon constitutive trafficking of G protein heterotrimers between the plasma membrane and internal membranes, it is proposed that the Gβγ dimer of some GPCR-activated G proteins (Gβγ dimers with Gγ 1,

9, 11, 13, 5, and 10) reversibly translocate to internal membranes, leaving their dissociated Gα in the plasma membrane [74]. In contrast, other G proteins (those with Gγ 2, 3, 4, 7, and 12), upon receptor activation, are proposed to remain associated with the plasma membrane [74]. The identity of the prenyl group on Gγ was not a direct determinant of this process because both farnesylated and geranylgeranylated Gβγ dimers traffic between membranes and because the behavior of different dimers could be changed by altering amino acid residues upstream of the Gγ CaaX motif [74]. The behavior of different Gβγ dimers was explained by the stability of the heterotrimer they make with Gα, along with the stability of this heterotrimer for its interaction with the receptor [70].

The idea that intact G proteins, or their dissociated subunits (Gβγ dimers), traffic between the plasma membrane and internal membranes as an integral part of their signaling function [70], although not entirely new [75], is a fundamental change in the way of thinking about G protein signaling in cells. There are still many questions to be resolved about this hypothesis, but one of these is the precise role of Gγ prenylation and whether this role is simply indirect, as one component of membrane targeting, or if it is more primary, possibly because of its role in mediating protein–protein interactions.

D. PROTEIN–PROTEIN INTERACTIONS

Early functional characterization of the prenylation reaction suggested that it is a determinant of protein–protein interactions [76–78]. Prenylation of Gγ is a primary determinant of fundamental properties of the Gβγ dimer, as well as the G protein heterotrimer. In biochemical studies, even in the absence of a membrane, prenylation of Gγ is a prerequisite for high-affinity interactions of Gβγ with Gα to form a heterotrimer [62,79–81], although farnesyl and geranylgeranyl moieties have been reported to have equivalent effects on Gβγ interactions with Gα [81]. Variation in the prenyl modification, as by farnesylation versus geranylgeranylation, or by variable carboxymethylation, is implicated in subcellular targeting of G proteins [82–84], receptor coupling to the heterotrimer [62,81,85–91], and the interactions of the Gβγ dimer with downstream effectors [62,80,92–95].

Although heterotrimer interactions with receptor, and Gβγ dimer interactions with effectors, require prenylation and are sensitive to the prenyl group attached to Gγ, a generalized summary of past studies, with rare exceptions [96], is that Gβγ interactions are of higher apparent affinity when Gγ contains a geranylgeranyl moiety than when it contains a farnesyl moiety. In the case of receptor–G protein interactions, this is true whether the receptor would most likely interact in a native environment with a geranylgeranylated Gγ, as in the case of the A1-adenosine receptor [97], or with a farnesylated Gγ, as in the case of rhodopsin [81]. In addition, multiple studies have demonstrated that C-terminal protein sequences of Gγ are also important, and might even be argued to be more primary, determinants of Gβγ specificity and properties [70,81,97,98]. However, the protein sequences that would confer receptor specificity for Gγ are also ambiguous because the same studies that show that geranylgeranylated Gγ has higher apparent affinity for rhodopsin than does its cognate farnesylated Gγ also show that Gγ2 C-terminal sequences confer higher apparent affinity for the receptor (rhodopsin) than do the analogous Gγ1 sequences [81].

Why would rhodopsin preferentially interact with the evolutionarily conserved farnesylated Gγ1 (Figure 6.2) to transmit intracellular signals when it could just as well use an even more highly conserved geranylgeranylated Gγ2 (Figure 6.2) that has higher affinity for receptors (at least rhodopsin) both because of its inherent C-terminal protein sequence and its conserved prenyl processing pattern? One answer to this conundrum is that affinity is one component of signaling specificity, but another may be the required kinetic interactions that determine on and off rates for protein–protein interactions. This is similar to the proposal to explain the trafficking patterns of G proteins at the plasma membrane, which may also be influenced by both the C-terminal protein sequence and the prenylation patterns of the Gγ subunits [70]. What this suggests is that the C-terminus of Gγ, both because of protein sequence, and because of prenyl processing pattern, generates a site of protein–protein interactions with both variable affinity and variable kinetics for the controlled interactions necessary for a variety of different signaling paradigms. This emphasizes also the need to understand what is the range of variation of the prenylation pattern on Gγ proteins that might generate different interaction sites in different environments or for different signaling mechanisms.

IV.   Variation in Prenyl Processing of Brain G Proteins

In most tissues, G proteins are a small fraction of a percent of total cellular protein. In contrast, in brain cortex, G proteins are on the order of 1% of particulate protein [99,100], allowing isolation of fairly large quantities of relatively pure protein (Figure 6.3A). An advantageous way of characterizing this protein is using matrix-assisted laser desorption ionization time-of-flight mass spectrometry (MALDI-TOF MS) [102]. This is a technique that can characterize the mass of proteins in biological solutions with high accuracy [103,104]. In spite of this, however, Gα and Gβ subunits of purified G proteins are not well observed by this technique. In contrast, Gγ subunits are readily observed (Figure 6.3B) [102]. Multiple prominent masses are routinely observed in MALDI spectra of purified G proteins that correspond to the predicted masses of modified forms of Gγ2, Gγ3, Gγ5, and Gγ7 [102]. Subsequent studies verified these structures by multiple techniques and characterized many others in these samples (see below).

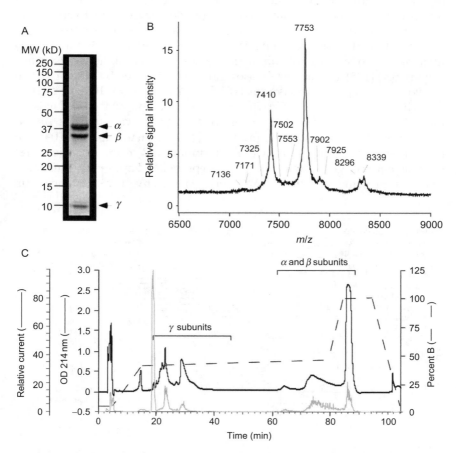

FIG. 6.3. Analysis of bovine brain G proteins. (A) Coomassie blue-stained 10–20% gradient SDS–polyacrylamide gel of purified bovine brain G protein. (B) MALDI MS in the Gγ subunit range of purified brain G protein. (C) Isolation of Gγ subunits by HPLC. Ion current was from electrospray ionization mass spectrometry on a Finnigan LCQ MS that detects Gα and Gγ subunits. Intact Gβ does not ionize well by standard techniques and has only been characterized after proteolysis. Adapted from Ref. [101].

Many studies have characterized the constituents of prenyl modifications using limited quantities of biological samples, using expressed proteins, often isolated through an epitope tag, or by following incorporation of radiolabeled precursors. The availability of large amounts of relatively pure protein from a (vertebrate) biological source (Figure 6.3A), and the small size and propensity of Gγ subunits to ionize by nearly all MS techniques (in contrast to Gα and Gβ), provided an opportunity to determine the variability at the prenyl modification site. This allowed exploration of a component of the proteome that is not accessible for many proteins—the degree of normal variation in the processing for a complex, multistep reaction such as prenylation.

Separation of G protein preparations by HPLC (Figure 6.3C) revealed a large number of consistently observed masses in the Gγ subunit mass range (Figure 6.4) [105]. Six distinct Gγ isoforms were observed by MS in purified brain G protein preparations after HPLC: Gγ2 [106], Gγ3 [105], Gγ5 [107], Gγ7 [105], Gγ10 [108], and Gγ12 [105]. Multiply modified forms of most of these Gγ were observed [105]. Some of these were variations at the N-terminus of the proteins, either predicted [105] or unanticipated [109], but most masses were variations at the C-terminus and of the canonical prenylation patterns normally associated with prenylated proteins (Figure 6.4). These represent variations in prenyl specificity (or presence), in

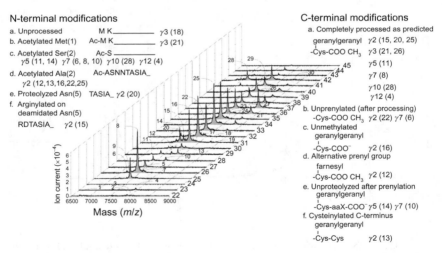

FIG. 6.4. Summary of the observed N-terminal and C-terminal modifications of bovine brain G proteins. Composite MALDI-TOF MS spectra for HPLC fractions from the Gγ range as in Figure 6.3. The number in parentheses after each Gγ subunit in the tables of N-terminal and C-terminal modifications refers to the corresponding numbered peaks in the composite of mass spectra from an HPLC run. Adapted from Ref. [105].

carboxymethylation or other C-terminal addition, and in Rce1 proteolytic processing after prenylation. Since these various forms are present in biological samples of purified, intact G protein heterotrimer, they are at least candidate modifications of biological significance. Often, the likely biological significance of these modifications can be inferred from extensive functional studies of Gγ subunit prenylation investigating the role of prenyl processing in G protein biology.

A. VARIATION OF THE PRENYL MOIETY

1. Geranylgeranylated Versus Farnesylated Gγ2

All six Gγ subunits routinely found in purified brain preparations (Gγ2, Gγ3, Gγ5, Gγ7, Gγ10, and Gγ12) have CaaX motifs that end in Leu and are predicted to be geranylgeranylated, which was found to be true of the major mass associated with each of them [105]. In addition, a MALDI signal at m/z 7684.5 (Figure 6.4; Mass 12) was subsequently characterized to be a farnesylated Gγ2 isoform [105]. Recombinant Gγ subunits [110] and expressed [111–113], and likely native [114], RhoB can be prenylated with geranylgeranyl, their predicted prenyl group, or farnesyl, albeit at substantially different levels under comparable conditions. Recently, in a survey of CaaL peptides that would normally be found to be geranylgeranylated [115], many, including CAIL of Gγ2, were also farnesylated. Another of these was CIIL, found on Gγ7 and Gγ12, which was not found in bovine brain, but may under other circumstances. It is a common observation that prenylated proteins such as Ras, predicted to be farnesylated, can also be geranylgeranylated in cells treated with a farnesyl transferase inhibitor (FTI) [116–119]. This appears to be due in part to acceptance of alternative substrates by geranylgeranyl transferase I and farnesyl transferase [120,121]. What all of these observations indicate is that prenyl substrate specificity of prenylating enzymes is not absolute, and under many different circumstances, alternatively prenylated proteins can be generated. The identification of farnesylated-Gγ2 in purified brain G proteins indicates that prenyl switching is also a normal occurrence with native proteins.

2. Unprenylated C-Terminus

Protein prenylation through a thioether linkage is a relatively stable bond. Studies of the metabolism of prenylcysteine degradation identified a lysosomal prenylcysteine lyase that is a FAD-dependent thioether oxidase [122,123], although this enzyme does not process prenylated peptides [122]. Other enzymes that would remove the prenyl group have not been described. This enzymology, notwithstanding, purified G proteins contain a

Gγ2 variant (Figure 6.4; Mass 22), as well as a Gγ7 protein (Figure 6.4; Mass 6), that are otherwise processed as prenylated proteins, but lack a prenyl group [105]. They were similar to other prenylated proteins in that they are carboxymethylated on a C-terminal Cys that corresponded to the prenylation site. In addition, because these proteins are isolated from G protein hetero-trimer, they are presumed to come from intact Gαβγ heterotrimers.

Although it is possible to generate unprenylated Gγ proteins during characterization by mass spectrometry, either by electrospray ionization (ESI) MS [110] or by collision-induced dissociation (CID) [124], this did not appear to explain the results from purified brain proteins. The prenylated and unprenylated proteins have different HPLC retention times [105], the unprenylated proteins eluting earlier, and the proteins are observed by multiple ionization techniques, not just ESI but, in particular, by MALDI [105].

Given what is generally thought to be the role of the prenyl group in membrane localization, such a reaction would produce cytosolic Gβγ dimers and would provide a potential intracellular signaling pattern in G protein responses. This would provide an irreversible mechanism for generating soluble Gβγ dimers in a cell. An alternative consequence is suggested by an interesting study recently published on the effects of expression of Gγ subunits that could not be prenylated in migratory cells of early developing zebrafish [125]. These studies suggested that unprenylated Gγ subunits are dominant negative inhibitors of Gβγ function, whereby the blocked effects could only be overcome by coexpression of wild-type Gγ, and not by over-expression of Gβ [125]. Another possibility is that unprenylated Gγ subunits would have novel effects not related to the actions of their membrane-localized precursors. For example, lack of prenylation of Gγ2 has been reported to promote nuclear localization of Gβγ dimers [126], where it may interact with the glucocorticoid receptor [127]. Interestingly, Gγ1 of the retina also contains a prominent unprenylated form [128], but one that is generated by proteolysis at the Gly immediately preceding the Cys of its CaaX motif [129,130]. Establishing the full significance of unprenylated Gγ products found in brain will require the isolation of an enzymatic activity that could account for their *in vivo* generation, along with identification of the biological functions associated with proteins produced by this activity.

B.   VARIATION IN CARBOXYMETHYLATION

1.   *Unmethylated Gγ*

Carboxymethylation is the final reaction in the progressive steps involved in prenylation of the CaaX motif proteins. It is a reversible modification [131] that has been identified as a possible regulated step affecting signaling for both small [39] and heterotrimeric [132] G proteins. Carboxymethylation influences protein turnover [133], membrane association of the Gβγ dimer [67], and heterotrimer stability [67]. Although expectations were that blocking the carboxymethylase would have modest effects on the function of Ras [38], this turned out not to be the case. Deletion of the carboxymethylase gene is early embryonic lethal [38] and appears to block the transforming activity of activated K-Ras and N-Ras [134].

Given the reversible and potentially regulated carboxymethylation of Gγ proteins, variation at this site in purified proteins might be expected. Although occasionally masses were seen that were compatible with low amounts of unmethylated versions of several Gγ isoforms, the only one consistently and prominently observed was an unmethylated Gγ2 (Figure 6.4; Mass 16) [105]. Signal intensity (and immunostaining of analogous fractions [101]) indicated that the unmethylated Gγ2 was nearly equally prominent with that for the primary methylated form of the protein (Figure 6.4; Mass 25). The presence of unmethylated Gγ2 was not due to nonspecific demethylation because other Gγ of nearly equivalent intensity for their methylated component were not commonly seen in an unmethylated form (e.g., Gγ7; Figure 6.4; Mass 8). Thus, Gγ2 is somewhat analogous to Gγ1 in the retina where a major component of it in purified preparations of transducin is also unmethylated [45]. These observations support the idea that some Gγ, such as Gγ2 [105], and perhaps Gγ1 [45], are prominently affected by processes regulating their carboxymethylation, whereas other Gγ, such as Gγ7, may not be so regulated, at least not under the conditions in which they were isolated for these studies [105]. This supports not only the idea that carboxymethylation is a functionally relevant regulatory step in G protein signaling but also that this is a process affecting some G protein-mediated responses, but perhaps not others.

2. C-Terminal Cysteinylation of Gγ Subunits

In addition to variations of carboxymethylation, a number of Gγ subunit variants including ones for Gγ3 and Gγ12, but most prominently Gγ2, contain a C-terminal modification that increases their predicted mass by 89 Da [105]. Progressive high-energy fragmentation of a C-terminal peptide of the Gγ2 variant generated fragments of a Cys residue attached to the C-terminus of the prenylcysteine that would have been the expected C-terminus of the peptide. The fragmentation pattern generated was compatible with a Cys residue attached either through a peptide bond or through a thioester linkage. Attempts to alkylate the Gγ2 variant with NEM or vinyl pyridine did not modify its C-terminus, supporting the thioester structure [105]. Although other proteins, such as serum albumin

[135] and PKC isoforms [136,137], are modified by cysteinylation via disulfide bond formation, attachment of Cys to the C-terminus of a prenylated protein via a thioester linkage would be expected to be more stable and would likely require enzymatic addition. Since Gγ, besides Gγ2 (Gγ3 and Gγ12), also had variants with a similar 89 Da increase from their predicted mass, this reaction may be a general phenomenon related to Gγ subunits, or perhaps generally to protein prenylation. Proteins with multiple Cys residues in their prenylation signal sequence (XCC or CXC) are often multiply prenylated. If the C-terminal Cys added to Gγ subunits is linked by a peptide bond, this would introduce into the protein a second prenylation site. If, as favored by the data at present, the Cys is added through a thioester linkage, this second site would not be available for prenylation. Masses associated with bovine G proteins compatible with multiply prenylated Gγ subunits were not observed [105]. Although the functional significance of these proteins has not been established, their existence suggests that the C-terminus of prenylated proteins is chemically diverse leading to the possibility that they are also functionally diverse.

C.   VARIATION IN PROTEOLYTIC PROCESSING BY RCE1

1.   Gγ Subunits not Processed by Rce1

The classical prenylation reaction of CaaX motif proteins includes prenylation of Cys at the -4 position, proteolysis of the C-terminal three amino acids, and carboxymethylation of the new C-terminus. The proteolysis step is mediated by one of two enzymes in yeast [36,138]. Although there are functional homologs of both yeast proteases in vertebrates [138], the proteolysis step in non-prelamin-related, prenylated proteins is thought to be carried out by a protease called Rce1 [31,138]. Absence (knockout) of this enzyme is embryonic lethal in mice [35], disrupts Ras localization [35], and leads to abnormal cardiac function, and eventual death, if disrupted cardioselectively [139]. These results indicate that the proteolysis reaction is essential, at least for some subset of prenylated proteins, and might suggest that those proteins that are targeted by it would be efficiently processed.

Of the six Gγ isoforms identified in purified brain G proteins, two of these, Gγ5 and Gγ7, had variants that retained their three terminal amino acids without carboxymethylation (Figure 6.4; Masses 14 and 10, respectively) [105]. For the Gγ7 variant, which was found to be prenylated while retaining its terminal-IIL sequence, this is a minor fraction of the protein observed. For Gγ5, by immunoreactivity and by MS signal intensity, the unprocessed Gγ5 is by far the more prevalent form observed in purified brain G proteins [107]. This is also true, based upon MS signal intensity, when $G\gamma5$ is expressed in cultured cell lines, including HEK293, NIH3T3, and Neuro2A cells [140]. Further, this is true of the endogenous $G\gamma5$ expressed in these cell lines when estimated by the ability of $G\gamma$ proteins to be captured by expressed, tagged $G\beta$ subunits [140]. Interestingly, data from insect sf9 cells indicate that the fully processed variant of $G\gamma5$ predominates in these cells unless the expressed protein is also modified at sites upstream of the CSFL signal [92]. This may suggest substrate specificity differences in the vertebrate and invertebrate Rce1 enzymes, but this may also suggest functional differences in the analogously processed proteins in different phyla.

2. Sequence Determinants for the Altered Prenyl Processing of $G\gamma5$

There have been reports of other prenylated proteins that lack proteolytic processing. The a and b subunits of farnesylated rabbit skeletal muscle glycogen phosphorylase kinase (GPK) have been shown to also retain their C-terminal residues after prenylation [141]. There is no obvious C-terminal motif common to $G\gamma5$, $G\gamma7$, and these GPK proteins that might direct this alternative processing pattern. Previous surveys of the specificity of, for example, the yeast Rce1-related enzymes would not adequately predict these observations in mammals either [36]. Nevertheless, the Ca_1a_2X motif of $G\gamma5$ (CSFL) is atypical in its inclusion of Ser at position a_1 and Phe at position a_2. That this atypical CaaX motif is the primary determinant of its resistance to Rce1 proteolysis could be shown by switching its CaaX motif for that of $G\gamma2$ (CAIL) and demonstrating that this altered their respective sensitivities to Rce1 (Figure 6.5) [140].

Further characterization of the requirements for Rce1 processing [140] showed that any aromatic amino acid at either a_1 or a_2 conferred relative insensitivity to Rce1 processing (Figure 6.6). In fact, many substitutions resulted in complete lack of processing, whereas Phe at a_2 resulted in substantial but not complete lack of processing.

It is interesting that this situation is different in sf9 insect cells, where a CSFL CaaX motif is processed by the Rce1 homolog, or not, depending upon upstream sequences [92]. In yeast, aromatic amino acids at a_1 or a_2 also often decrease prenylation, which complicates rigorous analysis of the effects of these residues on the proteolytic step. Nevertheless, several Ca_1a_2X sequences that contain aromatic amino acids do not appear to be substrates of either Afc1p/Ste24p or Rce1p [36], the two CaaX proteases in yeast. Other Ca_1a_2X sequences with aromatic amino acids, however, often do appear to be cleaved, but to the degree that they are, they appear to be preferentially substrates of Afc1p/Ste24p rather than Rce1p [36].

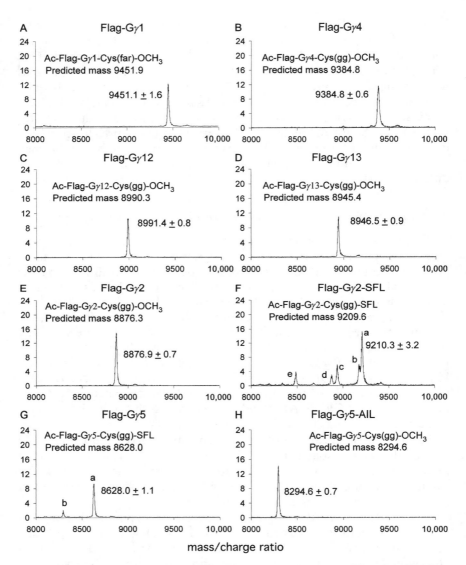

FIG. 6.5. Analysis by MALDI-TOF MS of Flag-Gγ subunits expressed in and purified from HEK293 cells. (A) Flag-Gγ1, (B) Flag-Gγ4, (C) Flag-Gγ12, (D) Flag-Gγ13, (E) Flag-Gγ2, (F) Flag-Gγ2-SFL, (G) Flag-Gγ5, (H) Flag-Gγ5-AIL. Adapted from Ref. [140].

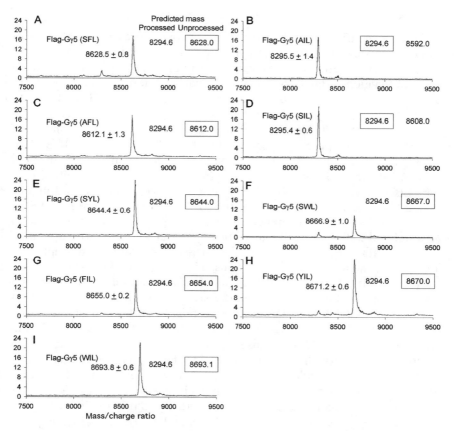

FIG. 6.6. Analysis by MALDI-TOF MS of the effect of changes in positions a1 and a2 of the Ca_1a_2X motif of Gγ5 expressed in HEK293 cells. A box is drawn around the mass corresponding to the processed (three terminal amino acids removed) or unprocessed (three terminal amino acids intact) form of the protein supported by the MS data. Adapted from Ref. [140].

The apparently greater role of Rce1 in targeting most prenyl substrates in vertebrates [35,142] would then be compatible with general lack of processing of Gγ5 and other prenyl proteins containing an aromatic amino acid in their Ca_1a_2X sequence. From this, it is interesting to speculate that the small but consistent amount of fully processed Gγ5 seen in cells could result from the activity of Zmpste24, the vertebrate functional homolog of yeast Afc1p/Ste24p [138], which seems to otherwise target somewhat selectively pre-lamin A [138].

3. Functional Role of Unprocessed CaaX Sequences

The full significance of the functional role of the presence or absence of Rce1 processing of prenyl proteins is unclear. A number of studies have been cited to suggest that this is of limited importance, or perhaps restricted to some functions related to protein prenylation, but not others. For example, farnesylation of Gγ1 is necessary and sufficient to allow Gβγ1 dimers to stimulate PLCβ even if the protein is not proteolytically processed or carboxymethylated [143]. In another case, studies of the relationship of Rheb1 and Rheb2 processing to functional activities found that Rce1 processing affected cellular localization but not downstream activation of the mTOR substrate p70 S6 kinase [144]. Finally, based upon modified Gγ5 subunits that varied in their relative proportion of processed and unprocessed Gγ5-CSFL, the retention of the -SFL sequence did not modify the ability of Gβγ dimers to form heterotrimers with Gαo and was presumed to be unrelated to the intracellular trafficking patterns of the protein [92].

Other data suggest that the Rce1 processing is related to multiple and important functions of the G proteins. First, absence (knockout) of this enzyme is embryonic lethal in mice and disrupts Ras localization [35] and leads to abnormal cardiac function if disrupted cardioselectively [134]. In another case, in characterization of G protein heterotrimers isolated from bovine brain, Gγ5 was associated equivalently with GoA, GoC, Gi1, and Gi2, but not with GoB [145]. This suggests specificity of Gγ5 association in heterotrimers, although this does not define the mechanism of this preferential association or indicate that it necessarily depends on lack of Rce1 processing. Perhaps, the strongest argument is that the CSFL CaaX motif of Gγ5 is rigorously conserved during vertebrate evolution (Figure 6.2). In fact, even the expression ratio of processed and unprocessed Gγ5 appears to be evolutionarily conserved, since this appears to be a property of Phe at the a_2 position. Even a conservative substitution with, for example, a Tyr (Figure 6.6E) would eliminate production of the minor form, and this does not appear to be evolutionarily observed. Thus, not only does Gγ5 represent an example of a prenylated substrate that is not efficiently processed by Rce1, but it also seems to be one designed to generate two forms of the protein with differential C-terminal processing, pointing to endogenous heterogeneity of prenyl substrates by design. One suggested function of this dual processing pattern for Gγ5 is that, while the typical form may be involved in traditional signaling mechanisms, the unprocessed form may account for Gγ5 localization in the nucleus in association with specific transcription factors [146]. Although this possibility should not be discounted, it is not clear that this function alone would account for the apparently widespread and prominent distribution of the unprocessed form of Gγ5 [105,140].

TABLE 6.1

Partial List of Proteins in the Human Genome with Aromatic Amino Acids in Their Prenylation CaaX Signal Sequence

Accession number	CaaX	GG/ Far	Description
Hs.523628	CAYL	GG	Death-associated protein 3
Hs.191540	CLYL	GG	Cohen syndrome 1 protein isoform 4
Hs.407709	CFKL	GG	DRE1 protein
Hs.195298	CLWS	Far	Sarcoglycan zeta; zeta-sarcoglycan
Hs.414300	CWPS	Far	Neuraminidase 4
Hs.515042	CYVM	Far	Lamin B2
Hs.351413	CSFM	Far	Rab37
Hs.546385	CFPS	Far	Mfa domain-containing protein p40
Hs.413801	CYYA	Far	Proteasome activator subunit 4
Hs.229988	CNFM	Far	GPI deacylase
Hs.521942	CWAS	Far	Zinc finger protein 517
Hs.13291	CFPS	Far	Cyclin G2
Hs.554795	CKWM	Far	Plexin C1; receptor for virally encoded semaphoring
Hs.455109	CTFS	Far	Beta-1,4-galactosyltransferase 7

Adapted from Ref. [140].

There are approximately 40 proteins or predicted proteins in the human genome that end in a putative prenylation signal "CaaX" sequence containing an aromatic amino acid at position a_1 or a_2 [140]. Most of these are of unknown function or protein family, but some recognized ones are identified in Table 6.1. These proteins belong to many different classes of proteins found in varying cellular compartments and organelles, suggesting that the functional role of this processing would relate to specific protein–protein interactions rather than general targeting to a common cellular site.

Although it may not yet be clear what specific function is associated with variable Rce1 processing of prenyl proteins, the evolutionary conservation of signals, as in $G\gamma 5$, that modify this enzymatic function, strongly argues for an important role both the processing of those prenyl proteins that are processed, as well as the importance of those that are not. The most likely role is the ability of this region of the protein to contribute to specific sites of protein–protein interaction.

V. Conclusion

The work reviewed here highlights some of the variation that can be observed for native proteins in each of the three constituent reactions associated with protein prenylation. A criticism of the significance of these observations is that these variant forms often represent minor components of the general population of G proteins observed, in this case, in brain. As a consequence, they may be simply the result of "sloppy" processing that generates otherwise irrelevant variants of an otherwise limited and well-defined set of biochemical reactions. The greatest argument against this criticism is the strong evolutionary conservation of the CaaX motif of many Gγ subunit isoforms indicated in Figure 6.2. Both the variable processing by Rce1 and the variable prenylation specificity of the prenyl transferases are dependent upon the CaaX signal sequence. The strong conservation of this signal sequence, especially when in all four positions, suggests evolutionary pressure to preserve even the subtle patterns of processing of these proteins.

What would be the functional effect, or biological role, of variation of the prenyl processing site on the Gγ subunits, or on any other prenyl protein substrates? The complex biological role of Gγ prenylation and the impact of this reaction on nearly all interactions and functions of the G proteins suggest a possible answer to this question. As sites of protein–protein interaction, in concert with amino acid sequences at the C-terminus of Gγ, these variations may lead to highly variable protein interaction sites that can interact with both upstream and downstream signaling partners in ways that are complex and, in the end, specific for each of the 12 Gγ isoforms found in the human genome. This argues that this variation is a fundamental component of the signaling diversity associated with the function of the heterotrimeric G proteins. If future research continues to support this idea, this is likely also true for many of the other prenylated proteins expressed in cells.

REFERENCES

1. Rodbell, M. (1995). Signal transduction: Evolution of an idea. *Biosci Rep* 15:117–133.
2. Bourne, H.R. (1986). GTP-binding proteins. One molecular machine can transduce diverse signals. *Nature* 321:814–816.
3. Gilman, A.G. (1995). G proteins and regulation of adenylyl cyclase. *Biosci Rep* 15:65–97.
4. Birnbaumer, L. (2007). Expansion of signal transduction by G proteins. The second 15 years or so: From 3 to 16 alpha subunits plus betagamma dimers. *Biochim Biophys Acta* 1768:772–793.
5. Katadae, M., *et al.* (2008). Interacting targets of the farnesyl of transducin gamma-subunit. *Biochemistry* 47:8424–8433.
6. Chisari, M., Saini, D.K., Cho, J.-H., Kalyanaraman, V., and Gautam, N. (2009). G Protein subunit dissociation and translocation regulate cellular response to receptor stimulation. *PLoS ONE* 4:e7797.
7. Magee, A.I., and Seabra, M.C. (2003). Are prenyl groups on proteins sticky fingers or greasy handles? *Biochem J* 2003:3e–34e.

8. Kuhn, H. (1980). Light and GTP regulated interaction of GTPase and other proteins with bovine photoreceptor membranes. *Nature* 283:587–589.

9. Fung, B.K., Hurley, J.B., and Stryer, L. (1981). Flow of information in the light-triggered cyclic nucleotide cascade of vision. *Proc Natl Acad Sci USA* 78:152–156.

10. Hildebrandt, J.D., Codina, J., Risinger, R., and Birnbaumer, L. (1984). Identification of a gamma subunit associated with the adenylyl cyclase regulatory proteins Ns and Ni. *J Biol Chem* 259:2039–2042.

11. Bokoch, G.M., Katada, T., Northup, J.K., Ui, M., and Gilman, A.G. (1984). Purification and properties of the inhibitory guanine nucleotide-binding regulatory component of adenylate cyclase. *J Biol Chem* 259:3560–3567.

12. Levitzki, A. (1986). Beta-adrenergic receptors and their mode of coupling to adenylate cyclase. *Physiol Rev* 66:819–854.

13. Rebois, R.V., Warner, D.R., and Basi, N.S. (1997). Does subunit dissociation necessarily accompany the activation of all heterotrimeric G proteins. *Cell Signal* 9:141–151.

14. Bünemann, M., Monika, F., and Lohse, M.J. (2003). Gi protein activation in intact cells involves subunit rearrangement rather than dissociation. *Proc Natl Acad Sci USA* 100:16077–16082.

15. Digby, G.J., Lober, R.M., Sethi, P.R., and Lambert, N.A. (2006). Some G protein heterotrimers physically dissociate in living cells. *Proc Natl Acad Sci USA* 103:17789–17894.

16. Oldham, W.M., and Hamm, H.E. (2008). Heterotrimeric G protein activation by G-protein-coupled receptors. *Nat Rev Mol Cell Biol* 9:60–71.

17. McIntire, W.E. (2009). Structural determinants involved in the formation and activation of G protein betagamma dimers. *Neurosignals* 17:82–99.

18. Bjarnadóttira, T.K., Gloriama, D.E., Hellstranda, S.H., Kristianssona, H., Fredrikssona, R., and Schiöth, H.B. (2006). Comprehensive repertoire and phylogenetic analysis of the G protein-coupled receptors in human and mouse. *Genomics* 88:263–273.

19. Hopkins, A.L., and Groom, C.R. (2002). The druggable genome. *Nat Rev Drug Discov* 1:727–730.

20. Hildebrandt, J.D. (1997). Role of subunit diversity in signaling by heterotrimeric G proteins. *Biochem Pharmacol* 54:325–339.

21. Casey, P.J. (1995). Protein lipidation in cell signaling. *Science* 268:221–225.

22. Milligan, G., and Grassie, M.A. (1997). How do G-proteins stay at the plasma membrane. *Essays Biochem* 32:49–60.

23. Resh, M.D. (2006). Trafficking and signaling by fatty-acylated and prenylated proteins. *Nat Chem Biol* 2:584–590.

24. Wedegaertner, P.B., Wilson, P.T., and Bourne, H.R. (1995). Lipid modifications of trimeric G proteins. *J Biol Chem* 270:503–506.

25. Yamane, H.K., and Fung, B.K. (1993). Covalent modifications of G Proteins. *Annu Rev Phamacol Toxicol* 32:201–241.

26. Clarke, S. (1992). Protein isoprenylation and methylation at carboxyl-terminal cysteine residues. *Annu Rev Biochem* 61:355–386.

27. Glomset, J.A., and Farnsworth, C.C. (1994). Role of protein modification reactions in programming interactions between ras-related GTPases and cell membranes. *Annu Rev Cell Biol* 10:181–205.

28. Zhang, F.L., and Casey, P. (1996). Protein prenylation: Molecular mechanisms and functional consequences. *Annu Rev Biochem* 65:241–269.

29. Sinensky, M. (2000). Recent advances in the study of prenylated proteins. *Biochim Biophys Acta* 1484:93–106.

30. Roskoski, R. (2003). Protein prenylation: A pivotal posttranslational process. *Biochem Biophys Res Commun* 303:1–7.
31. Winter-Vann, A.M., and Casey, P.J. (2005). Post-prenylation-processing enzymes as new targets in oncogenesis. *Nat Rev Cancer* 5:407–412.
32. Wright, L.P., and Philips, M.R. (2006). Thematic review series: Lipid posttranslational modifications. CAAX modification and membrane targeting of Ras. *J Lipid Res* 47:883–891.
33. Kato, K., Cox, A.D., Hiska, M.M., Graham, S.M., Buss, J.E., and Der, C.J. (1992). Isoprenoid addition to ras protein is the critical modification for its membrane association and transforming activity. *Proc Natl Acad Sci USA* 89:6403–6407.
34. Cox, A.D. (1995). Mutation and analysis of prenylation signal sequences. *Methods Enzymol* 250:105–121.
35. Kim, E., *et al.* (1999). Disruption of the mouse Rce1 gene results in defective Ras processing and mislocalization of Ras within cells. *J Biol Chem* 274:8383–8390.
36. Trueblood, C.E., Boyartchuk, V.L., Picologlou, E.A., Rozema, D., Poulter, D., and Rine, J. (2000). The CaaX proteases, Afc1p and Rce1p, have overlapping but distinct substrate specificities. *Mol Cell Biol* 20:4381–4392.
37. Clarke, S. (1985). Protein carboxyl methyltransferase: Two distinct classes of enzymes. *Annu Rev Biochem* 54:479–506.
38. Bergo, M.O., *et al.* (2001). Isoprenylcysteine carboxyl methyltransferase deficiency in mice. *J Biol Chem* 276:5841–5845.
39. Phillips, M.R., *et al.* (1993). Carboxyl methylation of ras-related proteins during signal transduction in neutrophils. *Science* 259:977–980.
40. Gautam, N., Baetscher, M., Aebersold, R., and Simon, M.I. (1989). A G protein gamma subunit shares homology with ras proteins. *Science* 244:971–974.
41. Casey, P.J., Solski, P.A., Der, C.J., and Buss, J.E. (1989). p21ras is modified by a farnesyl isoprenoid. *Proc Natl Acad Sci USA* 86:8323–8327.
42. Hancock, J.F., Magee, A.I., Childs, J.E., and Marshall, C.J. (1989). All ras proteins are polyisoprenylated but only some are palmitoylated. *Cell* 57:1167–1177.
43. Clarke, S., Vogel, J.P., Deschenes, R.J., and Stock, J. (1988). Posttranslational modification of the Ha-ras oncogene protein: Evidence for a third class of protein carboxyl methyltransferases. *Proc Natl Acad Sci USA* 85:4643–4647.
44. Backlund, P.S., Jr., Simonds, W.F., and Spiegel, A.M. (1990). Carboxyl methylation and COOH-terminal processing of the brain G-protein gamma-subunit. *J Biol Chem* 265:15572–15576.
45. Fukada, Y., Takao, T., Ohguro, H., Yoshizawa, T., Akino, T., and Shimonishi, Y. (1990). Farnesylated gamma-subunit of photoreceptor G protein indispensable for GTP-binding. *Nature* 346:658–660.
46. Fung, B.K., Yamane, H.K., Ota, I.M., and Clarke, S. (1990). The gamma subunit of brain G-proteins is methyl esterified at a C-terminal cysteine. *FEBS Lett* 260:313–317.
47. Lai, R.K., Perez-Sala, D., Canada, F.J., and Rando, R.R. (1990). The gamma subunit of transducin is farnesylated. *Proc Natl Acad Sci USA* 87:7673–7677.
48. Maltese, W.A., and Robishaw, J.D. (1990). Isoprenylation of C-terminal cysteine in a G-protein gamma subunit. *J Biol Chem* 265:18071–18074.
49. Mumby, S.M., Casey, P.J., Gilman, A.G., Gutowski, S., and Sternweis, P.C. (1990). G protein gamma subunits contain a 20-carbon isoprenoid. *Proc Natl Acad Sci USA* 87:5873–5877.
50. Sanford, J., Codina, J., and Birnbaumer, L. (1991). Gamma subunits of G proteins, but not their alpha or beta subunits, are polyisoprenylated. Studies on post-translational

modification using in vitro translation with rabbit reticulocyte lysates. *J Biol Chem* 266:9570–9579.

51. Yamane, H.K., *et al.* (1990). Brain G protein gamma subunits contain an all-trans-geranylgeranylcysteine methyl ester at their carboxyl termini. *Proc Natl Acad Sci USA* 87:5868–5872.

52. Casey, P.J., and Seabra, M.C. (1996). Protein prenyltransferases. *J Biol Chem* 271:5289–5292.

53. Sefton, B.M., Trowbridge, I.S., Cooper, J.A., and Scolnick, E.M. (1982). The transforming proteins of Rous sarcoma virus, Harvey sarcoma virus and Abelson virus contain tightly bound lipid. *Cell* 31:465–474.

54. Willumsen, B.M., Christensen, A., Hubbert, N.L., Papageorge, A.G., and Lowy, D.R. (1984). The p21 ras C-terminus is required for transformation and membrane association. *Nature* 310:583–586.

55. Glomset, J.A., Gelb, M.H., and Farnsworth, C.C. (1990). Prenyl proteins in eukaryotic cells: A new type of membrane anchor. *Trends Biochem Sci* 15:139–142.

56. Hori, Y., *et al.* (1991). Post-translational modifications of the C-terminal region of the rho protein are important for its interaction with membranes and the stimulatory and inhibitory GDP/GTP exchange proteins. *Oncogene* 6:515–522.

57. Silvius, J.R., and l'Heureux, F. (1994). Fluorimetric evaluation of the affinities of iso-prenylated peptides for lipid bilayers. *Biochemistry* 33:3012–3022.

58. Hancock, J.F., Paterson, H., and Marshall, C.J. (1989). A polybasic domain or palmitoylation is required in addition to the CAAX motif to localize p21ras to the plasma membrane. *Cell* 63:133–139.

59. Takida, S., and Wedegaertner, P.B. (2003). Heterotrimer formation, together with iso-prenylation, is required for plasma membrane targeting of Gbetagamma. *J Biol Chem* 278:17284–17290.

60. Simonds, W.F., Butrynski, J.E., Gautam, N., Unson, C.G., and Spiegel, A.M. (1991). G protein beta/gamma dimers. Membrane targeting requires subunit coexpression and intact gamma C-A-A-X domain. *J Biol Chem* 266:5363–5366.

61. Dietrich, A., Meister, M., Spicher, K., Schultz, G., Camps, M., and Gierschik, P. (1992). Expression, characterization and purification of soluble G protein beta/gamma dimers composed of defined subunits in baculovirus infected cells. *FEBS Lett* 313:220–224.

62. Matsuda, T., *et al.* (1998). Specific isoprenyl group linked to transducin gamma-subunit is a determinant of its unique signaling properties among G-proteins. *Biochemistry* 37:9843–9850.

63. Muntz, K.H., Sternweis, P.C., Gilman, A.G., and Mumby, S.M. (1993). Influence of gamma subunit prenylation on association of guanine nucleotide binding regulatory proteins with membranes. *Mol Biol Cell* 3:49–61.

64. Pomerantz, K.B., Lander, H.M., Summers, B., Robishaw, J.D., Balcueva, E., and Hajjar, D.P. (1997). G-protein-mediated signaling in cholesterol-enriched arterial smooth muscle cells. 1. Reduced membrane-associated G-protein content due to diminished isoprenylation of G-gamma subunits and p21ras. *Biochemistry* 36:9523–9531.

65. Pronin, A.N., and Gautam, N. (1993). Proper processing of a G protein gamma subunit depends on the complex formation with a beta subunit. *FEBS Lett* 328:89–93.

66. Schillo, S., *et al.* (2004). Targeted mutagenesis of the farnesylation site of Drosophila G {gamma}e disrupts membrane association of the G protein {beta}{gamma} complex and affects the light sensitivity of the visual system. *J Biol Chem* 279:36309–36316.

67. Fukada, Y., *et al.* (1994). Effects of carboxyl methylation of photoreceptor G protein gamma-subunit in visual transduction. *J Biol Chem* 269:5163–5170.

68. Choy, E., *et al.* (1999). Endomembrane trafficking of ras. The CAAX motif targets proteins to the ER and Golgi. *Cell* 98:69–80.
69. Higgins, J.B., and Casey, P.J. (1994). In vitro processing of recombinant G protein gamma subunits. Requirements for assembly of an active beta/gamma complex. *J Biol Chem* 269:9067–9073.
70. Saini, D.K., Chisari, M., and Gautam, N. (2009). Shuttling and translocation of heterotrimeric G proteins and ras. *Trends Pharmacol Sci* 30:278–286.
71. Chisari, M., Saini, D.K., Kalyanaraman, V., and Gautam, N. (2007). Shuttling of G protein subunits between the plasma membrane and intracellular membranes. *J Biol Chem* 282:24092–24098.
72. Goodwin, J.S., *et al.* (2005). Depalmitoylated Ras traffics to and from the Golgi complex via a nonvesicular pathway. *J Cell Biol* 170:261–272.
73. Rocks, O., *et al.* (2005). An acylation cycle regulates localization and activity of palmitoylated ras isoforms. *Science* 307:1746–1752.
74. Saini, D.K., Kalyanaraman, V., Chisari, M., and Gautam, N. (2007). A family of G protein betagamma subunits translocate reversibly from the plasma membrane to endomembranes on receptor activation. *J Biol Chem* 282:24099–24108.
75. Wedegaertner, P.B., Bourne, H.R., and von Zastrow, M. (1996). Activation-induced subcellular redistribution of Gs alpha. *Mol Biol Cell* 7:1225–1233.
76. Kuroda, Y., Suzuki, N., and Kataoka, T. (1993). The effect of posttranslational modifications on the interaction of ras2 with adenylyl cyclase. *Science* 259:683–686.
77. Marshall, C.J. (1993). Protein prenylation: A mediator of protein-protein interactions. *Science* 259:1865–1866.
78. Sinensky, M. (2000). Functional aspects of polyisoprenoid protein substituents: Roles in protein-protein interactions and trafficking. *Biochim Biophys Acta* 1529:203–209.
79. Linder, M.E., Pang, I.H., Duronio, R.J., Gordon, J.I., Sternweis, P.C., and Gilman, A.G. (1991). Lipid modification of G protein subunits. Myristolation of Go alpha increases its affinity for beta/gamma. *J Biol Chem* 266:4654–4659.
80. Iniguez-Lluhi, J.A., Simon, M.I., Robishaw, J.D., and Gilman, A.G. (1992). G protein beta/gamma subunits synthesized in sf9 cells. Functional characterization and the significance of prenylation of gamma. *J Biol Chem* 267:23409–23417.
81. Jian, X., Clark, W.A., Kowalak, J., Markey, S.P., Simonds, W.F., and Northup, J.K. (2001). Gbeta gamma affinity for bovine rhodopsin is determined by the carboxyl-terminal sequences of the gamma subunit. *J Biol Chem* 276:48518–48525.
82. Melkonian, K.A., Ostermeyer, A.G., Chen, J.Z., Roth, M.G., and Brown, D.A. (1999). Role of lipid modifications in targeting proteins to detergent-resistant membrane rafts. Many raft proteins are acylated, while few are prenylated. *J Biol Chem* 274:3910–3917.
83. Akgoz, M., Kalyanaraman, V., and Gautam, N. (2004). Receptor-mediated reversible translocation of the G protein betagamma complex from the plasma membrane to the Golgi complex. *J Biol Chem* 279:51541–51544.
84. Kubler, E., Dohlman, H.G., and Lisanti, M.P. (1996). Identification of triton X-100 insoluble membrane domains in the yeast Saccharomyces cerevisiae. Lipid requirements for targeting of heterotrimeric G protein subunits. *J Biol Chem* 271:32975–32980.
85. Azpiazu, I., Cruzblanca, H., Li, P., Linder, M., Zhuo, M., and Gautam, N. (1999). A G protein gamma subunit-specific peptide inhibits muscarinic receptor signaling. *J Biol Chem* 274:35305–35308.
86. Azpiazu, I., and Gautam, N. (2001). G protein gamma subunit interaction with a receptor regulates receptor-stimulated nucleotide exchange. *J Biol Chem* 276:41742–41747.
87. Figler, R.A., Lindorfer, M.A., Graber, S.G., Garrison, J.C., and Linden, J. (1997). Reconstitution of bovine A1 adenosine receptors and G proteins in phospholipid vesicles:

Betagamma-subunit composition influences guanine nucleotide exchange and agonist binding. *Biochemistry* 36:16288–16299.

88. Kisselev, O.G., Ermolaeva, M.V., and Gautam, N. (1994). A farnesylated domain in the G protein gamma subunit is a specific determinant of receptor coupling. *J Biol Chem* 269:21399–21402.

89. Kisselev, O., Ermolaeva, M., and Gautam, N. (1995). Efficient interaction with a receptor requires a specific type of prenyl group on the G protein gamma subunit. *J Biol Chem* 270:25356–25358.

90. Ohguro, H., Fulada, Y., Takao, T., Shimonishi, Y., Yoshizawa, T., and Akino, T. (1991). Carboxyl methylation and farnesylation of transducin gamma subunit synergistically enhance its coupling with metarhodopsin II. *EMBO J* 10:3669–3674.

91. Yasuda, H., Lindorfer, M.A., Woodfork, K.A., Fletcher, J.E., and Garrison, J.C. (1996). Role of the prenyl group on the G protein gamma subunit in coupling trimeric G proteins to A1 adenosine receptors. *J Biol Chem* 271:18588–18595.

92. Akgoz, M., Azpiazu, I., Kalyanaraman, V., and Gautam, N. (2002). Role of the G protein gamma subunit in beta gamma complex modulation of phospholipase Cbeta function. *J Biol Chem* 277:19573–19578.

93. Fogg, V.C., Azpiazu, I., Linder, M.E., Smrcka, A., Scarlata, S., and Gautam, N. (2001). Role of the gamma subunit prenyl moiety in G protein beta gamma complex interaction with phospholipase Cbeta. *J Biol Chem* 276:41797–41802.

94. Myung, C.S., Yasuda, H., Liu, W.W., Harden, T.K., and Garrison, J.C. (1999). Role of isoprenoid lipids on the heterotrimeric G protein gamma subunit in determining effector activation. *J Biol Chem* 274:16595–16603.

95. Kerchner, K.R., *et al.* (2004). Differential sensitivity of phosphatidylinositol 3-kinase p110 {gamma} to isoforms of G protein {beta}{gamma} dimers. *J Biol Chem* 279:44554–44562.

96. Kisselev, O., and Gautam, N. (1993). Specific interaction with rhodopsin is dependent on the gamma subunit type in a G protein. *J Biol Chem* 268:24519–24522.

97. Myung, C.-S., Lim, W.K., DeFilippo, J.M., Yasuda, H., Neubig, R.R., and Garrison, J.C. (2006). Regions in the G protein gamma subunit important for interaction with receptors and effectors. *Mol Pharmacol* 69:877–887.

98. Oldham, W.M., and Hamm, H.E. (2006). Structural basis of function in heterotrimeric G proteins. *Quart Rev Biophys* 39:117–166.

99. Sternweis, P.C., and Robishaw, J.D. (1984). Isolation of two proteins with high affinity for guanine nucleotides from membranes of bovine brain. *J Biol Chem* 259:13806–13813.

100. Neer, E.J., Lok, J.M., and Wolf, L.G. (1984). Purification and properties of the inhibitory guanine nucleotide regulatory unit of brain adenylate cyclase. *J Biol Chem* 259:14222–14229.

101. Cook, L.A., Wilcox, M.D., Dingus, J., Schey, K.L., and Hildebrandt, J.D. (2001). Separation and analysis of G Protein gamma subunits. *Methods Enzymol* 344:209–233.

102. Wilcox, M.D., *et al.* (1994). Analysis of G protein gamma subunit heterogeneity using mass spectrometry. *J Biol Chem* 269:12508–12513.

103. Beavis, R.C., and Chait, B.T. (1990). Rapid, sensitive analysis of protein mixtures by mass spectrometry. *Proc Natl Acad Sci USA* 87:6873–6877.

104. Biemann, K. (1992). Mass spectrometry of peptides and proteins. *Annu Rev Biochem* 61:977–1010.

105. Cook, L.A., *et al.* (2006). Proteomic analysis of bovine brain G protein gamma subunit processing heterogeneity. *Mol Cell Proteomics* 5:671–685.

106. Wilcox, M.D., Schey, K.L., Busman, M., and Hildebrandt, J.D. (1995). Determination of the complete covalent structure of the gamma-2 subunit of bovine brain G proteins by mass spectrometry. *Biochem Biophys Res Commun* 212:367–374.

107. Cook, L.A., Schey, K.L., Wilcox, M.D., Dingus, J., and Hildebrandt, J.D. (1998). Hetero-geneous processing of a G protein gamma subunit at a site critical for protein and membrane interactions. *Biochemistry* 37:12280–12286.
108. Cook, L.A., Schey, K.L., Cleator, J.H., Wilcox, M.D., Dingus, J., and Hildebrandt, J.D. (2002). Identification of a region in G protein gamma subunits conserved across species but hypervariable among subunit isoforms. *Protein Sci* 10:2548–2555.
109. Hamilton, M.H., Cook, L.A., McRackan, T.R., Schey, K.L., and Hildebrandt, J.D. (2003). gamma 2 subunit of G protein heterotrimer is an N-end rule ubiquitylation substrate. *Proc Natl Acad Sci USA* 100:5081–5086.
110. Lindorfer, M.A., Sherman, N.E., Woodfork, K.A., Fletcher, J.E., Hunt, D.F., and Garrison, J.C. (1996). G protein gamma subunits with altered prenylation sequences are properly modified when expressed in Sf9 cells. *J Biol Chem* 271:18582–18587.
111. Adamson, P., Marshall, C.J., Hall, A., and Tilbrook, P.A. (1992). Post-translational modifications of p21rho proteins. *J Biol Chem* 267:20033–20038.
112. Armstrong, S.A., Hannah, V.C., Goldstein, J.L., and Brown, M.S. (1995). CAAX ger-anylgeranyl transferase transfers farnesyl as efficiently as geranylgeranyl to RhoB. *J Biol Chem* 270:7864–7868.
113. Lebowitz, P.F., Casey, P.J., Prendergast, G.C., and Thissen, J.A. (1997). Farnesyltrasnfer-ase inhibitors alter prenylation and growth-stimulating function of RhoB. *J Biol Chem* 272:15591–15594.
114. Baron, R., *et al.* (2000). RhoB prenylation is driven by the three carboxyl-terminal amino acids of the protein: Evidenced in vivo by an anti-farnseyl cysteine antibody. *Proc Natl Acad Sci USA* 97:11626–11631.
115. Krzysiak, A.J., Aditya, A.V., Hougland, J.L., Fierke, C.A., and Gibbs, R.A. (2010). Synthesis and screening of a CaaL peptide library versus FTase reveals a surprising number of substrates. *Bioorg Med Chem Lett* 20:767–770.
116. James, G.L., Goldstein, J.L., and Brown, M.S. (1995). Polylysine and CVIM sequences of K-RasB dictate specificity of prenylation and confer resistance to benzodiazepine pepti-domimetic in vitro. *J Biol Chem* 270:6221–6226.
117. Rowell, C.A., Kowalczyk, J.J., Lewis, M.D., and Garcia, A.M. (1997). Direct demonstra-tion of geranylgeranylation and farnesylation of Ki-Ras in vivo. *J Biol Chem* 272:14093–14097.
118. Whyte, D.B., *et al.* (1997). K- and N-Ras are geranylgeranylated in cells treated with farnesyl protein transferase inhibitors. *J Biol Chem* 272:14459–14464.
119. Zhang, F.L., *et al.* (1997). Characterization of Ha-Ras, N-Ras, Ki-Ras4A, and Ki-Ras4B as in vitro substrates for farnesyl protein transferase and geranylgeranyl protein transfer-ase type I. *J Biol Chem* 272:10232–10239.
120. Yokoyama, K., Zimmerman, K., Scholten, J., and Gelb, M.H. (1997). Differential prenyl pyrophosphate binding to mammalian protein genranylgeranyltransferase-I and protein farnesyltransferase and its consequence on the specificity of protein prenylation. *J Biol Chem* 272:3944–3952.
121. Fiordalis, J.J., *et al.* (2003). High affinity for farnesyltransferase and alternative prenyla-tion contribute to K-Ras4B resistance to farnesyltransferase inhibitors. *J Biol Chem* 278:41718–41727.
122. Zhang, L., Tschantz, W.R., and Casey, P.J. (1997). Isolation and characterization of a prenylcysteine lyase from bovine brain. *J Biol Chem* 272:23354–23359.
123. Tschantz, W.R., Digits, J.A., Pyon, H.-J., Coates, R.M., and Casey, P. (2001). Lysosomal prenylcysteine lyase is a FAD-dependent thioether oxidase. *J Biol Chem* 276:2321–2324.

124. Schey, K.L., Busman, M., Cook, L.A., Hamm, H.E., and Hildebrandt, J.D. (2002). Structural characterization of intact G protein gamma subunits by mass spectrometry. *Methods Enzymol* 344:586–597.

125. Mulligan, T., Blaser, H., Raz, E., and Farber, S.A. (2010). Prenylation-deficient G protein gamma subunits disrupt GPCR signaling in the zebrafish. *Cell Signal* 22:221–233.

126. Kino, T., Kozasa, T., and Chrousos, G.P. (2005). Statin-induced blockade of prenylation alters nucleo-cytoplasmic shuttling of GTP binding proteins gamma2 and beta2 and enhances their suppressive effect on glucocorticoid receptor. *Eur J Clin Invest* 35:508–513.

127. Kino, T., Tiulpakov, A., Ichijo, T., and Kozasa, T. (2005). G Protein beta interacts with the glucocorticoid receptor and suppresses its transcriptional activity in the nucleus. *J Cell Biol* 169:885–896.

128. Fukada, Y., Ohguro, H., Saito, T., Yoshizawa, T., and Akino, T. (1989). Beta/gamma subunit of bovine transducin composed of two components with distinctive gamma subunits. *J Biol Chem* 264:5937–5943.

129. Ovchinnikov, Y.A., Lipkin, V.M., Shuvaeva, T.M., Bogachuk, A.P., and Shemyakin, V.V. (1985). Complete amino acid sequence of gamma subunit of the GTP binding protein from cattle retina. *FEBS Lett* 179:107–110.

130. McConnell, D.G., Kohnken, R.E., and Smith, A.J. (1984). Tentative identification of complete amino acid sequence for the smallest subunit of GTP-binding protein of bovine retinal rod outer segment. *Fed Proc* 43:1585.

131. Perez-Sala, D., Tan, E.W., Canada, F.J., and Rando, R.R. (1991). Methylation and demethylation reactions of guanine nucleotide-binding proteins of retinal rod outer segments. *Proc Natl Acad Sci USA* 88:3043–3046.

132. Philips, M.R., *et al.* (1995). Activation-dependent carboxyl methylation of neutrophil G-protein gamma subunit. *Proc Natl Acad Sci USA* 92:2283–2287.

133. Backlund, P.S. (1997). Post-translational processing of RhoA. *J Biol Chem* 272:33175–33180.

134. Bergo, M.O., *et al.* (2004). Inactivation of Icmt inhibits transformation by oncogenic K-Ras and B-Raf. *J Clin Invest* 113:539–545.

135. Andersson, L.O. (1966). The heterogeneity of bovine serum albumin. *Biochim Biophys Acta* 117:115–133.

136. Chu, F., Ward, N.E., and AO'Brian, C.A. (2001). Potent inactivation of representative members of each PKC isozyme subfamily and PKD via S-thiolation by the tumor-promotion/progression antagonist glutathione but not by its precursor cysteine. *Carcinogenesis* 22:1221–1229.

137. Chu, F., Ward, N.E., and AO'Brian, C.A. (2003). PKC isozyme S-cysteinylation by cysteine stimulates the pro-apoptotic isozyme PKC delta and inactivates the oncogenic PKC epsilon. *Carcinogenesis* 24:317–325.

138. Young, S.G., Fong, L.G., and Michaelis, S. (2005). Prelamin A, Zmpste24, misshapen cell nuclei, and progeria—New evidence suggesting that protein farnesylation could be important for disease pathogenesis. *J Lipid Res* 46:2531–2558.

139. Bergo, M.O., *et al.* (2004). On the physiological importance of endoproteolysis of CAAX proteins. Heart-specific RCE1 knockout mice develop a lethal cardiomyopathy. *J Biol Chem* 279:4729–4736.

140. Kilpatrick, E.L., and Hildebrandt, J.D. (2007). Sequence dependence and differential expression of Gγ5 subunit isoforms of the heterotrimeric G proteins variably processed after prenylation in mammalian cells. *J Biol Chem* 282:14038–14047.

141. Heilmeyer, L.M.G., Serwe, M., Weber, C., Metzger, J., Hoffmann-Posorske, E., and Meyer, H.E. (1992). Farnesylcysteine, a constituent of the alpha and beta subunits of

rabbit skeletal muscle phosphorylase kinase: Localization by conversion to S-ethylcysteine and by tandem mass spectrometry. *Proc Natl Acad Sci USA* 89:9554–9558.

142. Leung, G.K., *et al.* (2001). Biochemical studies of Zmpste24-deficient mice. *J Biol Chem* 276:29051–29058.

143. Dietrich, A., *et al.* (1996). Isoprenylation of the G protein gamma subunit is both necessary and sufficient for beta-gamma dimer-mediated stimulation of phospholipase C. *Biochemistry* 35:15174–15182.

144. Hanker, A.B., *et al.* (2010). Differential requirement of CAAX-mediated posttranslational processing for Rheb localization and signaling. *Oncogene* 29:380–391.

145. Wilcox, M.D., *et al.* (1995). Bovine brain Go isoforms have distinct gamma subunit compositions. *J Biol Chem* 270:4189–4192.

146. Park, J.-G., Muise, A., He, G.-P., Kim, S.-W., and Ro, H.-S. (1999). Transcriptional regulation by the gamma-5 subunit of a heterotrimeric G protein during adipogenesis. *EMBO J* 18:4004–4012.

7

Farnesylation Versus Geranylgeranylation in G-Protein-Mediated Light Signaling

HIDETOSHI KASSAI[a] • YOSHITAKA FUKADA[b]

[a]*Laboratory of Animal Resources, Center for Disease Biology and Integrative Medicine*
Graduate School of Medicine
The University of Tokyo
Tokyo, Japan

[b]*Department of Biophysics and Biochemistry, Graduate School of Science*
The University of Tokyo
Tokyo, Japan

I.   Abstract

Protein prenylation is a type of posttranslational lipid modification by either 15-carbon farnesyl or 20-carbon geranylgeranyl, which are found in signaling proteins such as G proteins. Farnesylation has a smaller and specific repertoire of modified proteins, when compared with geranylgeranylation. In the case of heterotrimeric G proteins, the γ subunit of retinal G protein, transducin (hereafter referred to as Gtα/Gt$\beta\gamma$), is selectively farnesylated, whereas the other γ subtypes are geranylgeranylated. The farnesylation of Gtγ is required for both transducin activity *in vitro* and light signaling *in vivo*. However, it remains poorly understood whether and how this particular prenylation regulates and contributes cellular functions. In this chapter, we discuss the biological significance of particular prenylation that we have pursued by *in vitro* and *in vivo* analyses. Farnesylation of transducin is interchangeable with geranylgeranylation as for transducin activity *in vitro*,

ISSN NO: 1874-6047
DOI: 10.1016/B978-0-12-381339-8.00007-X

while geranylgeranylated Gtγ augmented interaction of Gtβγ with Gtα and enhanced the rate of GTP-binding reaction to Gtα. The genetically engineered mice with farnesyl-to-geranylgeranyl replacement of Gtγ showed normal rod responses to dim flashes under dark-adapted conditions but exhibited impaired properties in light adaptation *in vivo*. Molecularly, farnesylation of Gtγ facilitates not only light-induced solubilization of Gtβγ from membranes but also its light-dependent translocation from the outer segment to the inner region, an intracellular event that decreases the light sensitivity of rod cells. In conclusion, the farnesylation of transducin is interchangeable with the geranylgeranylation in terms of the light signaling, while the particular prenylation plays an important role for visual sensitivity regulation by providing sufficient but not excessive membrane anchoring of Gtβγ that facilitates signal-dependent intracellular translocation of transducin.

II. Introduction

Transducin plays a pivotal role in retinal phototransduction. In 1980–1990s, the biochemical properties of transducin were intensively investigated by using a reconstitution system with rhodopsin and transducin subunits purified from bovine retinas. During purification of transducin, Gtβγ was chromatographically separated into two fractions with a striking difference in activities [1]. The difference was attributed to the heterogeneity of Gtγ, but Edman degradation analysis of Gtγ in both fractions revealed no difference in their amino acid sequences [1]. In 1990, high-accuracy mass spectrometry analysis of Gtγ unveiled that Gtγ in one of the fractions is modified with farnesyl group at its C-terminal cysteine residue, whereas Gtγ in the other fraction lacks the cysteine residue to be modified [2]. Importantly, Gtβγ with unmodified Gtγ showed extremely low activity, if any, to enhance the rate of GDP–GTP exchange reaction on Gtα in the presence of a photobleaching intermediate of rhodopsin (light-activated rhodopsin) [1]. This is the first case that prenylation of Gγ is a critical posttranslational modification indispensable for the activity of heterotrimeric G protein. At that time, small G-protein Ras had just been shown to be farnesylated [3–5]. Similar to the case of transducin, the oncogenic form of Ras protein was shown to require farnesylation for its transforming activity [3–5]. These discoveries together opened the research field of biological significance of protein prenylation in the intracellular signal transduction [6]. Especially, a particular type of protein prenylation, farnesylation, of transducin was predicted to play a specific, though yet unidentified, role in retinal phototransduction that can be characterized by extremely high sensitivity and dynamic adaptation.

III.   Characteristic Properties of Phototransduction in Retinal Rod Cells

Visual perception begins with the absorption of a photon by an opsin pigment in the retinal photoreceptors in vertebrates (Figure 7.1A and B) [7–9]. Two types of photoreceptors operate in response to discrete ranges of the ambient light intensity: rods for the scotopic (twilight) vision and cones for photopic (daylight) vision with color recognition (Figure 7.1C) [10]. These two types of photoreceptors are highly differentiated photoreceptive neurons, and each consists of an outer segment, inner segment, and synaptic terminal (Figure 7.1C). The rod photoreceptor shows extremely high photosensitivity, capable of detecting individual photons. Transducin, a member of heterotrimeric G protein family, plays a central role in the intensive amplification of the intracellular phototransduction process in the outer segment [11,12]. The GDP-bound form of Gtα and Gtβγ forms a transient ternary complex with light-activated rhodopsin (metarhodopsin II) that stimulates GDP-GTP-exchange reaction on Gtα. This reaction dissociates the ternary complex into Gtα-GTP and Gtβγ, both of which gain solubility due to reduction of their membrane affinities [13]. A single molecule of metarhodopsin II catalyzes the formation of several hundred molecules of Gtα-GTP, which in turn activates the effector enzyme cGMP–phosphodiesterase (PDE). Rapid hydrolysis of cGMP by PDE caused the closure of cyclic-nucleotide-gated cation channels in the rod plasma membrane and leads to hyperpolarization of the membrane potential of the rod cells (Figure 7.1D) [7,11]. This effective amplification of phototransduction depends greatly on the structural feature of the outer segment, which is densely packed with hundreds of disk membranes forming a lamellar structure with a hydrophobic milieu (Figure 7.1C).

While having high photosensitivity, the rod cell can adapt to ambient light condition by adjusting their photosensitivity to a broad range of light intensity of background illumination [14]: the rod cell can adapt over more than a 10^5-fold range of ambient illumination. Light adaptation occurs at many steps in the visual system from photoreceptors to central nervous system. Even when confined to the photoreceptor cells, the light adaptation is mediated by multiple molecular mechanisms such as the regulation by phosphorylation of visual pigment and intracellular concentrations of calcium ion and cyclic nucleotides [15]. In parallel with these mechanisms, the light adaptation of the rod cell has been shown to involve massive translocation of transducin from the outer segment to the inner region of the cell in a light-dependent manner [16]. We directed our attention to this regulatory mechanism because it was likely that the posttranslational lipid modification by farnesyl affects the localization and/ or translocation of transducin within the rod cells. We will summarize the biochemical properties of farnesylated transducin, especially from a viewpoint

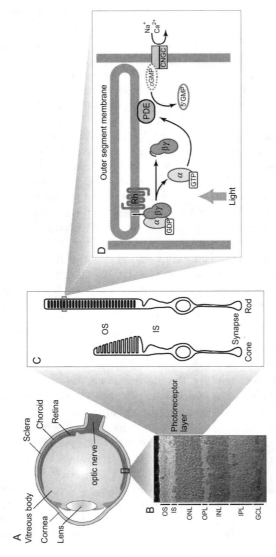

FIG. 7.1. Phototransduction in retinal rod photoreceptors. (A–C) Structures of mammalian eye, retina, and photoreceptors. In a rod cell, the outer segment consists of a stack of hundreds of disk membranes, which contain the light-absorbing photopigments, rhodopsin. The retinal image was merged with nuclear staining (B, *red*). OS, outer segment; IS, inner segment; ONL, outer nuclear layer; OPL, outer plexiform layer; INL, inner nuclear layer; IPL, inner plexiform layer; GCL, ganglion cell layer. (D) cGMP-mediated rod phototransduction cascade. Light-activated rhodopsin (Rh) activates hundreds of transducin (α/βγ) molecules by catalyzing the GDP–GTP exchange reaction on the α subunit. Each activated α subunit in turn binds to and activates cGMP–phosphodiesterase (PDE), resulting in rapid hydrolysis of cGMP and closure of cyclic-nucleotide-gated cation channel (CNGC). (See color plate section in the back of the book.)

of its contribution to the light-sensitivity regulation of the rod cells through translocation of transducin.

IV. Particular Lipidation of Photoreceptor Proteins

A. HETEROGENEOUS FATTY ACID MODIFICATION OF
TRANSDUCIN α SUBUNIT

Transducin α subunit is a member of Gi family of G protein and contains a consensus sequence for N-fatty acylation (myristoylation) at its N-terminus (NH$_2$MGXXXS). N-fatty acylation of Gα is essential not only for its membrane anchoring [17–19] but also for the functional coupling with Gβγ [20,21] and with effector proteins [22]. While Giα, Goα, and Gzα are uniformly myristoylated (C14:0), Tα is modified with a mixture of four types of fatty acids: laurate (C12:0), 5-*cis*-tetradecenoic acid (C14:1), 5-*cis*-, 8-*cis*-tetradecadienoic acid (C14:2), and myristate (Figure 7.2A) [20,23]. This heterogeneous N-fatty acylation has been found in other photoreceptor proteins, such as recoverin [24,25] and guanylyl cyclase-activating protein [26]. More than 90% of Gtα is modified by fatty acids other than myristic acid, and these acyl modifications are less hydrophobic than myristoylation. Such a less hydrophobic nature of Gtα modification may contribute to both prompt dissociation from the membranes upon receiving light stimulation and effective activation of PDE, supporting high photosensitivity of the rod cell. Although the physiological significance of heterogeneous acylation of Gtα has not been fully elucidated, *in vitro* reconstitution assays using Gtα–Giα chimaeric protein (Gt/iα) revealed that the steady-state GTP hydrolysis rate is significantly slower in the lauroylated form of Gt/iα than the myristoylated form [27]. Also, heterogeneously acylated Gtα with a diverged range of hydrophobicities appears to expand the range of light intensities over which the rod cell can generate detectable responses [28]. In addition, the less hydrophobic modification of Gtα should facilitate the light-dependent translocation of Gtα from the outer segment to the inner region of the rod cell. This effect should cooperate with that of Gtγ farnesylation which is less hydrophobic than geranylgeranylation of most Gγs in non-retinal tissues, and this issue will be discussed in the subsequent sections.

B. FARNESYLATION OF TRANSDUCIN γ SUBUNIT

Prenylated proteins have a C-terminal consensus sequence designated CAAX motif (C and A being cysteine and aliphatic amino acids, respectively), in which the prenyl group is linked via a thioether bond to the cysteine residue (Figure 7.2A). The X residue in the CAAX motif is the

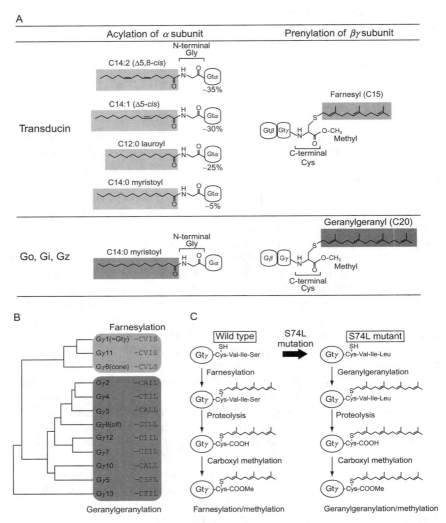

FIG. 7.2. Lipid modifications of heterotrimeric G protein subunits. (A) Comparison of lipid modifications of transducin subunits with those of other G proteins. (B) Phylogenic analysis of Gγ subtypes. Farnesylated and geranylgeranylated Gγs are clustered independently, suggesting the evolutionarily significant role of the particular prenylation, farnesylation. Amino acid sequences of mouse Gγs are aligned and UPGMA-based phylogenic tree is constructed by GENETYX software. (C) Schematic drawing of the C-terminal processing of transducin γ subunit. Wild-type Gtγ having C-terminal sequence of CVIS undergoes farnesylation, proteolysis of VIS residues, and carboxyl methylation of cysteine residue (*left*). Introduction of S74L mutation in Gtγ sequence provides the C-terminal CVIL sequence instead of CVIS, leading to geranylgeranylation without involving any alteration in the primary structure of mature Gtγ protein (*right*).

major determinant for the alternative prenylation, farnesylation (15-carbons), or geranylgeranylation (20-carbon) [29,30]. When X is serine, methionine, alanine, or glutamine, proteins are recognized by farnesyltransferase, whereas a leucine at this position leads to the modification by geranylgeranyltransferase I (Figure 7.2C). Following farnesylation or geranylgeranylation, the C-terminal tripeptide (AAX) are proteolyzed by the type II CAAX prenyl endopeptidase (RCE1) [31,32], and the newly exposed carboxyl group of the prenylcysteine residue is methylated by isoprenylcysteine carboxyl methyltransferase (ICMT; Figure 7.2C) [33,34]. Cooperatively with farnesylation, carboxyl methylation of Gtγ masks the negative charge of the C-terminal carboxyl group and promotes transient association of Gtβγ with the membranes, rhodopsin, and Gtα [35,36].

Among 12 subtypes of G protein γ subunits identified in mammalian cells, γ subunits of retinal rod transducin (Gtγ) [2,37] and cone transducin (Gγ8) have the C-terminal sequence of CVIS, which directs farnesylation. However, the other γ subtypes have leucine at the X position of CAAX motif and are geranylgeranylated (Figure 7.2B) [38]. Farnesylation and geranylgeranylation of Gγs are phylogenically clustered, indicating that the bifurcation of the prenylation (i.e., the switch of the amino acid X) is accompanied by the change in full-length sequence (Figure 7.2B). Indeed, the particular prenylation of Gγs is conserved across mammalian species, suggesting the evolutionarily significant role of selective farnesylation of transducin.

Interestingly, among members of the small G protein superfamily, only Ras family proteins are farnesylated, while Rho and Rab family proteins are geranylgeranylated, and farnesylation of Ras is indispensable for their transforming activity [4,39]. The transforming activity of Ras is enhanced not only by farnesyl but also by geranylgeranyl modification, demonstrating functional interchangeability between these two prenyl modifications [40]. In contrast to farnesylated Ras, however, expression of geranylgeranylated Ras exhibits potent growth inhibition of cultured NIH3T3 cells [40], indicating that selective farnesylation is required for Ras protein in terms of its function in normal cell growth. In this way, the particular modification of G protein has attracted considerable attention from the evolutionary and functional standpoints [30,41].

V. Biochemical Differences Between Farnesylation and Geranylgeranylation

Although a 15-carbon farnesylation is less hydrophobic than a 20-carbon geranylgeranylation, both of them are generally assumed to serve as a membrane anchor. However, the differences in chemical structure and hydrophobicity between farnesyl and geranylgeranyl are likely to impact on activities or functions of the modified proteins. This possibility can be examined by comparing the biochemical properties of farnesylated and geranylgeranylated transducin in *in vitro* reconstitution experiments [38,42]. Recombinant Gtβγ expressed in the baculovirus–insect cell system can be geranylgeranylated by coexpressing Gtβ and mutant Gtγ having the C-terminal sequence of CVIL (S74L mutation; Figure 7.2C) directing geranylgeranylation in place of CVIS for farnesylation [42]. The membrane affinity of Gtβγ was assessed by reconstituting recombinant Gtβγ with the rod outer segment membranes that had been stripped off the peripheral proteins. As expected from the hydrophobicity of the modifying lipids, geranylgeranylated Gtβγ showed much higher affinity for the membranes than the farnesylated form did, whereas unmodified Gtβγ that was neither farnesylated nor geranylgeranylated displayed extremely low affinity for the membranes. On the other hand, the degree of the subunit interaction between Gtα and modified Gtβγ in the absence of membranes was evaluated by the rate of ADP-ribosylation reaction of Gtα catalyzed by pertussis toxin, which ADP-ribosylates Gtα only when it associates with Gtβγ [43]. Interestingly, geranylgeranylated Gtβγ was far more effective in stimulation of the reaction than farnesylated Gtβγ, indicating that geranylgeranylated Gtβγ has a higher affinity to Gtα when compared with the farnesylated form. In contrast, unmodified Gtβγ showed substantially no affinity for Gtα. When GTP-binding activity was measured, transducin reconstituted with geranylgeranylated Gtβγ exhibited the rate constant of GTP-binding reaction much higher than that with farnesylated Gtβγ, whereas Gtα with unmodified Gtβγ exhibits essentially no GTP-binding activity [1,2], indicating that the function of Gtβγ depends absolutely on the lipid modification of Gtγ. The type of prenylation of Gβγ also affects the activation efficiency on effectors such as adenylyl cyclase and phospholipase C [44,45]. In this way, the prenyl modification *per se* is critical for the functions of G protein, but the "type" of prenylation is important as well for the G-protein signaling, and therefore it was predicted that particular prenylation, farnesylation, of transducin contributes to (some of) the characteristic features of light signaling in visual cells of the retina.

VI.   Physiological Significance of Farnesylation of Gtγ

A.   GENERATION AND MOLECULAR CHARACTERIZATION OF
       S74L KNOCKIN MICE

To study the physiological significance of the selective farnesylation of transducin, we have generated genetically-engineered mice, in which a gene encoding the farnesylation signal sequence CVIS at the C-terminus of Gtγ was mutated to encode CVIL (S74L mutation) that directs geranylgerany-lation [46]. Since the mutated amino-acid residue is cleaved off during the posttranslational processing (Figure 7.2C), the molecular and physiological phenotypes observed for the mutant mice hereafter is attributable solely to the change in the type of the C-terminal prenylation. The knockin mice homozygous for the S74L mutation (termed S74L knockin mice hereafter) are viable, fertile, and normal in appearance. The retinal morphology of the S74L knockin mouse was comparable to that of the wild type, with the thickness of each retinal layer unaltered and no symptom of retinal degen-eration even in the aged animals. Also, no significant differences were observed in the expression levels of retinal proteins such as transducin subunits and rhodopsin between wild-type and S74L knockin mice, indicat-ing that the S74L mutation does not affect the functional organization of the rod photoreceptors.

The molecular structure of Gtγ was examined by extraction of transdu-cin subunits from the rod outer segment membranes of wild-type and S74L knockin mice [13]. Extracted Gtγ was purified and subjected to mass spectrometric analysis [35]. The molecular mass of Gtγ matched the calcu-lated mass of the fully processed (i.e., farnesylated and methylated) form of mouse Gtγ, and top-down analysis of Gtγ using an electrospray ionization (ESI)-tandem mass spectrometry (MS/MS) [47] revealed farnesylation and methylation of wild-type Gtγ at the C-terminal region (Pro58-Cys71). On the other hand, similar analysis of Gtγ isolated from S74L knockin mice demonstrated geranylgeranylation and methylation of CAAX-mutated Gtγ at C-terminus.

The activities of Gtβγ were compared between farnesylated and gera-nylgeranylated forms by measuring the initial velocity of GTPγS-binding reaction in the retinal lysates prepared from wild-type and S74L knockin mice. Receiving a light flash, geranylgeranylated transducin in the retinal lysate from S74L knockin mice exhibited a higher initial velocity than farnesylated transducin in the wild-type lysate. This observation is consis-tent with the *in vitro* reconstitution experiments [42] (Figure 7.3C), sup-porting that farnesyl-to-geranylgeranyl replacement enhances the rate of GTP-binding activity reaction of transducin.

B. RETINAL ELECTROPHYSIOLOGY OF S74L KNOCKIN MICE

In vertebrates, a rod cell in the fully dark-adapted state has an extremely high photosensitivity, allowing it to respond to single photons [48,49], while being capable of regulating its sensitivity ranging over several orders of magnitude in response to background illumination (light adaptation) [7,11,14,15]. These physiological properties of rod cells were investigated

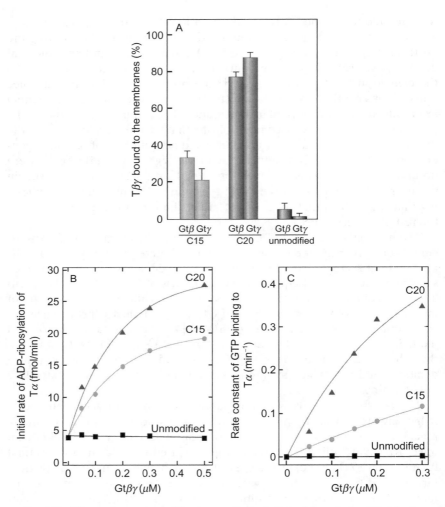

FIG. 7.3. *Effects of prenyl modification on the activities of transducin.* (A) Membrane affinity of Gtβγ. The rod outer segment membranes and Gtβγ was reconstituted *in vitro*, and the amounts of both Gtβ and Gtγ recovered in the soluble fraction after centrifugation were quantified by the immunoblot analysis. (B) Subunit interaction between Gtα and Gtβγ assessed by the rate of ADP-ribosylation reaction on Gtα. The ADP-ribosylation reaction catalyzed by pertussis toxin was carried out in the mixture containing Gtα (0.5 μM) and Gtβγ at various concentrations (0–0.5 μM). The initial rate of the reaction was estimated from the slope of the linear fitting of the reaction kinetics, and plotted against the concentration of Gtβγ. (C) Light-stimulated GTPγS-binding reaction of transducin. The GTPγS-binding reaction to constant amount of Gtα (1 μM) was measured by reconstitution with light-activated rhodopsin (30 nM), [^35S]GTPγS (10 μM) and various concentrations of Gtβγ (0–0.3 μM). The reaction kinetics was evaluated by measuring the radioactivity of Gtα-bound GTPγS, and then the data were fitted

by performing a couple of electrophysiological experiments: single-cell recordings and full-field electroretinogram (ERG) recordings.

The single-cell recording was performed by using the suction electrode method [50] to record photoresponses of the rods in the dark-adapted state. Light flashes of increasing intensities elicited similar sets of photoresponses from wild-type and S74L knockin mice (Figure 7.4A). Quantitative analysis of these traces demonstrated that there was no significant difference between the two genotypes both in the maximal response amplitude (r_{max}) and in light sensitivity calculated from the light intensity that evoked a half-saturating response. Their responses to a single photon were also indistinguishable in kinetics (Figure 7.4B). These results clearly indicate that the rod response properties are not affected by the type of prenylation of Gtγ under the dark-adapted condition, and this result apparently conflict with the *in vitro* observation that geranylgeranylated transducin has higher rate of GTP-binding reaction than farnesylated transducin (Figure 7.3C). Such a difference between the *in vivo* and *in vitro* experiments can be explained by the unique structure of the rod outer segments. In a dilute solution of the reconstitution experiments, lower membrane affinity of farnesylated Gtβγ may reduce the coupling efficiency between transducin and rhodopsin, resulting in the lower GTP-binding rate when compared with geranylgeranylated Gtβγ that is associated more strongly with the membranes. In contrast, light signaling in rod cells takes place in the outer segments, in which the signal-transducing proteins are densely packed in between the stacked disk membranes (Figure 7.1C) [7,51]. This unique structure of the outer segments maximizes the coupling efficiency between the signaling proteins irrespective of the modifying lipid, providing a minimal difference in the rate of GTP-binding reaction between farnesylated and geranylgeranylated transducin. This speculation is supported by *in vitro* experiments showing that the difference in the rate of GTP-binding reaction between farnesylated and geranylgeranylated transducin was reduced progressively when the concentration of the outer segment membrane suspension was increased [46].

---

with a equation, $B(t) = B_m(1 - \exp^{-kt})$, where $B(t)$ is the amount of GTPγS-bound Gtα at time t, B_m is the maximum binding, and k is the rate constant. The rate constant k was plotted against the concentration of Gtβγ in the reaction mixture. C15: farnesylated Gtβγ purified from bovine retinas (*blue bars* in (A) and *blue circles* in (B) and (C)), C20: S74L-mutated Gtβγ expressed in the baculovirus–insect cell system (the ratio of geranylgeranylation to farnesylation is 6:1, *red bars* in (A) and *red triangles* in (B) and (C)), unmodified: Gtβγ without farnesylation or geranylgeranylation (*black bars* in (A) and *black squares* in (B) and (C)). (For interpretation of the references to color in this figure legend, the reader is referred to the Web version of this chapter.)

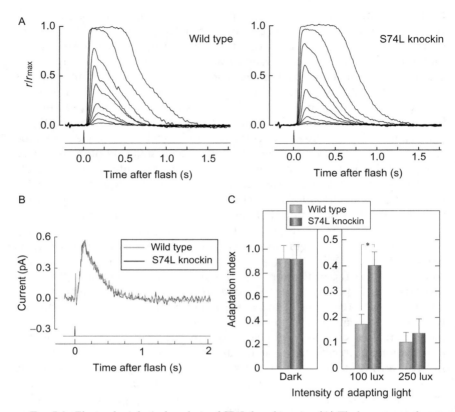

FIG. 7.4. *Electrophysiological analysis of S74L knockin mice.* (A) Flash responses from a single dark-adapted rod of wild-type (*left*) or S74L knockin (*right*) mouse to 500-nm flashes of increasing intensities (1.69, 3.45, 6.43, 13.13, 25.02, 51.08, 95.12, 194.21, and 770.35 photons μm^{-2}, respectively). Each trace is the averaged response from multiple flash trials. Stimulus monitor traces are shown below the records. (B) Single photon responses from a dark-adapted rod of wild-type (*blue*) or S74L/S74L (*red*) mouse. In both (A) and (B), the kinetics of the photoresponses from S74L knockin mice were indistinguishable from those from wild-type mice. (C) Light adaptation-dependent change in the light sensitivity. Mice were first subjected to the analysis of ERG in the dark-adapted condition, and then exposed to the adapting light of 0 lux (=dark), 100 lux or 250 lux before the second round of ERG recordings. Amplification constants were calculated from the analysis of the a-waves of the ERG responses from wild-type and S74L knockin mice. Adaptation index is the ratio of the amplification constant after the light adaptation to that of dark-adapted state. Error bars indicate standard deviations. *$p < 0.001$, Student's *t*-test wild-type versus S74L knockin mice. (For interpretation of the references to color in this figure legend, the reader is referred to the Web version of this chapter.)

The light-adapting property of the rod cells of S74L knockin mice was analyzed by measuring the ERG a-wave, which well reflects the light response from the photoreceptors (predominantly rods in mice). By analyzing the a-wave using the Lamb–Pugh model, the amplification of the photon signal can be quantified as an "amplification constant", a parameter that represents cellular light sensitivity of the rod [11,16,46,52]. Wild-type and S74L knockin mice were subjected to the analysis of ERG a-wave before and after 10-min light adaptation to 100- or 250-lux light, and the amplification constants were determined for the both genotypes. Under the dark-adapted condition, no significant difference in amplification constant was observed between wild-type and S74L knockin mice (Figure 7.4C, dark), being consistent with the results from the single-cell recordings (Figure 7.4A and B). After 10-min exposure to the 100-lux adapting light, the amplification constant of the wild-type mice was reduced to less than 20% of that of the dark-adapted state, indicating evident light adaptation. In S74L knockin mice, however, the reduction in the amplification constant after the light exposure was remarkably suppressed (Figure 7.4C, 100 lux). When the intensity of the adapting light was increased to 250 lux, then, the amplification constant of S74L knockin mice decreased sufficiently to a level almost identical to that observed for wild-type mice (Figure 7.4C, 250 lux). These results indicate that the farnesyl-to-geranylgeranyl replacement of Gtγ markedly impairs the light adaptation to a weaker (100 lux) light but does not to a brighter (250 lux) light, suggesting that the particular prenylation plays an important role in earlier phase of the light adaptation of rod cells in mice.

C. FARNESYLATION ESSENTIAL FOR LIGHT-DEPENDENT TRANSLOCATION OF TRANSDUCIN

The light-dependent translocation of various photoreceptor proteins has been reported to contribute to light adaptation on a relatively long time-scale [16,53–55]. In S74L knockin mice, higher hydrophobicity of geranylgeranylated Gtγ would inhibit the translocation of transducin, leading to the suppression of the light adaptation. We examined the light-dependent translocation of transducin under the light conditions identical to those used for the ERG recordings. In the dark-adapted state, both Gtα and Gtβγ were predominantly located in the outer segment layer of the retina in both wild-type and S74L knockin mice (Figure 7.5, dark). After receiving 10-min adaptation light (100 or 250 lux), Gtβγ in wild-type mice translocated to the inner region of the rod cell (Figure 7.5, left panel, 100 and 250 lux). The localization profiles of farnesylated Gtβγ display a striking contrast with those of geranylgeranylated Gtβγ, which remains in the outer

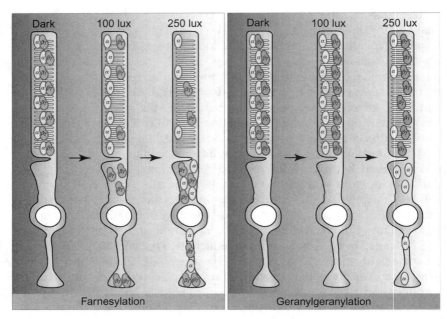

FIG. 7.5. Light-dependent translocation of transducin subunits in the rod cell during light adaptation. Gtα and Gtβγ localized predominantly in the outer segments in both wild-type and S74L knockin mice in the dark-adapted state (*dark in both panels*). After 10-min exposure to adaptation light of 100 or 250 lux, farnesylated Gtβγ was delocalized and distributed throughout the rod cell (*left panel*), whereas geranylgeranylated Gtβγ remained in the outer segment most probably due to the increased hydrophobicity (*right panel*). On the other hand, the translocation of Gtα was not affected by the type of prenylation of Gtβγ, that is, Gtα was detected mostly in the outer segment after 10-min exposure to the 100-lux light, while it translocated to the inner region after 10-min exposure to the 250-lux light. (See color plate section in the back of the book.)

segment even after the 10-min light exposure (Figure 7.5, right panel, 100 and 250 lux). These results indicate that geranylgeranylation of Gtγ inhibits intracellular translocation of Gtβγ and suppresses light adaptation, especially to the 100-lux weaker light. It is reasonable that the translocation of Gtβγ contributes to the decrease in light sensitivity because a reduced level of Gtβγ in the outer segment should decrease the coupling efficiency between transducin and rhodopsin. Very interestingly, Gtα translocated considerably to the inner region only when exposed to the 250-lux brighter light (for 10 min) but did not after 10-min exposure to the 100-lux weaker light in wild-type mice (Figure 7.5, left panel), as well as in S74L knockin mice (right panel). This property of Gtα provides consistent explanation for the observation that the decrease in light sensitivity was not significant between wild-type and S74L knockin mice after receiving 250-lux brighter light (Figure 7.4C); this phase of adaptation can be attributed to the translocation of Gtα, which occurs similarly in both genotypes (Figure 7.5). Together, the inhibition of the light-dependent translocation of Gtβγ by the S74L mutation was accompanied by a lower degree of light adaptation of the mutant rods to the weaker light, and we conclude that selective prenylation of Gtγ plays a critical role in the adaptation process. The light adaptation to a relatively weak ambient light is probably important for the nocturnal animals such as mice, because they are usually active in the dim light condition.

D. LIPID HYDROPHOBICITY DETERMINES THE INTRACELLULAR DYNAMICS OF TRANSDUCIN

Light-dependent translocation of various photoreceptor proteins is observed not only in vertebrates [56–58] but also in invertebrates [59]. Although the precise mechanisms for the translocation remains to be elucidated [54,60,61], the following three steps may be essential for the translocation of transducin subunits: (i) detachment of the subunits from the outer segment membranes upon light stimulation, (ii) translocation of transducin to the inner region of the rod cell, and (iii) return of the subunits back to the outer segment after light offset.

In the first step, farnesylated Gtβγ becomes soluble after dissociation of the ternary complex of light-activated rhodopsin and transducin subunits (Figure 7.6, left panel). However, a higher affinity of geranylgeranylated Gtβγ for the membrane may inhibit the solubilization, resulting in suppression of the light-dependent translocation of Gtβγ in S74L knockin mice (Figure 7.6, right panel). This idea is supported by our experimental evidence that farnesylated Gtβγ was dissociated from the outer segment membranes and solubilized *in vivo* in the light-dependent manner, whereas geranylgeranylated Gtβγ retained in the membranes even when activated Gtα became soluble in the mutant retina [46].

The second step is translocation of solubilized transducin from the outer segment to the inner region of the rod cell. We can speculate two general mechanisms that may underlie the translocation of transducin subunits: passive diffusion and active transport. Recent works suggest that the light-dependent dissociation and diffusion of transducin subunits are important for the intracellular translocation [54,62]. In the dark-adapted state, hydrophobicity of Gtβγ is augmented by the formation of the trimeric complex with Gtα, and hence it should be localized in the rod outer segment. This notion is also supported by the diffusely distributed pattern of farnesylated Gtβγ in the rod cells of Gtα-knockout mice not only in the

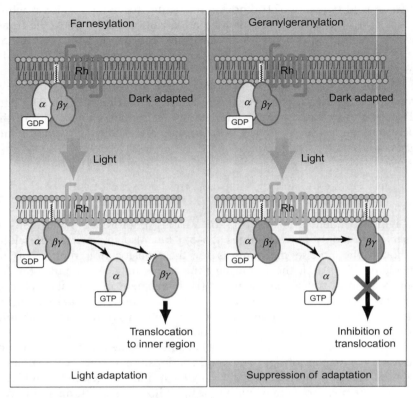

Fɪɢ. 7.6. Light-dependent transition of transducin subunits from the membrane to the soluble (cytosol) fraction. In the dark-adapted state, transducin subunits are tethered to the outer segment membranes. Upon exposure to the adapting light, light-activated rhodopsin and transducin subunits transiently form a ternary complex, which rapidly dissociates after GDP–GTP exchange reaction on Gtα (*left and right panels*). Farnesylated Gtβγ dissociates from the membranes, and translocates to the inner region of the rod cell (*left panel*). However, geranylgeranylated Gtβγ remains tethered to the outer segment membranes after the light exposure, and the membrane tethering inhibits intracellular translocation of Gtβγ and suppresses the light adaptation (*right panel*). In both cases, Gtα becomes soluble after dissociation of the ternary complex. (See color plate section in the back of the book.)

light but also in the dark [57]. More generally, the distribution pattern of photoreceptor protein seems to be governed by its balance between hydrophobicity and hydrophilicity. The rod outer segment consists of hundreds of densely packed membrane structures, leaving small and hydrophobic intracellular space, while the inner segment reserves relatively large hydrophilic space in the cytosol. Here, it should be noticed that transducin, cGMP–

PDE, rhodopsin kinase, and recoverin localized in the outer segment are all modified with the hydrophobic prenyl group or fatty acyl group, whereas hydrophilic proteins such as phosducin and arrestin are distributed in the inner region of the rod cells.

In the third step, translocated transducin subunits need to return back to the outer segment. In the light condition, translocated Gtβγ may be trapped in the inner region of the rod cells and be prevented from backward movement toward the outer segment by being associated with phosducin that is known to decrease hydrophobicity of Gtβγ [63–66]. In the dark, phosphorylation of phosducin by protein kinase A reduces the stability of the phosducin–Gtβγ complex [67], allowing Gtβγ to reassociate promptly with Gtα that has already hydrolyzed bound GTP and stays in the GDP-bound form showing higher affinity for Gtβγ. The trimer formation gains hydrophobicity, which should enhance the relocalization of transducin in the outer segment.

VII.   Biological Significance and Conclusion

The farnesyl-to-geranylgeranyl replacement in transducin provides unequivocally a new role of lipid modification in proteins; in this case, the physiological importance of selective farnesylation over geranylgeranylation in regulation of the G protein-mediated signal transduction. The selective farnesylation of Gtγ provides Gtβγ with a "proper" (sufficient but not excessive) membrane affinity, allowing Gtβγ (i) to stay with Gtα in the outer segment in the dark and (ii) to translocate to the inner region during light adaptation for modulation of light sensitivity. Generally, membrane anchoring of the prenylated proteins is important for effective signal transduction by accumulating relevant proteins at the membrane surface. In this respect, geranylgeranylation must be more advantageous due to its higher membrane affinity compared with farnesylation. As a consequence of such evolutionary advantage, most of G proteins may adopt geranylgeranylation for their membrane localization. In the photoreceptor cell, however, highly differentiated unique structure of the outer segment should support the compartmentalization and enable effective signaling of less hydrophobic G protein transducin, relieving the evolutionary constraint of its lipid modification for membrane anchoring. Rather, less hydrophobic nature of farnesylation can be advantageous for Gtβγ translocation to expand the cellular receptive sensitivity to ambient light intensities. This speculation may be applicable to Gtα and other photoreceptor proteins that are modified by unique lipids.

REFERENCES

1. Fukada, Y., Ohguro, H., Saito, T., Yoshizawa, T., and Akino, T. (1989). βγ subunit of bovine transducin composed of two components with distinctive γ subunits. *J Biol Chem* 264:5937–5943.
2. Fukada, Y., Takao, T., Ohguro, H., Yoshizawa, T., Akino, T., and Shimonishi, Y. (1990). Farnesylated γ subunit of photoreceptor G protein indispensable for GTP-binding. *Nature* 346:658–660.
3. Schafer, W., Kim, R., Sterne, R., Thorner, J., Kim, S., and Rine, J. (1989). Genetic and pharmacological suppression of oncogenic mutations in ras genes of yeast and humans. *Science* 245:379–385.
4. Casey, P.J., Solski, P.A., Der, C.J., and Buss, J.E. (1989). p21ras is modified by a farnesyl isoprenoid. *Proc Natl Acad Sci USA* 86:8323–8327.
5. Hancock, J., Magee, A., Childs, J., and Marshall, C. (1989). All ras proteins are poly-isoprenylated but only some are palmitoylated. *Cell* 57:1167–1177.
6. Finegold, A.A., Schafer, W.R., Rine, J., Whiteway, M., and Tamanoi, F. (1990). Common modifications of trimeric G proteins and ras protein: involvement of polyisoprenylation. *Science* 249:165–169.
7. Burns, M.E., and Baylor, D.A. (2001). Activation, deactivation, and adaptation in vertebrate photoreceptor cells. *Annu Rev Neurosci* 24:779–805.
8. Lamb, T. (2009). Evolution of vertebrate retinal photoreception. *Philos Trans R Soc Lond B Biol Sci* 364:2911–2924.
9. Lamb, T., Arendt, D., and Collin, S. (2009). The evolution of phototransduction and eyes. *Philos Trans R Soc Lond B Biol Sci* 364:2791–2793.
10. Fu, Y., and Yau, K. (2007). Phototransduction in mouse rods and cones. *Pflugers Arch* 454:805–819.
11. Arshavsky, V.Y., Lamb, T.D., and Pugh, E.N., Jr. (2002). G proteins and phototransduction. *Annu Rev Physiol* 64:153–187.
12. Stryer, L. (1986). Cyclic GMP cascade of vision. *Annu Rev Neurosci* 9:87–119.
13. Kuhn, H. (1980). Light- and GTP-regulated interaction of GTPase and other proteins with bovine photoreceptor membranes. *Nature* 283:587–589.
14. Fain, G.L., Matthews, H.R., Cornwall, M.C., and Koutalos, Y. (2001). Adaptation in vertebrate photoreceptors. *Physiol Rev* 81:117–151.
15. Pugh, E.N., Jr., Nikonov, S., and Lamb, T.D. (1999). Molecular mechanisms of vertebrate photoreceptor light adaptation. *Curr Opin Neurobiol* 9:410–418.
16. Sokolov, M., *et al.* (2002). Massive light-driven translocation of transducin between the two major compartments of rod cells: a novel mechanism of light adaptation. *Neuron* 34:95–106.
17. Bigay, J., Faurobert, E., Franco, M., and Chabre, M. (1994). Roles of lipid modifications of transducin subunits in their GDP-dependent association and membrane binding. *Biochemistry* 33:14081–14090.
18. Jones, T., Simonds, W., Merendino, J.J., Brann, M., and Spiegel, A. (1990). Myristoylation of an inhibitory GTP-binding protein α subunit is essential for its membrane attachment. *Proc Natl Acad Sci USA* 87:568–572.
19. Mumby, S., Heukeroth, R., Gordon, J., and Gilman, A. (1990). G-protein α subunit expression, myristoylation, and membrane association in COS cells. *Proc Natl Acad Sci USA* 87:728–732.
20. Kokame, K., Fukada, Y., Yoshizawa, T., Takao, T., and Shimonishi, Y. (1992). Lipid modification at the N terminus of photoreceptor G-protein α subunit. *Nature* 359:749–752.

21. Linder, M., Pang, I., Duronio, R., Gordon, J., Sternweis, P., and Gilman, A. (1991). Lipid modifications of G protein subunits. Myristoylation of Goα increases its affinity for βγ. *J Biol Chem* 266:4654–4659.
22. Taussig, R., Iñiguez-Lluhi, J., and Gilman, A. (1993). Inhibition of adenylyl cyclase by Giα. *Science* 261:218–221.
23. Neubert, T., Johnson, R., Hurley, J., and Walsh, K. (1992). The rod transducin α subunit amino terminus is heterogeneously fatty acylated. *J Biol Chem* 267:18274–18277.
24. Dizhoor, A., *et al.* (1992). The NH_2 terminus of retinal recoverin is acylated by a small family of fatty acids. *J Biol Chem* 267:16033–16036.
25. Sanada, K., Kokame, K., Yoshizawa, T., Takao, T., Shimonishi, Y., and Fukada, Y. (1995). Role of heterogeneous N-terminal acylation of recoverin in rhodopsin phosphorylation. *J Biol Chem* 270:15459–15462.
26. Palczewski, K., *et al.* (1994). Molecular cloning and characterization of retinal photoreceptor guanylyl cyclase-activating protein. *Neuron* 13:395–404.
27. Hashimoto, Y., Matsuda, T., Matsuura, Y., Haga, T., and Fukada, Y. (2004). Production of N-lauroylated G protein α subunit in Sf9 insect cells: the type of N-acyl group of Gα influences G protein-mediated signal transduction. *J Biochem* 135:319–329.
28. Neubert, T.A., and Hurley, J.B. (1998). Functional heterogeneity of transducin α subunits. *FEBS Lett* 422:343–345.
29. Casey, P.J., and Seabra, M.C. (1996). Protein prenyltransferases. *J Biol Chem* 271:5289–5292.
30. Fu, H.W., and Casey, P.J. (1999). Enzymology and biology of CaaX protein prenylation. *Recent Prog Horm Res* 54:315–342.
31. Kim, E., *et al.* (1999). Disruption of the mouse Rce1 gene results in defective Ras processing and mislocalization of Ras within cells. *J Biol Chem* 274:8383–8390.
32. Otto, J.C., Kim, E., Young, S.G., and Casey, P.J. (1999). Cloning and characterization of a mammalian prenyl protein-specific protease. *J Biol Chem* 274:8379–8382.
33. Dai, Q., *et al.* (1998). Mammalian prenylcysteine carboxyl methyltransferase is in the endoplasmic reticulum. *J Biol Chem* 273:15030–15034.
34. Bergo, M.O., *et al.* (2001). Isoprenylcysteine carboxyl methyltransferase deficiency in mice. *J Biol Chem* 276:5841–5845.
35. Fukada, Y., *et al.* (1994). Effects of carboxyl methylation of photoreceptor G protein γ subunit in visual transduction. *J Biol Chem* 269:5163–5170.
36. Ohguro, H., Fukada, Y., Takao, T., Shimonishi, Y., Yoshizawa, T., and Akino, T. (1991). Carboxyl methylation and farnesylation of transducin γ subunit synergistically enhance its coupling with metarhodopsin II. *EMBO J* 10:3669–3674.
37. Lai, R.K., Perez-Sala, D., Canada, F.J., and Rando, R.R. (1990). The γ subunit of transducin is farnesylated. *Proc Natl Acad Sci USA* 87:7673–7677.
38. Matsuda, T., and Fukada, Y. (2000). Functional analysis of farnesylation and methylation of transducin. *Methods Enzymol* 316:465–481.
39. Jackson, J.H., Cochrane, C.G., Bourne, J.R., Solski, P.A., Buss, J.E., and Der, C.J. (1990). Farnesol modification of Kirsten-ras exon 4B protein is essential for transformation. *Proc Natl Acad Sci USA* 87:3042–3046.
40. Cox, A.D., Hisaka, M.M., Buss, J.E., and Der, C.J. (1992). Specific isoprenoid modification is required for function of normal, but not oncogenic, Ras protein. *Mol Cell Biol* 12:2606–2615.
41. Sebti, S.M., and Der, C.J. (2003). Opinion: searching for the elusive targets of farnesyltransferase inhibitors. *Nat Rev Cancer* 3:945–951.

42. Matsuda, T., *et al.* (1998). Specific isoprenyl group linked to transducin γ subunit is a determinant of its unique signaling properties among G-proteins. *Biochemistry* 37:9843–9850.

43. Ui, M., and Katada, T. (1990). Bacterial toxins as probe for receptor-Gi coupling. *Adv Second Messenger Phosphoprotein Res* 24:63–69.

44. Myung, C.S., Yasuda, H., Liu, W.W., Harden, T.K., and Garrison, J.C. (1999). Role of isoprenoid lipids on the heterotrimeric G protein γ subunit in determining effector activation. *J Biol Chem* 274:16595–16603.

45. Fogg, V.C., Azpiazu, I., Linder, M.E., Smrcka, A., Scarlata, S., and Gautam, N. (2001). Role of the γ subunit prenyl moiety in G protein βγ complex interaction with phospholipase Cβ. *J Biol Chem* 276:41797–41802.

46. Kassai, H., *et al.* (2005). Farnesylation of retinal transducin underlies its translocation during light adaptation. *Neuron* 47:529–539.

47. Kassai, H., Satomi, Y., Fukada, Y., and Takao, T. (2005). Top-down analysis of protein isoprenylation by electrospray ionization hybrid quadrupole time-of-flight tandem mass spectrometry; the mouse Tγ protein. *Rapid Commun Mass Spectrom* 19:269–274.

48. Baylor, D.A., Lamb, T.D., and Yau, K.W. (1979). Responses of retinal rods to single photons. *J Physiol* 288:613–634.

49. Baylor, D.A., Lamb, T.D., and Yau, K.W. (1979). The membrane current of single rod outer segments. *J Physiol* 288:589–611.

50. Sung, C.H., Makino, C., Baylor, D., and Nathans, J. (1994). A rhodopsin gene mutation responsible for autosomal dominant retinitis pigmentosa results in a protein that is defective in localization to the photoreceptor outer segment. *J Neurosci* 14:5818–5833.

51. Calvert, P.D., Govardovskii, V.I., Krasnoperova, N., Anderson, R.E., Lem, J., and Makino, C.L. (2001). Membrane protein diffusion sets the speed of rod phototransduction. *Nature* 411:90–94.

52. Lamb, T.D., and Pugh, E.N., Jr. (1992). A quantitative account of the activation steps involved in phototransduction in amphibian photoreceptors. *J Physiol* 449:719–758.

53. Hardie, R. (2002). Adaptation through translocation. *Neuron* 34:3–5.

54. Calvert, P., Strissel, K., Schiesser, W., Pugh, E.J., and Arshavsky, V. (2006). Light-driven translocation of signaling proteins in vertebrate photoreceptors. *Trends Cell Biol* 16:560–568.

55. Artemyev, N. (2008). Light-dependent compartmentalization of transducin in rod photoreceptors. *Mol Neurobiol* 37:44–51.

56. Mendez, A., Lem, J., Simon, M., and Chen, J. (2003). Light-dependent translocation of arrestin in the absence of rhodopsin phosphorylation and transducin signaling. *J Neurosci* 23:3124–3129.

57. Zhang, H., Huang, W., Zhu, X., Craft, C.M., Baehr, W., and Chen, C.K. (2003). Light-dependent redistribution of visual arrestins and transducin subunits in mice with defective phototransduction. *Mol Vis* 9:231–237.

58. Rajala, A., *et al.* (2009). Growth factor receptor-bound protein 14 undergoes light-dependent intracellular translocation in rod photoreceptors: functional role in retinal insulin receptor activation. *Biochemistry* 48:5563–5572.

59. Bahner, M., Frechter, S., Da Silva, N., Minke, B., Paulsen, R., and Huber, A. (2002). Light-regulated subcellular translocation of Drosophila TRPL channels induces long-term adaptation and modifies the light-induced current. *Neuron* 34:83–93.

60. Giessl, A., *et al.* (2004). Differential expression and interaction with the visual G-protein transducin of centrin isoforms in mammalian photoreceptor cells. *J Biol Chem* 279:51472–51481.

61. Lee, S.J., and Montell, C. (2004). Light-dependent translocation of visual arrestin regulated by the NINAC myosin III. *Neuron* 43:95–103.
62. Rosenzweig, D., *et al.* (2007). Subunit dissociation and diffusion determine the subcellular localization of rod and cone transducins. *J Neurosci* 27:5484–5494.
63. Lee, R.H., Lieberman, B.S., and Lolley, R.N. (1987). A novel complex from bovine visual cells of a 33,000-dalton phosphoprotein with β- and γ-transducin: purification and subunit structure. *Biochemistry* 26:3983–3990.
64. Tanaka, H., *et al.* (1996). MEKA/phosducin attenuates hydrophobicity of transducin βγ subunits without binding to farnesyl moiety. *Biochem Biophys Res Commun* 223:587–591.
65. Sokolov, M., Strissel, K.J., Leskov, I.B., Michaud, N.A., Govardovskii, V.I., and Arshavsky, V.Y. (2004). Phosducin facilitates light-driven transducin translocation in Rod photoreceptors: evidence from the phosducin knockout mouse. *J Biol Chem* 279:19149–19156.
66. Gaudet, R., Savage, J., McLaughlin, J., Willardson, B., and Sigler, P. (1999). A molecular mechanism for the phosphorylation-dependent regulation of heterotrimeric G proteins by phosducin. *Mol Cell* 3:649–660.
67. Yoshida, T., Willardson, B.M., Wilkins, J.F., Jensen, G.J., Thornton, B.D., and Bitensky, M.W. (1994). The phosphorylation state of phosducin determines its ability to block transducin subunit interactions and inhibit transducin binding to activated rhodopsin. *J Biol Chem* 269:24050–24057.

8

Organization and Function of the Rab Prenylation and Recycling Machinery

KIRILL ALEXANDROV[a] • YAOWEN WU[b] •
WULF BLANKENFELDT[c] • HERBERT WALDMANN[d] •
ROGER S. GOODY[b]

[a]*Institute for Molecular Bioscience*
The University of Queensland, St. Lucia
Queensland, Australia

[b]*Max-Planck-Institute for Molecular Physiology*
Department of Physical Biochemistry, Otto-Hahn-Straße
Dortmund, Germany

[c]*Department of Biochemistry University of Bayreuth, Universitätsstraße*
Bayreuth, Germany

[d]*Max-Planck-Institute for Molecular Physiology Department of Chemical Biology*
Otto-Hahn-Straße Dortmund, Germany

I. Abstract

The Rab proteins form the largest subgroup of the Ras superfamily and control multiple steps of intracellular vesicular transport. Similar to many other small GTPases, Rab proteins are posttranslationally prenylated. Prenylation by Rab Geranylgeranyl transferase increases hydrophobicity of the RabGTPases and enables them to reversibly associate with their target membrane. Due to the vectoriality of vesicular transport, RabGTPases are recycled upon completion of their functional cycle and returned to a

ISSN NO: 1874-6047
DOI: 10.1016/B978-0-12-381339-8.00008-1

cytosolic intermediate. The cytosolic form of Rab proteins is stabilized by the tightly binding chaperon termed GDP dissociation inhibitor (GDI). GDI also plays a central role in loading of prenylated RabGTPases onto the target membranes. In this chapter, we discuss available structural and biochemical data that shed light on the molecular mechanism of Rab prenylation, membrane delivery, and recycling. We discuss the important functional differences between GDI and a structurally related RabGGTase accessory factor termed Rab escort protein. We summarize the available data on the identity of factors mediating dissociation of the Rab:GDI complex prior to membrane loading and provide arguments for the critical role of guanine nucleotide exchange factors in this process.

II. Rab GTPases

The Rab proteins form the largest subgroup of the Ras superfamily and control multiple steps of intracellular vesicular transport. Like other members of the Ras superfamily, Rab GTPases cycle between a GTP- and a GDP-bound forms. This cycle is regulated by GTP/GDP exchange factors (GEFs), which accelerate the otherwise slow rate of GDP dissociation and promote binding of more abundant cytosolic GTP to the GTPases. The GTPase-activating proteins (GAPs) accelerate the intrinsic GTPase activity thus promoting deactivation of the Rab protein. The activation of Rab GTPases coincides with their localization to the target membranes and the GTP-bound form of Rab proteins displays increased affinity for other proteins commonly referred to as "effectors." Interaction with effector molecules initiates a cascade of events resulting in execution of critical steps of vesicular transport, including budding, transport, and tethering/docking of transport vesicles to target membranes. GTP hydrolysis allows extraction of Rab from the membrane by a recycling factor termed GDP dissociation inhibitor (GDI), which forms a soluble complex with the otherwise hydrophobic Rab GTPases and makes them available for new rounds of recruitment. The number of Rab proteins in eukaryotic organisms varies significantly, with *Cryptosporidium parvum*'s genome containing only 8 Rab genes and *Trichomonas vaginalis* possessing the largest Rab complement of nearly 300 genes. Combined with the finding that mammalian organisms have only ca. 60 Rab GTPases, this raises the question about the relationship of RabGTPase family size and organization of the intracellular membrane compartments (see Ref. [1] and references therein).

III. Pathways of Rab Prenylation

Similar to many other small GTPases, Rab proteins are posttranslationally prenylated. This modification allows the reversible attachment with membranes as an essential feature of their functional cycle (Figure 8.1).

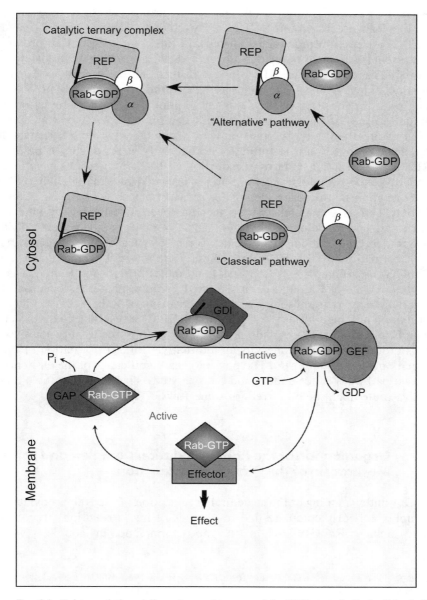

FIG. 8.1. Rab prenylation, delivery to membranes, and the GTPase cycle. In the "classical" pathway, the GDP form of Rab is recognized by REP (Rab escort protein) and presented to RabGGTase (α- and β-subunits), resulting in geranylgeranylation at one or two C-terminal cysteines. In the alternative pathway, the Rab molecule is recruited by the phosphoisoprenoid containing complex of REP and RabGGTase. Prenylated Rab proteins are escorted to their specific membranes and anchored by insertion of the lipid residues. They are activated by the action of GEF (guanine nucleotide exchange factor) molecules, which results in replacement of

After their synthesis on cytosolic ribosomes, Rab proteins are recognized by the Rab escort protein (REP), of which two isoforms (REP-1 and REP-2) are present in mammalian cells. In this complex, the Rab protein is presented to the prenylating enzyme, Rab geranylgeranyltransferase (RabGGTase or GGTase II), which transfers one or in most cases two geranylgeranyl groups to cysteines at the C-terminus of the Rab molecule. The need for the REP molecule as an additional factor in prenylation is unique to the Rab family; proteins with C-terminal CAAX box are pre-nylated directly by farnesyl transferase or geranylgeranyl transferase type I [2]. Since prenylation is essential for Rab activity, it is not surprising that REP is also essential, and loss or reduction of this activity results in various phenotypes or pathological conditions, as discussed later in this chapter [3].

Formation of the ternary complex between a Rab protein, REP, and RabGGTase is a complex process that has been investigated by kinetic, structural, and modeling methods [4,5]. The complex can be generated via two routes. One of them is referred to as the "classical" route, which starts with formation of the REP:Rab complex, followed by binding of the RabGGTase (Figure 8.1). REP has micromolar affinity for RabGGTase in the absence of Rab and geranylgeranyl pyrophosphate. Association of geranylgeranyl pyrophosphate with the active site of RabGGTase induces formation of a high-affinity complex with REP. The resulting complex can bind Rab protein with low nanomolar affinity to form the ternary complex. This order of binding represents an alternative route to formation of the ternary protein complex [6]. The physiological significance of the existence of two pathways is not understood but one study suggested that the alternative route might be preferred for some RabGTPases [7].

IV.   Organization of the Rab Prenylation Complex and Mechanism of the Prenylation Reaction

Our understanding of the structural organization of the Rab prenylation complex is nearly complete due to a series of high-resolution structures that include RabGGTase in the apo form and in complex with

---

GDP by GTP. In their active GTP bound form, Rabs can interact with effector molecules setting off the signaling cascade leading to vesicle budding, transport, and fusion. Upon completion of the functional cycle, the slow intrinsic rate of GTP hydrolysis is accelerated by GAP (GTPase activat-ing protein) activity, after which the Rab molecule can be extracted from the membrane by GDI (GDP dissociation inhibitor), which returns the Rab molecule to the cytosolic recruitable pool. (See color plate section in the back of the book.)

phosphoisoprenoids and prenylated peptides together with a Rab7:REP-1 and a REP-1:RabGGTase complexes [8–11]. The structure of the catalytic ternary complex is still not available, but a biochemically validated computational model explains the molecular details of the Rab prenylation process [5]. Since the structures of subcomplexes were reviewed elsewhere, we will focus here on the final structural model of the ternary complex and the current view of the prenylation process [12].

In the catalytic ternary complex, Rab and RabGGTase bind to distinct sides of the REP molecule (Figure 8.2). Rab contacts domain I of REP-1 via two binding sites: the GTPase domain of Rab7 is bound to the Rab binding platform (RBP) of REP-1, while the visible part of the extended C-terminus of Rab7 is coordinated by the C-terminus binding region (CBR) of REP-1. The former interaction involves the Switch I and, particularly, Switch II regions of Rab molecule. These regions undergo structural changes on nucleotide exchange and hydrolysis. In the activated (GTP-bound) form, the switch regions adopt conformations less favorable for formation of the Rab:REP complex [8]. This explains the higher affinity of REP for GDP-bound forms of Rab molecule [13,14].

The second interaction involves formation of the interface between the CBR of REP and the hydrophobic CBR interacting motif (CIM) of Rab GTPases. This is driven by binding of a cryptic hydrophobic sequence on the Rab C-terminus to a hydrophobic cavity on the REP molecule. The role of this interaction appears to be twofold: it increases the affinity of the Rab:REP complex by an order of magnitude and also directs the disordered C-terminus to the active site of RabGGTase bound to domain 2 of REP (Figure 8.2). The affinity of the ternary complex is further increased by the weak and largely unspecific interactions of the C-terminus with the active site of RabGGTase. According to the current view, the prenylation machinery assembles sequentially and cooperatively, initially utilizing high affinity and high specificity interactions (Rab with RBP of REP) followed by progressively smaller, weaker, and less specific interactions of Rab C-terminal region first with REP and then with RabGGTase [15].

This arrangement enables RabGGTase to process C-terminal sequences that have nothing in common except the presence of one or mostly two cysteine residues at or close to the C-terminus. Thus, the RabGGTase-mediated prenylation appears to be purely concentration-driven once the ternary protein complex is established, and as a consequence, any cysteine-containing peptide concentrated in the active center undergoes modification [10,16]. Thus, in contrast to other protein prenyltransferases that harbor the substrate specificity in their active sites, RabGGTase "outsources" substrate selection to the accessory protein REP [2]. The cost of this feature for RabGGTase is a 5- to 100-fold reduction in reaction rate

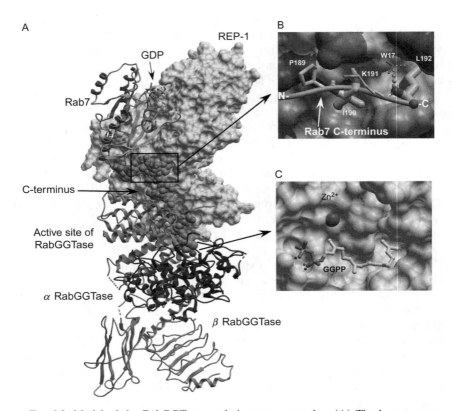

FIG. 8.2. Model of the RabGGTase catalytic ternary complex. (A) The lowest energy conformation of the Rab7 C-terminus docked onto the model of the ternary complex. REP-1 is displayed in surface representation and colored in gray. Rab7 is displayed as ribbons and colored according to secondary structure. The C-terminus of Rab7 is displayed in CPK representation and colored in green. RabGGTase is displayed in ribbon representation with the α-subunit colored in orange and the β subunit in blue. (B) Structural basis of the REP-1 CBR interaction with the CIM of Rab7. The hydrophobic CBR patch of REP-1 is displayed in surface representation, hydrophobic residues are colored yellow, and polar and charged residues are colored in pink, blue (+), and red (−). The main chain atoms involved in hydrogen bonding are displayed in atomic colors. The C-terminus of Rab7 is displayed as an orange worm, and the residues RabP189–RabL192 are displayed in stick format. N- and C- denote the termini. Atoms involved in hydrophobic interactions are colored in green. The thickness of the stick representation is an indication of the degree of interaction with REP. (C) Structure of the active site of RabGGTase with GGPP bound to it. The active site of RabGGTase is displayed in surface representation, and the ball-and-stick representation of GGPP is displayed in atomic colors. Zn^{2+} ion is displayed as magenta CPK spheres. (See color plate section in the back of the book.)

compared to that of FTase and GGTase-I. In return, the arrangement allows RabGGTase be active toward Rab GTPases with hypervariable C-terminal sequences.

RabGGTase catalyzes one or two prenyl transfer reactions, and in case of double prenylation, the reaction proceeds without dissociation of the singly prenylated intermediate [17]. Structural analysis of RabGGTase in complex with prenylated peptides revealed no defined position in the active site of RabGGTase that would "store" the isoprenoid of the monoprenylated intermediate when the second GGPP binds to RabGGTase [10]. Following the second transfer reaction, binding of GGPP to the lipid binding site dislodges the prenylated peptide and promotes its relocation to the lipid binding site on the domain II of REP. The lipid binding site is located in the vicinity of and is allosterically connected to the RabGGTase: REP interface. Binding of the prenylated Rab C-terminus to the lipid-binding site leads to disturbance of the residues involved in interaction with RabGGTase. In particular, the essential REP residue F279 is moved to a different position, disrupting the binding interface and reducing the affinity to RabGGTase [6,8]. It is interesting to note that F279 is missing in GDI, which is otherwise highly homologous to REP, explaining the different physiological functions of these two proteins [8,12]. The prenylated Rab:REP complex mediates delivery of the prenylated Rab protein to its target membrane [18].

V.   Structure of the Rab:GDI Complex and Its Functional Segregation from Rab:REP Complex

The structure of the monoprenylated Ypt1:GDI complex (the yeast homolog of a Rab:GDI complex) was determined at 1.5 Å resolution [19]. Although its structure is very similar to that of the Rab:REP complex, it displays several functionally important differences. The main binding interface is conserved between REP and GDI, but it is both smaller and involves fewer residues in the latter. As a consequence, the strength of the interaction with GDI is more influenced by the prenylation status of the GTPase. As discussed above, REP binds prenylated and unprenylated Rabs with low nanomolar affinity and in the unprenylated state only modestly responds to changes in the nucleotide bound state [13]. In contrast, incorporation of prenyl group to Rab leads to increase in the affinity of GDI for Rab from micromolar to low nanomolar [20,21]. Moreover, the state of the nucleotide bound to Rab dramatically affects the affinity of the Rab:GDI interaction. GTP-bound Rab binds to GDI with very low affinity, while the GDP-bound Rab displays low nanomolar affinity to GDI [15]. Similar to the arrangement in the Rab:REP complex, the C-terminus of the Rab molecule is anchored to the GDI molecule via hydrophobic interactions between the CBR of GDI and the hydrophobic CIM of the GTPase. The structure of the Ypt1:GDI complex provides an unambiguous explanation of why the lipid binding sites of GDI could not be identified despite a large number of genetic and biochemical studies and the availability of high-resolution structures of GDI [22,23]. The lipid binding site is formed by helices D, E, H, and F of domain II by an outward movement of helix D. This conformational change results in formation of a deep cavity penetrating the hydrophobic core of domain II that is occupied by the geranylgeranyl [19]. The subsequently determined structure of the doubly prenylated Ypt1:GDI complex demonstrated that the second lipid is more exposed to solvent and is skewed across several atoms of the first lipid moiety [24]. Interestingly, double prenylation slightly decreases the affinity of the Rab:REP and Rab:GDI complex [15,20,25].

VI.   Extraction of Rab Proteins from Membranes by GDI and REP

Based on the studies described briefly above, a model for the extraction of Rabs from membranes can be constructed. After initial interaction of the RBP of REP/GDI with the globular domain of the Rab molecule, the flexible C-terminus of Rab docks via its hydrophobic CIM region to the hydrophobic CBR of GDI/REP. Domain II of the GDI/REP molecule is now close to the membrane-anchored prenyl groups, and this leads to extraction of the prenyl groups from the membrane and binding to the site on GDI/REP described above. Although details of mechanism of these steps are not available, based on interaction analysis, we can make qualified statements about the thermodynamic driving force for these events, and use these to elucidate the differences between the relative efficiencies of REP and GDI in Rab extraction. REP is less efficient than GDI in Rab extraction, in accordance with their biological roles, that is, REP probably is only involved in delivery of Rabs to membranes. In contrast, GDI must also be able to extract Rabs from membranes. These properties can be explained at the molecular level by comparing the affinities of REP/GDI for unmodified and prenylated forms of Rab, respectively. REP binds with high affinity to both unprenylated and prenylated Rabs. The property of binding tightly to unprenylated Rabs is a requirement for its role in supporting Rab prenylation. In contrast, GDI binds only prenylated Rabs with high affinity. The very large increase in affinity of GDI to Rab on docking of the C-terminus and the prenyl groups into their respective binding sites provides the driving force for Rab extraction [20]. Thus, the high efficiency of GDI in extracting Rabs from membranes is a consequence of the large difference in binding energy when only the GTPase domain interacts with GDI and the situation where the C-terminus and prenyl groups are also docked. The difference in interaction energies is the thermodynamic driving force for Rab extraction. In the case of REP, most of the binding energy comes from the interaction with the GTPase domain and there is only very little driving force for extraction arising from the interaction of the C-terminus.

The discussed properties of REP and GDI are in line with their cell biological functions. The first role played by the REP interaction with Rab proteins is to present them to RabGGTase. The second role is to solubilize the prenylated Rab molecule. REP must then be ready to release Rab to its specific membrane. REP appears to be able to achieve this without strong fixation of the conjugated isoprenoids. At this point, the relatively weak affinity of the prenyl groups for REP is probably an advantage. An aspect that is not understood at present is the fact that REP would still be bound with relatively high affinity to the membrane-bound Rab. Additional accessory molecules that have the ability to displace GDI (GDFs, or GDI displacement factors; discussed below) might contribute to the resolution of this thermodynamic (and kinetic) problem, assuming they have a similar activity toward REP:Rab complexes. As also discussed below, the problem remains that the GDF would have to interact strongly with either the REP or the Rab molecule to disturb the Rab:REP interaction, leaving a similar energetic problem to be resolved to allow either REP or Rab to dissociate from the GDF. The ultimate driving force for the separation of prenylated Rab and REP is the weakening of the interaction when GDP is replaced by GTP, as elaborated on in other sections.

GDI does not have a high affinity for unprenylated Rabs. Because of this, a dramatic increase in affinity when prenyl groups are present does not result in nondissociable prenylated Rab:GDI that would otherwise fail to release prenylated protein in the appropriate part of the Rab cycle. However, this would be the case for prenylated Rab:REP complex if the increase in affinity on docking of the prenyl groups were as large as for GDI.

The arguments discussed here offer an explanation for the fact that there are two similar but distinct interaction partners for Rab required for prenylation (in addition to RabGGTase), solubilization, membrane delivery, and membrane extraction. A combination of all necessary properties in one molecule would probably not be possible for the reasons discussed here and elsewhere [12,20]. It is therefore not surprising that attempts to generate molecules with the combined properties of REP and GDI have not been significantly successful [26].

VII.   Targeting of Rabs to Specific Membranes

A still unresolved question in vesicular transport concerns the targeting of Rab proteins to specific membranes or membrane domains. An attractive model, for which experimental evidence was provided, postulates that the hypervariable C-terminus contains targeting information, that is, that the exact sequence of this 30- to 40-residue long stretch determines Rab localization [27]. Later work suggested that this model does not hold, at least not without the additional involvement of other regions of the Rab structure. Indeed, Rab orthologs from distant organisms with quite divergent C-termini were found to locate correctly, and replacement of the Rab27A C-terminus with the corresponding region from Rab5A did not affect its localization [28]. Some of the results of this work were interpreted to indicate that effector interactions might be critical in Rab localization, and evidence has been presented suggesting Rab9 targeting is mediated by interaction with its putative effector TIP47 [29].

A class of proteins that have been postulated as being important in delivery of Rabs to membranes are termed GDFs (GDI displacement factors) and evidence for their ability to disrupt the very stable Rab:GDI complexes has been presented. However, so far only one protein with an established activity of this kind has been described (PRA-1/Yip3) [30,31], and its properties suggest that it does not play the role of a specific receptor for individual Rab proteins; it has in fact been described as an endosomal GDF and interacts with a wide spectrum of prenylated GTPases [32]. The mechanism of action of PRA1/Yip3 is unclear, since it has been reported to interact strongly with Rab proteins and weakly with GDI, which are somewhat unexpected properties for a putative GDF. Thus, a factor with high affinity to a binding site on GDI which does not overlap with the Rab binding site but which can negatively affect the Rab affinity would appear more likely from first principles (Figure 8.3).

Quantitative studies on PRA1/Yip3 are hampered by the fact that it is an integral membrane protein. For this reason, reports on a soluble bacterial protein that appeared to have both GDF and GEF properties [35,36] generated expectations that the mechanism of GDF activity might be more easily investigated in this case. The protein in question is DrrA (or SidM), one of a large number of so-called effector proteins injected into the cytoplasm of *Legionella pneumophila* infected cells by its Dot-Icm protein type IV secretion system. DrrA localizes to the cytoplasmic face of *Legionella* replication vacuoles and is involved in the recruitment of ER-derived vesicles by interaction with the ER Rab protein Rab1. The conclusion that DrrA harbors both GEF and GDF activity toward Rab1 was subsequently shown to be incorrect [37]: the apparent GDF activity is solely

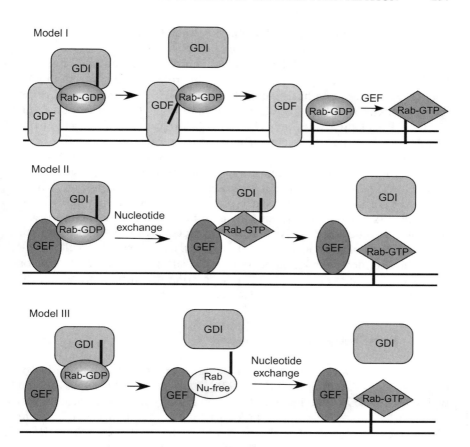

FIG. 8.3. Models of modulation of Rab targeting of Rabs to membranes by the state of bound nucleotide. In model I, GDF facilitated Rab–GDI dissociation, resulting in membrane attachment followed by GEF-mediated nucleotide exchange [33, 34]. The other models describe GEF-mediated Rab insertion to membranes. In model II, GEF acts directly on the Rab–GDI complex, leading to nucleotide exchange and Rab dissociation. In model III, spontaneous dissociation of Rab–GDI complex is rendered effectively irreversible by GEF-mediated nucleotide exchange, leading to membrane attachment [15]. (See color plate section in the back of the book.)

due to the GEF activity of DrrA and is a direct consequence of the GTP/ GDP-dependent interaction of Rab proteins with GDI discussed in a previous section. Thus, the dramatic drop in affinity between Rab and GDI when GDP is replaced by GTP [15] is the driving force for dissociation of the Rab:GDI complex, but there is no active disruption of the Rab:GDI complex by formation of a ternary complex with DrrA. This effect is to be expected for any Rab/GEF cognate pair, since it arises solely from the change in affinity of the Rab:GDI complex when GDP is replaced by GTP and may explain why reports of explicit GDFs are so scarce.

Although these studies did not shed light onto the mechanism of putative GDFs, they did draw attention to the possible role of GEFs in Rab targeting. The working hypothesis would be that a GEF localized to a specific membrane can stabilize the interaction of its cognate Rab by catalyzing GDP/GTP exchange, since the GTP form of Rabs cannot be extracted from membranes because of the low affinity of its interaction with GDI. Evidence obtained with DrrA strongly supports this model. Transfection of cells with DrrA results in localization of the protein to the plasma membrane. Intriguingly, Rab1 then also locates to the plasma membrane, where it is never normally seen. This suggests strongly that GEFs are likely to be major factors in the specificity of Rab–membrane interactions. However, it is likely that other factors also play a role, especially in the light of studies using purified complexes and membranes or in a permeabilized cell model. Here, nucleotide exchange appears to occur after the release of REP or GDI from the membrane [18,33,34], suggesting initial membrane attachment with Rab in the GDP-bound state.

VIII.   Rab Prenylation in Disease

Lack of or impaired GDI and REP activities are involved in a number of diseases [3]. A mutation in αGDI (L92P) has been found to be associated with nonspecific X-linked mental retardation [38]. The mutant protein was shown to be less efficient than wild-type GDI in extracting Rab3A from membranes and thus reduce the pool of cytoplasmic Rab3A. Since Rab3 cycling is an essential aspect of neurotransmitter release, the possible connection to mental retardation is apparent. The mutation is on the edge of the CBR patch of GDI and is likely to change its configuration and thereby the overall affinity of the complex. However, this has not been confirmed experimentally either in yeast or in a mammalian system.

A different X-chromosome-linked disease, choroideremia, leads to progressive loss of sight due to chorioretinal degradation, an effect arising from loss of function of REP-1 [39]. It appears that REP-2 can take over the prenylation and be functional for most Rabs, but Rab27A, which is expressed in the cell types affected in the disease (the retinal pigment epithelium and the choroid), appears to accumulate in the unprenylated form [40]. This was assumed to be due to the specific inability of REP-2 to support Rab27A prenylation at the required level in these cells. This proposal has been reexamined by several subsequent studies. One study suggested that several factors contribute to Rab27 underprenylation including low affinity of the Rab27:REP-2 complex for RabGGTase [41]. An alternative explanation was provided by the study that used semisynthetic fluorescent Rab27A to measure both its affinity to prenylation machinery and the rate of its prenylation [8]. The study found that REP-2 bound several tested Rabs ca. fivefold less strongly than REP-1 [8]. More significantly, Rab27A was found to have a generally very low affinity to both REP-1 and REP-2. This means that if Rabs compete for binding to REP, then limitation of REP activity will affect the weakly bound Rabs most. The simulation of this situation *in vitro* showed that Rab7 could potently inhibit prenylation of Rab27A. The suggestion was therefore made that loss of REP-1 activity in choroideremia reduces the total REP activity in the cell, and that this will affect weakly binding Rabs such as Rab27A more profoundly than others, leading to specific deficits in cells for which Rab27a plays an essential role. According to this model, overexpression of REP-2 should reduce the amount of unprenylated Rab27A and possibly revert the choroidermia phenotype. The *in vivo* proof of this model is still forthcoming.

A recent study pointed to previously undetected differences in membrane trafficking of choroidermia patients' peripheral cells. Fibroblasts of CHM patients displayed increased lysosomal pH, reduced rates of proteolysis, and altered secretion of cytokines [42]. The observed pleiotropic effect is difficult to reconcile with the model of single Rab underprenylation and may point to additive minor effects of a larger spectrum of unprenylated Rab GTPases.

IX.   Conclusions

The organization of the machinery involved in prenylation and recycling of Rab proteins is at the very heart of the process of intracellular vesicular transport. This process is a distinguishing feature of eukaryotic organisms, since the requirement for such a mechanism, or some variant of it, arises logically from the need for transport between subcellular compartments within the cell. It is an essential component of the organization of the endomembrane system, which is intricately involved in the generalized processes of exocytosis and endocytosis. The endomembrane system is considered to be an essential feature of what is referred to as the last eukaryotic common ancestor (LECA) [1], and Rabs as well as other proteins involved in vesicular transport such as SNAREs are considered part of the LECA. The development of a relatively complex endomembrane system in the LECA appears to have occurred rapidly after the initial diversion of a proto-eukaryote from other prokaryotes [43]. In view of the essential nature of the processes regulated by Rab proteins, it is therefore of utmost interest to understand the mechanisms involved, and the work discussed in this contribution describes a central aspect of their cycling between membranes and the cytosol. Further progress in this field can be expected to lead to significant advances in the overall understanding of cellular organization and dynamics, and to shed light on disease mechanisms involving processes in vesicular transport.

REFERENCES

[1] Brighouse, A., Dacks, J.B., and Field, M.C. (2010). Rab protein evolution and the history of the eukaryotic endomembrane system. *Cell Mol Life Sci* 67:3449–3465.
[2] Nguyen, U.T., Goody, R.S., and Alexandrov, K. (2010). Understanding and exploiting protein prenyltransferases. *Chembiochem* 11:1194–1201.
[3] Seabra, M.C., Mules, E.H., and Hume, A.N. (2002). Rab GTPases, intracellular traffic and disease. *Trends Mol Med* 8:23–30.
[4] Anant, J.S., *et al.* (1998). Mechanism of Rab geranylgeranylation: formation of the catalytic ternary complex. *Biochemistry* 37:12559–12568.
[5] Wu, Y.W., Goody, R.S., Abagyan, R., and Alexandrov, K. (2009). Structure of the disordered C terminus of Rab7 GTPase induced by binding to the Rab geranylgeranyl transferase catalytic complex reveals the mechanism of Rab prenylation. *J Biol Chem* 284:13185–13192.
[6] Thoma, N.H., Iakovenko, A., Kalinin, A., Waldmann, H., Goody, R.S., and Alexandrov, K. (2001). Allosteric regulation of substrate binding and product release in geranylgeranyltransferase type II. *Biochemistry* 40:268–274.
[7] Baron, R.A., and Seabra, M.C. (2008). Rab geranylgeranylation occurs preferentially via the pre-formed REP-RGGT complex and is regulated by geranylgeranyl pyrophosphate. *Biochem J* 415:67–75.
[8] Rak, A., Pylypenko, O., Niculae, A., Pyatkov, K., Goody, R.S., and Alexandrov, K. (2004). Structure of the Rab7:REP-1 complex: insights into the mechanism of Rab prenylation and choroideremia disease. *Cell* 117:749–760.
[9] Pylypenko, O., *et al.* (2003). Structure of rab escort protein-1 in complex with rab geranylgeranyltransferase. *MolCell* 11:483–494.
[10] Guo, Z., *et al.* (2008). Structures of RabGGTase-substrate/product complexes provide insights into the evolution of protein prenylation. *EMBO J* 27:2444–2456.
[11] Zhang, H., Seabra, M.C., and Deisenhofer, J. (2000). Crystal structure of Rab geranylgeranyltransferase at 2.0 Å resolution. *Structure* 8:241–251.
[12] Goody, R.S., Rak, A., and Alexandrov, K. (2005). The structural and mechanistic basis for recycling of Rab proteins between membrane compartments. *Cell Mol Life Sci* 62:1657–1670.
[13] Alexandrov, K., Simon, I., Iakovenko, A., Holz, B., Goody, R.S., and Scheidig, A.J. (1998). Moderate discrimination of REP-1 between Rab7 × GDP and Rab7 × GTP arises from a difference of an order of magnitude in dissociation rates. *FEBS Lett* 425:460–464.
[14] Seabra, M.C. (1996). Nucleotide dependence of Rab geranylgeranylation. Rab escort protein interacts preferentially with GDP-bound Rab. *J Biol Chem* 271:14398–14404.

[15] Wu, Y.W., Oesterlin, L.K., Tan, K.T., Waldmann, H., Alexandrov, K., and Goody, R.S. (2010). Membrane targeting mechanism of Rab GTPases elucidated by semisynthetic protein probes. *Nat Chem Biol* 6:534–540.

[16] Guo, Z., *et al.* (2008). Development of selective RabGGTase inhibitors and crystal structure of a RabGGTase-inhibitor complex. *Angew Chem Int Ed Engl* 47:3747–3750.

[17] Thoma, N.H., Niculae, A., Goody, R.S., and Alexandrov, K. (2001). Double prenylation by RabGGTase can proceed without dissociation of the mono-prenylated intermediate. *J Biol Chem* 276:48631–48636.

[18] Alexandrov, K., Horiuchi, H., Steele-Mortimer, O., Seabra, M.C., and Zerial, M. (1994). Rab escort protein-1 is a multifunctional protein that accompanies newly prenylated rab proteins to their target membranes. *EMBO J* 13:5262–5273.

[19] Rak, A., *et al.* (2003). Structure of Rab GDP-dissociation inhibitor in complex with prenylated YPT1 GTPase. *Science* 302:646–650.

[20] Wu, Y.W., Tan, K.T., Waldmann, H., Goody, R.S., and Alexandrov, K. (2007). Interaction analysis of prenylated Rab GTPase with Rab escort protein and GDP dissociation inhibitor explains the need for both regulators. *Proc Natl Acad Sci USA* 104:12294–12299.

[21] Ignatev, A., Kravchenko, S., Rak, A., Goody, R.S., and Pylypenko, O. (2008). A structural model of the GDP dissociation inhibitor rab membrane extraction mechanism. *J Biol Chem* 283:18377–18384.

[22] Luan, P., Balch, W.E., Emr, S.D., and Burd, C.G. (1999). Molecular dissection of guanine nucleotide dissociation inhibitor function in vivo. Rab-independent binding to membranes and role of Rab recycling factors. *J Biol Chem* 274:14806–14817.

[23] Schalk, I., *et al.* (1996). Structure and mutational analysis of Rab GDP-dissociation inhibitor. *Nature* 381:42–48.

[24] Pylypenko, O., *et al.* (2006). Structure of doubly prenylated Ypt1:GDI complex and the mechanism of GDI-mediated Rab recycling. *EMBO J* 25:13–23.

[25] Shen, F., and Seabra, M.C. (1996). Mechanism of digeranylgeranylation of Rab proteins. Formation of a complex between monogeranylgeranyl-Rab and Rab escort protein. *J Biol Chem* 271:3692–3698.

[26] Luan, P., *et al.* (2000). A new functional domain of guanine nucleotide dissociation inhibitor (alpha-GDI) involved in Rab recycling. *Traffic* 1:270–281.

[27] Chavrier, P., Gorvel, J.P., Stelzer, E., Simons, K., Gruenberg, J., and Zerial, M. (1991). Hypervariable C-terminal domain of rab proteins acts as a targeting signal. *Nature* 353:769–772.

[28] Ali, B.R., Wasmeier, C., Lamoreux, L., Strom, M., and Seabra, M.C. (2004). Multiple regions contribute to membrane targeting of Rab GTPases. *J Cell Sci* 117:6401–6412.

[29] Aivazian, D., Serrano, R.L., and Pfeffer, S. (2006). TIP47 is a key effector for Rab9 localization. *J Cell Biol* 173:917–926.

[30] Sivars, U., Aivazian, D., and Pfeffer, S. (2005). Purification and properties of Yip3/PRA1 as a Rab GDI displacement factor. *Methods Enzymol* 403:348–356.

[31] Sivars, U., Aivazian, D., and Pfeffer, S.R. (2003). Yip3 catalyses the dissociation of endosomal Rab-GDI complexes. *Nature* 425(6960):856–859.

[32] Figueroa, C., Taylor, J., and Vojtek, A.B. (2001). Prenylated Rab acceptor protein is a receptor for prenylated small GTPases. *J Biol Chem* 276:28219–28225.

[33] Soldati, T., Shapiro, A.D., Svejstrup, A.B., and Pfeffer, S.R. (1994). Membrane targeting of the small GTPase Rab9 is accompanied by nucleotide exchange. *Nature* 369:76–78.

[34] Ullrich, O., Horiuchi, H., Bucci, C., and Zerial, M. (1994). Membrane association of Rab5 mediated by GDP-dissociation inhibitor and accompanied by GDP/GTP exchange. *Nature* 368:157–160.

[35] Ingmundson, A., Delprato, A., Lambright, D.G., and Roy, C.R. (2007). Legionella pneumophila proteins that regulate Rab1 membrane cycling. *Nature* 450:365–369.

[36] Machner, M.P., and Isberg, R.R. (2007). A bifunctional bacterial protein links GDI displacement to Rab1 activation. *Science* 318:974–977.

[37] Schoebel, S., Oesterlin, L.K., Blankenfeldt, W., Goody, R.S., and Itzen, A. (2009). RabGDI displacement by DrrA from Legionella is a consequence of its guanine nucleotide exchange activity. *Mol Cell* 36:1060–1072.

[38] D'Adamo, P., *et al.* (1998). Mutations in GDI1 are responsible for X-linked non-specific mental retardation. *Nat Genet* 19:134–139.

[39] Seabra, M.C. (1996). New insights into the pathogenesis of choroideremia: a tale of two REPs. *Ophthalmic Genet* 17:43–46.

[40] Seabra, M.C., Ho, Y.K., and Anant, J.S. (1995). Deficient geranylgeranylation of Ram/Rab27 in choroideremia. *J Biol Chem* 270:24420–24427.

[41] Larijani, B., Hume, A.N., Tarafder, A.K., and Seabra, M.C. (2003). Multiple factors contribute to inefficient prenylation of Rab27a in Rab prenylation diseases. *J Biol Chem* 278:46798–46804.

[42] Strunnikova, N.V., *et al.* (2009). Loss-of-function mutations in Rab escort protein 1 (REP-1) affect intracellular transport in fibroblasts and monocytes of choroideremia patients. *PLoS ONE* 4:e8402.

[43] Dacks, J.B., and Field, M.C. (2007). Evolution of the eukaryotic membrane-trafficking system: origin, tempo and mode. *J Cell Sci* 120:2977–2985.

9

Protein Prenylation
CaaX Processing in Plants

SHAUL YALOVSKY

Department of Molecular Biology and Ecology of Plants
Tel Aviv University
Tel Aviv, Israel

I. Abstract

Plants have evolutionarily conserved prenyl transferases and CaaX processing enzymes. Their study has illuminated important evolutionary and
mechanistic aspects relevant to the study of protein prenylation and CaaX
processing. In the model plant species, *Arabidopsis thaliana* (*Arabidopsis*)
mutants in the farnesyltransferase (FT), geranylgeranyltransferase-I (GGT-
I) α and β subunits, and Rab geranylgeranyltransferase (RGGT) β subunit
have been identified and characterized. A mutant, *RNAi*-silenced plants and
overexpressing plant of isoprenyl carboxy methyl transferases (ICMTs)
have been created and characterized. Characterization of these mutant
prenyltransferases and ICMT plants has shown the important role of protein
prenylation and CaaX processing during plant development and in responses
to the environment, in particular to abiotic stress and light signaling. The
phenotypes of the *Arabidopsis* prenyltransferases and ICMT mutants have
prompted studies directed for their utilization for biotechnological purposes
of producing agronomically superior plant species. The identification and
characterization of FT and GGT-I substrates have shown that plants posses
both conserved and unique prenylation substrates. More recent genome
analyses have revealed that some of these protein substrates only exist in
certain plant family/ies. Importantly, plants produce isoprenoids via two

DOI: 10.1016/B978-0-12-381339-8.00009-3

pathways: the cytoplasmic mevalonate (MVA) pathway and the plastidial methyl-D-erythritol-4-phosphate (MEP) pathway. This fact may have important implications to the regulatory function of protein prenylation in coordinating signaling and metabolic pathways. In the present chapter, I will cover these topics with the hope that it would be both a useful source for experts in the field of protein prenylation and for drawing more research to this understudied area of protein lipidation.

II.   The Plant Protein Prenyltransferase Enzymes

The plant FT, GGT-I, and RGGT are structurally and functionally conserved with their homologs from yeast and animals. Sequence and phylogenetic analysis show that the α and β subunits of FT, GGT-I, and RGGT and the Rab Escort Protein (REP) are more closely conserved between plant and animal than either one with yeast [1–4]. However, plants have diverged as a distinct group prior to the divergence of fungi and animals. Hence, the greater conservation between the plant and animal subunits implies that, in ancestral eukaryotic cells, prenyltransferase subunits and REP resembled more the current plant and animal proteins. Often in plants, proteins exist in multimember families. FT and GGT-I α and β subunits are exceptional and are encoded by single gene *loci* in different plant species. In *Arabidopsis* and other plant species, the RGGT α and β subunits, designated RGGTA and RGGTB, are each encoded by two genes and REP is encoded by a single gene [2,4]. RGGTA1 and RGGTB1 are expressed in different tissues and developmental stages. In contrast, expression of RGGTA2 and RGGTB2 is mainly restricted to developing pollen [2]. Pollen tubes elongate at high rate in a highly polar fashion known as tip growth. Tip growth is characterized by extensive vesicle secretion toward the growing cell tip [5]. It could be that the RGGTA2–RGGTB2 complex might have specific function in facilitating Rab function during this highly specialized process.

A.   BIOCHEMICAL AND MOLECULAR PROPERTIES OF
      THE PLANT FT AND GGT-I

The plant FT and GGT-I are soluble enzymes that form heterodimers comprising α and β subunits in the cytoplasm [1,3,6]. When expressed in baculovirus-infected insect cells, expression of each of the FT/GGT-I α and β subunits by itself affected protein stability. Furthermore, protein purification succeeded only when the α and β subunits of either FT or GGT-I were coexpressed together [1,3]. RNA and protein expression analysis indicated that both FT and GGT-I are expressed to more or less the same levels in different tissues, indicating their role as housekeeping enzymes.

B.   THE PLANT FT

Coexpression of tomato FT α and β subunits, SlFTA and SlFTB, in yeast *ram1Δ* FT β-subunit mutant cells restored Ras1p and Ras2p membrane localization, growth at restrictive temperatures, and mating. Pheromone diffusion halo and *in vitro* prenylation assays showed that farnesylation of the **a**-factor by plant FT was less efficient than by the yeast enzyme indicating differences in substrate specificity [3]. The plant FT β-subunit contains an insert of 66 amino acids starting at position 335 that interrupts a highly conserved region, which is implicated in interaction with the CaaX peptide and prenyl substrate [7–9]. Complementation assays in yeast showed that this domain required for full activity of plant FT [3].

1.   The Plant GGT-I

Biochemical characterization demonstrated that *Arabidopsis* GGT-I preferentially prenylates proteins terminating with a CaaL rather than CaaX box sequence motifs. Similar to yeast and animal GGT-I, the plant enzyme can utilize both geranylgeranyl pyrophosphate (GGPP) and farnesyl pyrophosphate (FPP) as prenyl group donors but prenylation with GGPP is more efficient. As with mammalian GGT-I [10], a polybasic domain proximal to the prenylated Cys residue increases prenylation efficiency by an order of magnitude [1]. For example, the K_m value for prenylation of the petunia calmodulin CaM53 and the *Arabidopsis* Rho of Plants 6 (ROP6) was in the range of 0.1–0.3 μM whereas the K_m value for prenylation of GST–CTIL fusion protein, corresponding to the CaM53 CaaL box fused to GST, was 2–3 μM [1,11]. The k_{cat} of ROP6 geranylgeranylation by GGT-I is approximately 40-fold higher than the k_{cat} of its farnesylation by FT (28.9 vs. 0.71 min^{-1}) [11]. Similarly, it has been reported that geranylgeranylation of *Arabidopsis* Gγ subunit proteins AGG1 and AGG2 is more efficient than their farnesylation by FT [12]. In line with the *in vitro* data, chemical analysis of prenylation *in vivo*, using gas chromatography coupled mass spectrometry (GC–MS), revealed that both ROP6 and AGG1 are primarily geranylgeranylated in *Arabidopsis* and tobacco. Surprisingly, the *Arabidopsis* GGT-I β-subunit, *ggt-Ib*, mutants have mild developmental and physiological phenotypes [12,13]. In these plants, both ROP6 and AGG1 are farnesylated and functional [11]. Hence, regardless of the lower efficiency of farnesylation by FT of GGT-I substrates, in plants, FT can compensate for the loss of GGT-I function.

Taken together the structural and functional analysis of plant, FT and GGT-I have indicated that they share many common properties with their yeast and mammalian homologs. *In vivo* analysis of mutants, however, showed that in comparison with yeast in which GGT-I β subunit mutant cells are nonviable [14,15], under most conditions the plant mutant display only mild phenotype.

C. THE PLANT RGGT

Early biochemical analysis using protein extracts and recombinant tomato Rab1 proteins indicated that Rab prenylation takes place by a conserved mechanism [16,17]. The early characterization of Rab prenylation in plants has been confirmed by recent studies [2,4]. Recombinant *Arabidopsis* REP (AtREP) can interact with nonprenylated Rabs and enhance Rab prenylation when added to plant protein extracts but not to yeast *mrs6^{ts}*, REP mutant, protein extracts. Sequence alignment showed that a conserved Arg residue (R195 in yeast Mrs6/R290 in rat REP1), which is required for REP function [18], is occupied by an Asp or Thr in plant REPs. An AtrepN188R mutant in which Asp was changed to Arg restored Rab prenylation in *mrs6^{ts}* protein extracts and *mrs6^{ts}* cell growth at restrictive temperatures [4]. The report by Hala *et al.* showed that AtREP is a bona fide REP with plant-specific features. Further support for the conserved mechanism of RGGT in plants came from a study showing that Rab prenylation is reduced in *rggtb1* mutant *Arabidopsis* plants [2].

Based on functional analysis of arbitrary point mutants in REP and RabGDI, it has been proposed that RabGDI has evolved from REP [18]. This suggestion was challenged by a phylogenetic analysis of REPs and RABGDIs, which suggests that REP and RabGDI have evolved from a common ancestor that presumably had both GDI and REP functions [4].

III.   FT and GGT-I Substrates in Plants

According to the Prenbase database of prenylated proteins (http://mendel.imp.ac.at/sat/PrePS/PRENbase/; [19]), there are 327 predicted prenylated plant proteins that are organized into 36 clusters. Some of the listed proteins have homologs in yeast and animals and some are unique to plants. Among the proteins that have homologs in other organisms are the ROPs, Gγ subunits, the Nucleosome Assembly Protein 1 (NAP1) and DnaJ. Among the plant unique proteins there are Ca^{2+}-calmodulin and calmodulin-related proteins, *m*embrane-bound *U*biquitin fold (MUB) proteins, E2 ubiquitin ligases, putative SNAREs, the floral transcriptional regulator

APETALA1, and a group of copper-binding proteins with unknown functions. Unfortunately, only few of the listed proteins were experimentally tested, some of the proteins predicted to be prenylated could not be verified experimentally and several proteins predicted not to be prenylated were found to be prenylated both *in vitro* and *in vivo*. Therefore, I have expanded the discussion to several proteins, which were experimentally proven to be prenylated and for which the role of prenylation in their function or subcellular localization was tested.

A. FARNESYLATED PROTEINS

1. NAP1

The Nucleosome Assembly Protein 1 (NAP1;1) promotes cell proliferation or cell expansion in a developmental context, depending on farnesylation status of the protein. During the cell-proliferation stage of leaf development, NAP1;1 was farnesylated and accumulated in the nucleus. During the cell-expansion stage, nonfarnesylated NAP1;1 accumulated in the cytoplasm. Ectopic expression of a nonfarnesylated mutant disrupted cell proliferation [20].

2. IPT3

Cytokinin is a hormone regulating diverse processes in plants. The cytokinin biosynthesis enzyme adenosine phosphate Isopentenyltransferase 3 (IPT3) is farnesylated. The farnesylated form of IPT3 was found in the cytoplasm and nuclei, and the nonfarnesylated form in plastids. The cytoplasmic/nucleus-localized and plastid-localized IPT3 forms were associated with synthesis of isopentenyl-type or zeatine-type cytokinins, respectively [21].

3. MUBs

MUBs are thought to function as modifiers in diverse signaling cascades. *Arabidopsis* has six MUBs and five of them terminate with a CaaX box motif. It has been shown that farnesylation and geranylgeranylation of AtMUB1 and AtMUB6, respectively, facilitate their attachment to the plasma membrane. AtMUB2 lacks a CaaX box and its attachment to the plasma membrane requires at least two C-terminal cysteine residues that may be *S*-acylated. In addition to the CaaX box, all prenylated MUBs contain a conserved C-terminal cysteine that could be *S*-acylated. The function of MUBs in plants has not yet been described and thus it is not yet known how prenylation affects their function [22].

4. APETALA1

The floral transcription factor APETALA1 (AP1) was shown to be farne-sylated both *in vitro* and *in vivo*. Functional analysis showed that prenylation was required for its function in regulation of flower development.

B. GERANYLGERANYLATED PROTEINS

1. CaM53

One of the earliest and most detailed studies that exemplified the role of protein prenylation in plants was on the petunia Ca^{2+}-calmodulin CaM53 [23]. CaM53 is geranylgeranylated and localizes to the plasma membrane. Nonprenylated CaM53 mutant accumulated in the nucleus due to a C-terminal polybasic domain. Similarly, inhibition of isoprenoid biosynthesis by the hydroxymethylglutaryl CoA reductase (HMGR) inhibitor mevinolin (also known as lovastatin) caused accumulation of the protein in the nucleus. *Nicotiana benthamiana* plants that overexpressed the wild-type (WT) CaM53 were stunted, while plants that overexpressed a nonprenylated CaM53 mutant were smaller compared to WT plants but not stunted [23]. These results suggested that CaM53 might function in two distinctive signaling cascades in the membrane and nucleus. Ca^{2+}-Calmodulins and calmodulin-like proteins that terminate with a CaaL box motif can be found in other plant species including *Arabidopsis*, rice, and sorghum.

2. Gγ Subunits

Both *Arabidopsis* Gγ homologs AGG1 and AGG2 are prenylated, and prenylation is essential but not sufficient for their proper plasma membrane association [11,12]. *S*-acylation functions as an important second signal to efficiently target AGG2 to the plasma membrane. When *S*-acylation was inhibited, AGG2 accumulated in the endoplasmic reticulum (ER) and Golgi [12]. Reversible *S*-acylation may thus provide the means to shuttle AGG2 between the plasma and internal membranes. Analysis of protein lipid modification by GC–MS showed that AGG1 is only geranylgerany-lated but not *S*-acylated. Further, in *ggt-Ib* mutant plants, AGG1 was farnesylated, indicating that FT can function redundantly to GGT-I.

3. ROP/RAC GTPases

ROPs also known as RACs are master regulators of cell polarity and signaling [24–27]. A classification method based on the amino acid sequence divided ROPs into two major subgroups designated type I and type II [28] and will be used hereon for clarity. All type I ROPs terminate with a CaaL box motif and are preferentially geranylgeranylated [11,29]. Prenylation is required for plasma membrane attachment and function of ROPs (reviewed in Ref. [26]). Type II ROPs are not prenylated and attach to the membrane by *S*-acylation of cysteines that are part of a conserved domain designated the GC-CG box, in which the acylated cysteines are separated by a stretch of five or six aliphatic residues and flanked by glycines [30,31]. It has been shown that the composition and structure of the GC-CG box is required for stable association with the membrane [31].

Both type I and type II ROPs contain polybasic regions (PBRs), comprised of multiple lysines and/or arginines, proximal to the CaaX or GC-CG boxes. The PBRs enhance prenylation by PGGT-I [1,10] as well as protein interaction with negatively charged phosphatidylinositol 3,4,5 triphosphate (PtdIns 3,4,5-P3) and phosphatidylinositol 4,5 bisphosphate (PtdIns 4,5-P2) in the membrane [32]. A PBR deletion mutant of the type II ROP AtROP10 failed to associate with the membrane and remained in the cytoplasm [31]. Likely, PBRs have similar effect on membrane localization of type I ROPs. The PBR also facilitated nuclear localization of type I and type II ROPs CaaX and GC-CG box mutants [11,29–31]. Thus, the PBRs appear to have a dual role in shuttling proteins between membrane and nuclei, depending on whether proteins are lipid modified or not.

The viability of the *Arabidopsis ggt-Ib* mutant plants enabled studies aimed to reveal the effect of prenyl group type on steady-state distribution between the plasma membrane and the cytoplasm, membrane interaction dynamics, and ROP function in regulation of cell polarity [11]. Analysis of prenylation of ROP6 and the Gγ subunit AGG1 by GC–MS demonstrated that both proteins were farnesylated in the *ggt-Ib* background. Both confocal imaging and immunoblots with anti-ROP-specific antibodies showed that relative to *wt* plants, increased amounts of ROPs could be detected in the cytoplasm in steady state. Furthermore, the type of prenyl group had no effect on the distribution of ROPs in the membrane between detergent soluble membranes and lipid rafts. Interestingly, recombinant ROP6, which was purified from the soluble protein extracts of *ggt-Ib* plants, was farnesylated and *S*-acylated. Hence, unlike H-ras, where transient *S*-acylation of hypervariable domain cysteines promotes stable interaction with the plasma membrane [33], the transient *S*-acylation of G-domain cysteines in ROP6 only promotes stable membrane association in conjunction with geranylgeranylation.

The interaction dynamics of ROPs with the membrane was determined by FRAP beam-size analysis [34] that enables calculating the contribution of lateral diffusion in the membrane versus exchange with the cytoplasm to fluorescence recovery. It was found that the type of prenyl group had little effect on membrane interaction dynamics and on ROP signaling in cell polarity [11]. The minor differences in cell polarity that were observed likely resulted from the smaller fraction of ROPs in the plasma membrane rather than direct effect on their function. In contrast, mutations that abolished ROP G-domain transient *S*-acylation strongly affected ROP membrane interaction dynamics and compromised ROP function in regulation of cell polarity [35].

In *Arabidopsis*, the FT GGT-I common α subunit mutant called *pluripetala* (*plp*) is viable although membrane targeting of CaaX prenylated proteins is lost and they become unstable [12,36]. Cell polarity is strongly compromised in *plp* plants, likely resulting from loss of type I ROP function. These data, together with the results described above, indicate that prenylation is required for membrane targeting of ROPs and is therefore crucial for their function. Farnesylated ROPs are still targeted to the plasma membrane and functional, although their interaction with the membrane is weaker. The recovery of farnesylated and *S*-acylated ROPs from the soluble protein extracts prepared from *ggt-Ib* mutant plants is compatible with two alternative scenarios. One scenario is that the prenyl group is embedded in the lipid bilayer and that the greater hydrophobicity of geranylgeranyl in the C-terminal hypervariable domain together with the proximal polybasic domain is required for stable association with the membrane. An alternative explanation is that the prenyl moiety is associated with the plasma membrane through an adaptor with higher affinity for geranylgeranyl compared to farnesyl. Existence of such factor/s is compatible with the identification of Galectin-1 as adaptor for H-ras membrane anchoring [37]. Given that cells have proteins that specifically recognize the isoprenylated cysteine including; CaaX proteases, isoprenyl cysteine carboxy methyl transferase (ICMT), and RhoGDI, existence of membrane adaptor for geranylgeranylated ROPs is not unlikely.

Geranylgeranylation of Rho GTPases is required for their interaction with RhoGDI [38]. It is believed that RhoGDIs deanchor inactive GDP-bound Rhos from the plasma membrane and facilitate their recycling back to the membrane [38,39]. In tobacco pollen tubes, polar localization of RAC5 at the tip was facilitated by RhoGDI2-dependent recycling [40]. *Arabidopsis* RhoGDI *scn1* mutant plants develop root hairs with multiple tips, supporting the role of RhoGDI in polarized tip growth [41]. Palmitoylation was shown to inhibit the interaction of Rhos with RhoGDI [42]. Hence, it is likely that activation-dependent transient *S*-acylation of type I ROPs [35] inhibits their interaction with RhoGDI. Further, type II ROPs, which are not prenylated very likely, do not interact with RhoGDI regardless of their activation status.

C.   PRENYL GROUP TYPE AFFECT ON PROTEIN FUNCTION

Geranylgeranylated proteins are mostly found associated with the plasma membrane, for example, ROPs, CaM53, AGG1, AGG2, AtMUB6, [11,12,22,23,29,35] and most Rho GTPases in yeast and mammals [42].

This does not seem to be the case with farnesylated proteins, several of which are not localized to membranes, for example, NAP1;1, IPT3 [20,21], APETALA1 (AP1) [43], nuclear lamins [44], and the kinetochore-associated CENP-E and CENP-F proteins [45]. Often farnesylated proteins that are attached to the plasma membrane are also S-acylated, for example, H-ras, N-ras [46], and possibly AtMUB1 and AtMUB4 [22]. Consistent with this, association of AGG1, AGG2, and type-I ROPs with the plasma membrane was weaker in *pggt-Ib* mutant background [11,12], presumably since they were farnesylated rather than geranylgeranylated. Collectively, prenylation and the type of prenyl groups affect proteins in plant yeast and animal cells in a similar fashion. The mild phenotype of *Arabidopsis ggt-Ib* mutants possibly results from G-domain S-acylation of type I ROPs, which stabilizes their interaction with the membrane and the S-acylation of AGG2, which promotes its transport from the endomembrane system to the plasma membrane. Interestingly, *ggt-Ib* mutant are viable in *S. pombe* [47] and several mouse cell types remained viable in the absence of GGT-I activity. It would be now interesting to examine whether Rho proteins in animal and yeast also undergo activation-dependent transient S-acylation in the G domain, which may stabilize their membrane interaction, enabling them to function.

IV.   Protein Prenylation—A Crossroad Between Signaling and Metabolism

Both prenyl group donors FPP and GGPP are early intermediates in metabolic pathways that produce myriad of compounds, including the plant hormones abscisic acid (ABA), gibberellins (GA), cytokinins, and brassinosteroids (BRs) as well as important metabolites such as the phytyl side chain of chlorophylls, carotenoids, and all membrane sterols. Thus, protein prenylation may link between central metabolic pathways and diverse signaling cascades. Plants possess two distinct isoprenoid biosynthetic pathways: the cytosolic MVA pathway and the plastidial 2-C-methyl-D-erythritol-4-phosphate (MEP) pathway [48–52]. Recent studies demonstrated that, in tobacco BY-2 cells, the plastidial MEP pathway provides the isoprenyl moiety for protein geranylgeranylation [53]. Thus, a delicate balance may exist between the MVA and MEP pathways and G protein signaling in plants. Future studies on the MVA and MEP pathways will be required to elucidate such mode of regulation.

V.   Prenyltransferase Mutants in Plants

In *Arabidopsis*, mutants in FT, GGT-I, RGGT β-subunits, and the FT and GGT-I common α subunit have been characterized. The analysis of these mutants implicated protein prenylation in regulation of plant development, response to abiotic stress, light, and gravitropic stimuli.

A. THE *ARABIDOPSIS* FT β-SUBUNIT MUTANT *era1*

The first prenyltransferase mutant, which was identified and analyzed, was enhanced response to ABA (*era1*) FT β-subunit mutant. ABA regulates seed germination, and plant responses to drought stress. The *era1* mutant plants are hypersensitive to the ABA, resulting in inhibition of seed germination and increased resistance to drought stress [54,55]. Water loss and gas exchange in plants occurs primarily via specialized pores in leaves called stomata. The opening and closing of the stomata pores are regulated by dynamic changes in structure of two cells that surround the stomatal pore that are called the stomata guard cells. Specific responses to ABA in the stomata guard cells lead to change in their structure and closing of the stomata pores. The drought tolerance of *era1* plants results from closing of stomata in response to low ABA concentrations that do not affect *wt* cells [55]. Drought-tolerant crops have significant agricultural and environmental benefits since they reduce water consumption and enable growing plants under unfavorable hot and dry conditions. The ABA hypersensitivity of the FT β-subunit mutant in *Arabidopsis* has led to the development of drought-tolerant canola varieties. Canola seeds are used for production of cooking oil. Silencing the expression of *FTB* or *FTA* in canola leaves by *RNAi*, using a heat-shock inducible promoter, resulted development of drought-tolerant plants that had greater seed yields [56,57]. The success of utilizing FTA and FTB silencing in canola for biotechnological purposes, which stemmed from basic studies in *Arabidopsis*, highlights the importance of studying protein prenylation in plants and may lead to attempts to utilize a similar approach in other plant species.

Binding of ABA to its receptors in the cytoplasm promotes their interaction with TYPE 2 C PROTEIN PHOSPHATASEs (PP2C) and inhibition of their function. The inactivation of PP2C leads to the activation of SNF1-RELATED PROTEIN KINASE 2S (SnRK2S), which phosphorylates and activates transcriptional regulators and ion channels that mediate the ABA response [58,59]. Genetic analysis showed the FTB (ERA1) functions downstream of the PP2C proteins ABA Insensitive 1 and 2 (ABI1 and ABI2) and upstream of the transcriptional regulator ABI3. It was also shown that ABI3 expression is enhanced in *era1* mutant plants. ABA levels quickly increase in response to drought leading to transient increases in cytoplasmic Ca^{2+} concentrations in guard cells. In *era1* mutants, these Ca^{2+} transients appear following treatment with concentrations of ABA that do not cause similar affect in *wt* guard cells [60], suggesting that protein prenylation acts upstream or close to the ABA-associated Ca^{2+} response. ABI1, ABI2, and SnRKS2 do not terminate with a CaaX, box and it is currently unknown which farnesylated protein affects ABA signaling.

Analysis of *era1* plants also implicated protein farnesylation in plant development. In plants, the shoot and root apical meristems are specialized groups of cells located at the tips of shoots and roots that maintain stem cell identity. The shoot apical meristem (SAM) is enlarged in *era1* mutants leading to the formation of larger leaves, flowers with increased organ number and abnormal organ organization [61–64]. In addition, the *era1* mutants are partially male sterile, due to abnormal cell division of cells, which form the male gametophyte [61,63]. It has also been found that *era1* plants are late flowering under both long (16 h light/8 h dark) and short (8 h light/16 h dark) day conditions [63]. It is not known which farnesylated proteins regulate cell differentiation, cell division, and transition to flowering.

B.   The *GGT-Ib* and FT/GGT-I α Subunit Mutants

In contrast to *era1*, the *ggt-1b* mutant plants have mild phenotype [13]. The *plp*, FT, and GGT-I α-subunit mutant has a severe developmental phenotype. The SAM of *plp* is even more enlarged than in *era1*, the plants are small, have slow growth rate, are very late flowering and are almost completely male sterile [36]. *era1 ggt-1b* double mutant plants are phenotypically inseparable from *plp*, and overexpression of GGT-IB in *era1* partially restores the mutant phenotype [13]. Taken together, the genetic analyses of FT and GGT-I mutants indicate that FT and GGT-I have partially overlapping functions, FT can almost fully compensate for the loss of GGT-I function, but GGT-I can only partially compensate for the loss of FT function. These conclusions are compatible with the biochemical analysis showing that the GGT-I substrates ROP6 and AGG1 are farnesylated in *ggt-1b* mutants [11] but that prenylation of FT substrates by GGT-I is inefficient [1].

"The green revolution", which modernized agriculture, affecting the life of billions of people, was marked by the identification of mutants that changed plant architecture [65–67]. Furthermore, precocious flowering due to global warming has become a serious problem in several agricultural crops. Hence, the involvement of protein prenylation in regulation of plant development and transition to flowering offers interesting venues for improvement of crop growth and yields.

C. The *RGGTB1* Mutant

Both vesicle secretion and endocytosis are compromised in *rggtb1* mutant *Arabidopsis*. The *rggt1b* plants have pleotropic phenotype including altered light-regulated development, response to gravitropic stimuli in the shoot, and overall small size [2]. There are two RGGTB subunits in *Arabidopsis*, and expression studies suggest that RGGTB1 is more highly and widely expressed compared to RGGTB2 and was therefore suggested to play a major role in Rab prenylation. Rab function in *rggtb1* is likely maintained by residual RGGT activity enabled by RGGTB2 [2].

VI. CaaX Processing

Homologs of all CaaX processing enzymes have been identified and characterized in *Arabidopsis* [68–73]. Similar to their animal and yeast homologs, the *Arabidopsis* CaaX proteases and ICMTs are localized at the ER [68,69,73], suggesting that following prenylation in the cytoplasm prenylated proteins are targeted to the ER.

A. STE24

When expressed in *ste24Δ rce1Δ* mutant yeast cells *Arabidopsis* AtSET24 facilitated processing of yeast **a**-factor [68]. The growth inhibition halos formed around *ste24Δ rce1Δ* cells expressing AtTE24 from either high or low copy number plasmids were, however, smaller compared to halos formed around *Ste24 Rce1* cells, indicating that the plant and yeast enzymes likely have different substrate specificities. Additional analyses of AtSTE24 function *in vitro* and in yeast indicated that the plant enzyme has different and wider substrate specificity compared to its yeast counterpart. *In vitro* CaaX processing coupled methylation assays showed that AtSTE24 can cleave prenylated CaM53, which terminates with a CTIL CaaX box [68], whereas a threonine at the CaaX a_1 position inhibited processing by yeast Ste24 [74]. When expressed in yeast, AtSTE24 enabled plasma membrane localization of a Rac protein with a CLLM CaaX box, whereas, yeast Rce1 but not Ste24 promoted plasma membrane localization of this protein [68].

Homologs of the yeast **a**-factor and nuclear lamins have not been identified in plant genomes. A search for *Arabidopsis* proteins, which terminate with a CaaX box and contain an upstream the prelamin A-like STE24 cleavage sequence, yielded one protein called AAT1 (accession # *AT4G21120*). AAT1 terminates with a CSAT sequence motif and contains a putative internal YVL endoproteolytic cleavage site, which is identical to the prelamin A Ste24 internal cleavage site. In AAT1 the YVL motif is located 37 residues upstream of the putative the CSAT Cys residue [75]. However, AAT1 is a highly hydrophobic transmembrane amino acid transporter. Furthermore, according to the "Prenbase" (http://mendel.imp.ac.at/PrePS/PRENbase/) AAT1 is not predicted to be prenylated. Hence, the likelihood that AAT1 is a substrate for AtSTE24 is questionable. Interestingly, when expressed fused to GFP in either yeast or plants, AtSTE24 cleaved itself as well as the GFP moiety resulting in loss of its activity as well as the GFP fluorescence [75]. This suggests that under certain conditions, AtSTE24 endoproteolytic activity is independent form isoprenyl cysteine.

B. RCE1

The *Arabidopsis* AtRCE1 homolog has been identified and characterized biochemically [70]. When expressed in *rce1Δ ste24Δ* mutant cells, AtRCE1 was able to restore **a**-factor production, demonstrating its functionality as a CaaX protease in yeast [76]. The yeast Ste24 and Rce1 enzymes are known to have partially overlapping substrate specificity [74]. It was not known whether this specificity pattern extends to other organisms. The results indicate that this may be the case. It was found that the substrate specificity of AtRCE1 is partially overlapping with that of AtSTE24 as demonstrated by their ability to process both the *Arabidopsis* AtROP9 and the yeast **a**-factor [76]. The halos that formed around the AtRCE1 complemented *ste24Δ rce1Δ* cells were smaller than the halos formed around cells that were complemented with AtSTE24. The differences in halo sizes indicate **a**-factor processing by AtRCE1 was less efficient than by AtSTE24, likely owing the dual role of AtSTE24 in **a**-factor processing.

The subcellular localization of the farnesylated GFP-AtROP9 in *wt, rce1Δ ste24Δ,* and *rce1Δ ste24Δ* cells complemented with *AtRCE1* revealed a role for CaaX proteolysis in targeting of prenylated protein to the plasma membrane. Consistently, in fibroblasts derived from *rce1* or *icmt* knockout mice farnesylated Ras accumulated in the endomembrane [77]. Likewise, the geranylgeranylated petunia calmodulin CaM53 [1, 23] accumulated in the endomembranes following transient expression *N. benthamiana* leaf epidermal cells treated with the carboxymethylation inhibitor acetyl farnesyl cysteine (AFC) [73].

C. ICMT

Querying of plant sequence databases revealed small families of ICMT proteins in monocot and dicot of plant species [69]. In *Arabidopsis,* there are two ICMTs, designated ICMTA and ICMTB. AtICMTB exhibits lower K_m and a higher catalytic activity compared to AtICMTA in *in vitro* carboxymethylation assays [72]. In agreement, larger growth inhibition halos were formed around AtICMTB expressing *ste14Δ* cells [69]. Analysis of the sequences differences between AtICMTA and AtICMTB revealed five amino acids conserved between AtICMTB, the moss *Physomitrella patens,* yeast Ste14p, and human and ICMT proteins that differ in ICMTA. It was found that these five amino acids are responsible for the different activities of AtICMTA and AtICMTB [69].

Hydropathy plots suggest that ICMTA and ICMTB contain between 6 and 8 transmembrane helixes and have similar topology to that of Ste14p and other members of the ICMT family [69,78]. Sequence analysis revealed that four of the five amino acids unique to AtICMTA (N^{111}, Y^{112}, Q^{165}, and S^{188}) introduce neutral or hydrophobic residues into cytosolic spans, which may disrupt the topology or structure of AtICMTA. It is noteworthy that Q^{165} is located in region close to the end of the transmembrane span and the beginning of the cytosolic part. The charged amino acid E^{187} in AtICMTA is represented by the neutral residues Q^{187} and N^{230} in AtICMTB and Ste14p, respectively. The sequence alignment of ICMT protein from *Arabidopsis,* poplar, rice, *Physomitrella,* yeast, and human shows that while AtICMTA contain a serine residue at position 188, all the ICMT protein have positively charged lysine or arginine residues at this position [69]. In contrast, the charged residues at positions 111 and 112 are less conserved. Notably, the activity of AticmtAm^{E165} single mutant and the AticmtAm^{R111R112E165} triple mutant was similar to *wt* AtICMTA. Taken together, these data suggest that a positively charged residue at position 188 is critical for catalytic activity of ICMT. The findings that activity of AtICMTA is reduced compared to AtICMTB and that substituting all five residues to AtICMTB-like restore activity in yeast [69] are consistent with the hypothesis that residues in the C-terminal hydrophilic domains play a critical role in the function of ICMT [78]. The existence of small ICMT protein families in different plant species and identity between AtICMTA and poplar ICMT at positions 111 (N) and 165 (Q) and between AtICMTA and rice ICMT protein at position 112 (Y) may indicate that the ICMT proteins have diverged early during the evolution of higher plants.

ICMTA T-DNA knockout mutants have no visible phenotype. However, silencing of both *ICMTA* and *ICMTB* by *RNAi* induced subtle developmental phenotypes that were similar to *era1* (FTB) and *plp* (FTA/

GGTA) mutants [69]. Thus, prenylation and CaaX processing are required for proper function of the SAM. The relatively mild phenotype of the *ICMT* silenced plants could have been a result of incomplete repression of At*ICMTA* and At*ICMTB* expression. Alternatively, the mild phenotype could be due to the partial function of prenylated but nonmethylated proteins as was shown in yeast and mammals [74, 77]. Taken together, the genetic analysis of ICMT function [69], the significantly lower expression level, activity, and sequence conservation of ICMTA compared to ICMTB [69,72] indicate that ICMTB is the major ICMT enzyme in *Arabidopsis*.

α-carboxyl methylation of prenylated proteins is the final and the only potentially reversible step during prenylation and CaaX processing. Studies on isoprenylcysteine methyl esterase (ICME) and farnesylcysteine lyase (FCLY) in *Arabidopsis* highlight potential role of reversible methylation in regulation of prenylated proteins [79–81]. It was shown that overexpression of ICMTB results in ABA insensitivity during seed germination and stomata closure while ectopic expression of ICME had the opposite effect leading to ABA hypersensitivity [81]. *fcly* mutants accumulate higher levels of farnesyl-cysteine (FC) and are ABA hypersensitive. It was suggested that the ABA hypersensitivity was due to accumulation of higher levels of FC, which in turn inhibit ICMT [80]. These data implicate CaaX processing in regulation of ABA signaling and suggest that manipulation ICMT, ICME, and FCLY function might be used to enhance plant resistance to drought stress.

VII.   Summary

Plant protein prenyltransferase and CaaX processing enzymes are structurally and functionally conserved with their counterparts in yeast and animals. Identification of FT and GGT-I substrates showed, however, that they are often unique to plants. Some of these substrates such as the Ca^{2+}-calmodulin CaM53, MUBs, and IPT3 belong to larger protein families in which other members are not prenylated. Evolvement of these proteins has therefore occurred later in evolution and sometimes only in specific plant species. Thus, while the mechanisms of prenylation and CaaX processing are conserved between plant, yeast, and animals, they mostly serve different signaling cascades in these kingdoms of living organisms. The involvement of protein prenylation and CaaX processing in the ABA response and developmental regulation make signaling processes regulated by prenylation and CaaX processing important targets for the purpose of producing superior crop plants. Hopefully, the essential role of prenylation and CaaX processing for plant development, physiology, and biotechnology would attract more research effort to this field.

ACKNOWLEDGMENTS

Research in my lab is supported grants from the Israel Science Foundation (ISF 312/07) and the US—Israel Binational Science Foundation (BSF 2009309).

REFERENCES

1. Caldelari, D., Sternberg, H., Rodriguez-Concepcion, M., Gruissem, W., and Yalovsky, S. (2001). Efficient prenylation by a plant geranylgeranyltransferase-I requires a functional CaaL box motif and a proximal polybasic domain. *Plant Physiol* 126:1416–1429.
2. Hala, M., Soukupova, H., Synek, L., and Zarsky, V. (2010). Arabidopsis RAB geranyl-geranyl transferase beta-subunit mutant is constitutively photomorphogenic, and has shoot growth and gravitropic defects. *Plant J* 62:615–627.
3. Yalovsky, S., Trueblood, C.E., Callan, K.L., Narita, J.O., Jenkins, S.M., Rine, J., and Gruissem, W. (1997). Plant farnesyltransferase can restore yeast ras signaling and mating. *Mol Cell Biol* 17:1986–1994.
4. Hala, M., Elias, M., and Zarsky, V. (2005). A specific feature of the angiosperm Rab escort protein (REP) and evolution of the REP/GDI superfamily. *J Mol Biol* 348:1299–1313.
5. Cheung, A.Y., and Wu, H.M. (2008). Structural and signaling networks for the polar cell growth machinery in pollen tubes. *Annu Rev Plant Biol* 59:547–572.
6. Bracha-Drori, K., Shichrur, K., Katz, A., Oliva, M., Angelovici, R., Yalovsky, S., and Ohad, N. (2004). Detection of protein–protein interactions in plants using bimolecular fluorescence complementation. *Plant J* 40:419–427.
7. Long, S.B., Casey, P.J., and Beese, L.S. (1998). Cocrystal structure of protein farnesyl-transferase complexed with a farnesyl diphosphate substrate. *Biochemistry* 37:9612–9618.
8. Park, H.W., Boduluri, S.R., Moomaw, J.F., Casey, P.J., and Beese, L.S. (1997). Crystal structure of protein farnesyltransferase at 2.25 angstrom resolution. *Science* 275:1800–1804.
9. Strickland, C.L., Windsor, W.T., Syto, R., Wang, L., Bond, R., Wu, Z., Schwartz, J., Le Hung, V., Beese, L.S., and Weber, P.C. (1998). Crystal structure of farnesyl protein transferase complexed with a CaaX peptide and farnesyl diphosphate analogue. *Biochemistry* 37:16601–16611.
10. James, G.L., Goldstein, J.L., and Brown, M.S. (1995). Polylysine and CVIM sequences of K-RasB dictate specificity of prenylation and confer resistance to benzodiazepine pepti-domimetic *in vitro*. *J Biol Chem* 270:6221–6226.
11. Sorek, N., Gutman, O., Bar, E., Abu-Abied, M., Feng, X., Running, M.P., Lewinsohn, E., Ori, N., Sadot, E., Henis, Y.I., *et al.* (2011). Differential effects of prenylation and S-acylation on type I and II ROPS membrane interaction and function. *Plant Physiol* 155:706–720.
12. Zeng, Q., Wang, X., and Running, M.P. (2007). Dual lipid modification of Arabidopsis Ggamma-subunits is required for efficient plasma membrane targeting. *Plant Physiol* 143:1119–1131.
13. Johnson, C.D., Chary, S.N., Chernoff, E.A., Zeng, Q., Running, M.P., and Crowell, D.N. (2005). Protein geranylgeranyltransferase I is involved in specific aspects of abscisic acid and auxin signaling in Arabidopsis. *Plant Physiol* 139:722–733.
14. Ohya, Y., Goebl, M., Goodman, L.E., Peterson-Bjorn, S., Friesen, J.D., Tamanoi, F., and Anraku, Y. (1991). Yeast CAL1 is a structural and functional homologue to the DPR1 (RAM) gene involved in ras processing. *J Biol Chem* 266:12356–12360.

15. Trueblood, C.E., Ohya, Y., and Rine, J. (1993). Genetic evidence for *in vivo* cross-specificity of the CaaX-box protein prenyltransferases farnesyltransferase and geranylgeranyltransferase-I in Saccharomyces cerevisiae. *Mol Cell Biol* 13:4260–4275.

16. Loraine, A.E., Yalovsky, S., Fabry, S., and Gruissem, W. (1996). Tomato Rab1A homologs as molecular tools for studying Rab geranylgeranyl transferase in plant cells. *Plant Physiol* 110:1337–1347.

17. Yalovsky, S., Loraine, A.E., and Gruissem, W. (1996). Specific prenylation of tomato Rab proteins by geranylgeranyl type-II transferase requires a conserved cysteine-cysteine motif. *Plant Physiol* 110:1349–1359.

18. Alory, C., and Balch, W.E. (2003). Molecular evolution of the rab-escort-protein/guanine-nucleotide-dissociation-inhibitor superfamily. *Mol Biol Cell* 14:3857–3867.

19. Maurer-Stroh, S., Koranda, M., Benetka, W., Schneider, G., Sirota, F.L., and Eisenhaber, F. (2007). Towards complete sets of farnesylated and geranylgeranylated proteins. *PLoS Comput Biol* 3:e66.

20. Galichet, A., and Gruissem, W. (2006). Developmentally controlled farnesylation modulates AtNAP1;1 function in cell proliferation and cell expansion during Arabidopsis leaf development. *Plant Physiol* 142:1412–1426.

21. Galichet, A., Hoyerova, K., Kaminek, M., and Gruissem, W. (2008). Farnesylation directs AtIPT3 subcellular localization and modulates cytokinin biosynthesis in Arabidopsis. *Plant Physiol* 146:1155–1164.

22. Downes, B.P., Saracco, S.A., Lee, S.S., Crowell, D.N., and Vierstra, R.D. (2006). MUBs, a family of ubiquitin-fold proteins that are plasma membrane-anchored by prenylation. *J Biol Chem* 281:27145–27157.

23. Rodrigues-Concepcion, M., Yalovsky, S., Zik, M., Fromm, H., and Gruissem, W. (1999). The prenylation status of a novel plant calmodulin directs plasma membrane or nuclear localization of the protein. *EMBO J* 18:1996–2007.

24. Mucha, E., Fricke, I., Schaefer, A., Wittinghofer, A., and Berken, A. (2011). Rho proteins of plants—Functional cycle and regulation of cytoskeletal dynamics. *Eur J cell biol* 10.1016/j.ejcb.2010.11.009.

25. Nibau, C., Wu, H.M., and Cheung, A.Y. (2006). RAC/ROP GTPases: "Hubs" for signal integration and diversification in plants *Trends Plant Sci* 11:309–315.

26. Yalovsky, S., Bloch, D., Sorek, N., and Kost, B. (2008). Regulation of membrane trafficking, cytoskeleton dynamics, and cell polarity by ROP/RAC GTPases. *Plant Physiol* 147:1527–1543.

27. Yang, Z. (2008). Cell polarity signaling in Arabidopsis. *Annu Rev Cell Dev Biol* 24:551–575.

28. Winge, P., Brembu, T., and Bones, A.M. (1997). Cloning and characterization of rac-like cDNAs from Arabidopsis thaliana. *Plant Mol Biol* 35:483–495.

29. Sorek, N., Poraty, L., Sternberg, H., Bar, E., Lewinsohn, E., and Yalovsky, S. (2007). Activation status-coupled transient S acylation determines membrane partitioning of a plant rho-related GTPase. *Mol Cell Biol* 27:2144–2154.

30. Lavy, M., Bracha-Drori, K., Sternberg, H., and Yalovsky, S. (2002). A cell-specific, prenylation-independent mechanism regulates targeting of type II RACs. *Plant Cell* 14:2431–2450.

31. Lavy, M., and Yalovsky, S. (2006). Association of Arabidopsis type-II ROPs with the plasma membrane requires a conserved C-terminal sequence motif and a proximal polybasic domain. *Plant J* 46:934–947.

32. Heo, W.D., Inoue, T., Park, W.S., Kim, M.L., Park, B.O., Wandless, T.J., and Meyer, T. (2006). PI(3,4,5)P3 and PI(4,5)P2 lipids target proteins with polybasic clusters to the plasma membrane. *Science* 314:1458–1461.

33. Rotblat, B., Prior, I.A., Muncke, C., Parton, R.G., Kloog, Y., Henis, Y.I., and Hancock, J.F. (2004). Three separable domains regulate GTP-dependent association of H-ras with the plasma membrane. *Mol Cell Biol* 24:6799–6810.

34. Henis, Y.I., Rotblat, B., and Kloog, Y. (2006). FRAP beam-size analysis to measure palmitoylation-dependent membrane association dynamics and microdomain partitioning of Ras proteins. *Methods* 40:183–190.

35. Sorek, N., Segev, O., Gutman, O., Bar, E., Richter, S., Poraty, L., Hirsch, J.A., Henis, Y.I., Lewinsohn, E., Jurgens, G., *et al.* (2010). An S-acylation switch of conserved G-domain cysteines is required for polarity signaling by ROP GTPases. *Curr Biol* 20:914–920.

36. Running, M.P., Lavy, M., Sternberg, H., Galichet, A., Gruissem, W., Hake, S., Ori, N., and Yalovsky, S. (2004). Enlarged meristems and delayed growth in plp mutants result from lack of CaaX prenyltransferases. *Proc Natl Acad Sci USA* 101:7815–7820.

37. Belanis, L., Plowman, S.J., Rotblat, B., Hancock, J.F., and Kloog, Y. (2008). Galectin-1 is a novel structural component and a major regulator of h-ras nanoclusters. *Mol Biol Cell* 19:1404–1414.

38. Scheffzek, K., Stephan, I., Jensen, O.N., Illenberger, D., and Gierschik, P. (2000). The Rac-RhoGDI complex and the structural basis for the regulation of Rho proteins by RhoGDI. *Nat Struct Biol* 7:122–126.

39. DerMardirossian, C., and Bokoch, G.M. (2005). GDIs: Central regulatory molecules in Rho GTPase activation. *Trends Cell Biol* 15:356–363.

40. Klahre, U., Becker, C., Schmitt, A.C., and Kost, B. (2006). Nt-RhoGDI2 regulates Rac/Rop signaling and polar cell growth in tobacco pollen tubes. *Plant J* 46:1018–1031.

41. Carol, R.J., Takeda, S., Linstead, P., Durrant, M.C., Kakesova, H., Derbyshire, P., Drea, S., Zarsky, V., and Dolan, L. (2005). A RhoGDP dissociation inhibitor spatially regulates growth in root hair cells. *Nature* 438:1013–1016.

42. Michaelson, D., Silletti, J., Murphy, G., D'Eustachio, P., Rush, M., and Philips, M.R. (2001). Differential localization of Rho GTPases in live cells: Regulation by hypervariable regions and RhoGDI binding. *J Cell Biol* 152:111–126.

43. Yalovsky, S., Rodríguez-Concepción, M., Bracha, K., Toledo-Ortiz, G., and Gruissem, W. (2000). Prenylation of the floral transcription factor APETALA1 modulates its function. *Plant Cell,* in this issue.

44. Young, S.G., Fong, L.G., and Michaelis, S. (2005). Prelamin A, Zmpste24, misshapen cell nuclei, and progeria—New evidence suggesting that protein farnesylation could be important for disease pathogenesis. *J Lipid Res* 46:2531–2558.

45. Ashar, H.R., James, L., Gray, K., Carr, D., Black, S., Armstrong, L., Bishop, W.R., and Kirschmeier, P. (2000). Farnesyl transferase inhibitors block the farnesylation of CENP-E and CENP-F and alter the association of CENP-E with the microtubules. *J Biol Chem* 275:30451–30457.

46. Hancock, J.F., Paterson, H., and Marshall, C.J. (1990). A polybasic domain or palmitoylation is required in addition to the CAAX motif to localize p21ras to the plasma membrane. *Cell* 63:133–139.

47. Diaz, M., Sanchez, Y., Bennet, T., Sun, C.R., Godoy, C., Tamanoi, F., Duran, A., and Perez, P. (1993). The Schizosaccharomyces pombe cwg2+ gene codes for the b subunit of a geranylgeranyltransferase type I required for b-glucan synthesis. *EMBO J* 12:5245–5254.

48. Eisenreich, W., Rohdich, F., and Bacher, A. (2001). Deoxyxylulose phosphate pathway to terpenoids. *Trends Plant Sci* 6:78–84.

49. Kuzuyama, T., and Seto, H. (2003). Diversity of the biosynthesis of the isoprene units. *Nat Prod Rep* 20:171–183.

50. Lichtenthaler, H.K. (2000). Non-mevalonate isoprenoid biosynthesis: Enzymes, genes, and inhibitors. *Biochem Soc Trans* 28:785–789.
51. Rodriguez-Concepcion, M., and Boronat, A. (2002). Elucidation of the methylerythritol phosphate pathway for isoprenoid biosynthesis in bacteria and plastids. A metabolic milestone achieved through genomics. *Plant Physiol* 130:1079–1089.
52. Rohmer, M. (1999). The discovery of a mevalonate-independent pathway for isoprenoid biosynthesis in bacteria, algae, and higher plants. *Nat Prod Rep* 16:565–574.
53. Gerber, E., Hemmerlin, A., Hartmann, M., Heintz, D., Hartmann, M.A., Mutterer, J., Rodriguez-Concepcion, M., Boronat, A., Van Dorsselaer, A., Rohmer, M., *et al.* (2009). The plastidial 2-C-methyl-D-erythritol 4-phosphate pathway provides the isoprenyl moiety for protein geranylgeranylation in tobacco BY-2 cells. *Plant Cell* 21:285–300.
54. Cutler, S., Ghassemian, M., Bonetta, D., Cooney, S., and McCourt, P. (1996). A protein farnesyl transferase involved in abscisic acid signal transduction in Arabidopsis. *Science* 273:1239–1241.
55. Pei, Z.M., Ghassemian, M., Kwak, C.M., McCourt, P., and Schroeder, J.I. (1998). Role of farnesyltransferase in ABA regulation of guard cell anion channels and plant water loss. *Science* 282:287–290.
56. Wang, Y., Beaith, M., Chalifoux, M., Ying, J., Uchacz, T., Sarvas, C., Griffiths, R., Kuzma, M., Wan, J., and Huang, Y. (2009). Shoot-specific down-regulation of protein farnesyltransferase (alpha-subunit) for yield protection against drought in Canola. *Mol Plant* 2:191–200.
57. Wang, Y., Ying, J., Kuzma, M., Chalifoux, M., Sample, A., McArthur, C., Uchacz, T., Sarvas, C., Wan, J., Dennis, D.T., *et al.* (2005). Molecular tailoring of farnesylation for plant drought tolerance and yield protection. *Plant J* 43:413–424.
58. Hubbard, K.E., Nishimura, N., Hitomi, K., Getzoff, E.D., and Schroeder, J.I. (2010). Early abscisic acid signal transduction mechanisms: Newly discovered components and newly emerging questions. *Genes Dev* 24:1695–1708.
59. Weiner, J.J., Peterson, F.C., Volkman, B.F., and Cutler, S.R. (2010). Structural and functional insights into core ABA signaling. *Curr Opin Plant Biol* 13:495–502.
60. Allen, G.J., Murata, Y., Chu, S.P., Nafisi, M., and Schroeder, J.I. (2002). Hypersensitivity of abscisic acid-induced cytosolic calcium increases in the Arabidopsis farnesyltransferase mutant era1-2. *Plant Cell* 14:1649–1662.
61. Bonetta, D., Bayliss, P., Sun, S., Sage, T., and McCourt, P. (2000). Farnesylation is involved in meristem organization in Arabidopsis. *Planta* 211:182–190.
62. Running, M.P., Fletcher, J.C., and Meyerowitz, E.M. (1998). The WIGGUM gene is required for proper regulation of floral meristem size in Arabidopsis. *Development* 125:2545–2553.
63. Yalovsky, S., Kulukian, A., Rodriguez-Concepcion, M., Young, C.A., and Gruissem, W. (2000). Functional requirement of plant farnesyltransferase during development in Arabidopsis. *Plant Cell* 12:1267–1278.
64. Ziegelhoffer, E.C., Medrano, L.J., and Meyerowitz, E.M. (2000). Cloning of the Arabidopsis WIGGUM gene identifies a role for farnesylation in meristem development. *Proc Natl Acad Sci USA* 97:7633–7638.
65. Peng, J., Richards, D.E., Hartley, N.M., Murphy, G.P., Devos, K.M., Flintham, J.E., Beales, J., Fish, L.J., Worland, A.J., Pelica, F., *et al.* (1999). "Green revolution" genes encode mutant gibberellin response modulators *Nature* 400:256–261.
66. Salamini, F. (2003). Plant Biology. Hormones and the green revolution. *Science* 302:71–72.

67. Sasaki, A., Ashikari, M., Ueguchi-Tanaka, M., Itoh, H., Nishimura, A., Swapan, D., Ishiyama, K., Saito, T., Kobayashi, M., Khush, G.S., *et al.* (2002). Green revolution: A mutant gibberellin-synthesis gene in rice. *Nature* 416:701–702.

68. Bracha, K., Lavy, M., and Yalovsky, S. (2002). The Arabidopsis AtSTE24 is a CAAX protease with broad substrate specificity. *J Biol Chem* 277:29856–29864.

69. Bracha-Drori, K., Shichrur, K., Lubetzky, T.C., and Yalovsky, S. (2008). Functional analysis of Arabidopsis post-prenylation CaaX processing enzymes and their function in subcellular protein targeting. *Plant Physiol* 148:119–131.

70. Cadinanos, J., Varela, I., Mandel, D.A., Schmidt, W.K., Diaz-Perales, A., Lopez-Otin, C., and Freije, J.M. (2003). AtFACE-2, a functional prenylated protein protease from Arabidopsis thaliana related to mammalian Ras-converting enzymes. *J Biol Chem* 278:42091–42097.

71. Crowell, D.N., and Kennedy, M. (2001). Identification and functional expression in yeast of a prenylcysteine alpha-carboxyl methyltransferase gene from Arabidopsis thaliana. *Plant Mol Biol* 45:469–476.

72. Narasimha Chary, S., Bultema, R.L., Packard, C.E., and Crowell, D.N. (2002). Prenylcysteine alpha-carboxyl methyltransferase expression and function in Arabidopsis thaliana. *Plant J* 32:735–747.

73. Rodriguez-Concepcion, M., Toledo-Ortiz, G., Yalovsky, S., Caldelari, D., and Gruissem, W. (2000). Carboxyl-methylation of prenylated calmodulin CaM53 is required for efficient plasma membrane targeting of the protein. *Plant J* 24:775–784.

74. Trueblood, C.E., Boyartchuk, V.L., Picologlou, E.A., Rozema, D., Poulter, C.D., and Rine, J. (2000). The CaaX proteases, Afc1p and Rce1p, have overlapping but distinct substrate specificities. *Mol Cell Biol* 20:4381–4392.

75. Bracha-Drori, K. (2005). Cloning and characterization of the CaaX-protease AtSTE24. PhD Dissertation Volume Ph.D. (Tel Aviv: Tel Aviv University).

76. Bracha-Drori, K., Shichrur, K., Lubetzky, T.C., and Yalovsky, S. (2008). Functional analysis of Arabidopsis postprenylation CaaX processing enzymes and their function in subcellular protein targeting. *Plant Physiol* 148:119–131.

77. Michaelson, D., Ali, W., Chiu, V.K., Bergo, M., Silletti, J., Wright, L., Young, S.G., and Philips, M. (2005). Postprenylation CAAX processing is required for proper localization of Ras but not Rho GTPases. *Mol Biol Cell* 16:1606–1616.

78. Romano, J.D., and Michaelis, S. (2001). Topological and mutational analysis of Saccharomyces cerevisiae Ste14p, founding member of the isoprenylcysteine carboxyl methyltransferase family. *Mol Biol Cell* 12:1957–1971.

79. Deem, A.K., Bultema, R.L., and Crowell, D.N. (2006). Prenylcysteine methylesterase in Arabidopsis thaliana. *Gene* 380:159–166.

80. Huizinga, D.H., Denton, R., Koehler, K.G., Tomasello, A., Wood, L., Sen, S.E., and Crowell, D.N. (2010). Farnesylcysteine lyase is involved in negative regulation of abscisic acid signaling in Arabidopsis. *Mol Plant* 3:143–155.

81. Huizinga, D.H., Omosegbon, O., Omery, B., and Crowell, D.N. (2008). Isoprenylcysteine methylation and demethylation regulate abscisic acid signaling in Arabidopsis. *Plant Cell* 20:2714–2728.

10

Posttranslational Isoprenylation of Tryptophan Residues in Bacillus subtilis

MASAHIRO OKADA[a] • FUMITADA TSUJI[b] •
YOUJI SAKAGAMI[b]

[a]*College of Bioscience and Biotechnology Chubu University*
Matsumoto-cho, Kasugai, Japan

[b]*Graduate School of Bioagricultural Sciences*
Nagoya University
Chikusa-ku, Nagoya, Japan

I.   Abstract

Bacillus subtilis and related bacilli produce a posttranslationally modified oligopeptide, termed ComX pheromone, which stimulates natural genetic competence controlled by quorum sensing. ComX pheromones are modified with either geranyl or farnesyl groups on a tryptophan residue at the 3-position of its indole ring, resulting in the formation of a tricyclic structure, including a newly formed five-membered ring. ComX pheromone is the first example of not only isoprenoidal modification of tryptophan residues in living organisms but also posttranslational isoprenylation in prokaryotes. As the geranylation of peptides, including monoterpene metabolites, is unusual in prokaryotes, posttranslational geranylation is unprecedented in nature. The posttranslational modification of ComX pheromone with isoprenoid plays an essential role for a functional pheromone inducing genetic competence in *B. subtilis* and is more significant

THE ENZYMES, Vol. XXIX   183   ISSN NO: 1874-6047
© 2011 Elsevier Inc. All rights reserved.   DOI: 10.1016/B978-0-12-381339-8.00010-X

than the ComX amino acid sequence. We speculate that the posttranslational isoprenylation of tryptophan will soon be discovered in other prokaryotic peptides and proteins.

II. Quorum Sensing in Bacteria

For microorganisms, cell population density is one of the most crucial factors affecting viability in natural environments. Prokaryotic bacteria continually coordinate their behavior to regulate gene expression in response to cell density, in a process known as quorum sensing [1–4]. The responses governed by quorum sensing are varied and include acquisition of virulence, biofilm formation, bioluminescence, conjugation, sporulation, antibiotics production, and genetic competence [5]. The strategy developed by bacteria to collect information on cell density involves the constant secretion of specific extracellular signaling molecules, termed quorum-sensing pheromones. These pheromones increase in concentration with increasing cell density and, upon reaching threshold levels, trigger various signal transduction pathways through binding to specific transmembrane receptors (Figure 10.1). Thus, bacteria detect their own population density based on the concentration of secreted pheromones.

The quorum-sensing pheromones of Gram-negative bacteria are generally low molecular weight secondary metabolites, such as *N*-acylhomoserine lactones, while Gram-positive bacteria typically produce oligopeptide pheromones [6]. Various classes of pheromones are produced by bacteria, with each type controlling distinct responses expressing distinct gene expression, and each quorum-sensing pheromone generally shows species (or group)-specific

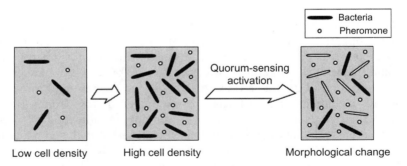

Low cell density   High cell density   Morphological change

Fig. 10.1. Schematic illustration of bacterial quorum sensing. At low bacterial cell density, the concentration of pheromone is also low. However, at high cell density, the pheromone reaches threshold levels, resulting in activation of specific quorum-sensing systems that leads to morphological changes, such as genetic competence.

bioactivity. In other words, these secreted pheromones function as languages for cell–cell conversation among bacterial groups [7].

III.   Genetic Competence in *B. subtilis*

B. subtilis is a nonpathogenic, Gram-positive bacterium that has several important industrial roles. A number of biosynthetic products from *B. subtilis* are commercially prepared, including *B. subtilis* itself, while the closely related species *Bacillus natto* is widely used for the fermentation of soybean in eastern Asia. *B. subtilis* exhibits two characteristic phenotypes, which are both controlled by quorum sensing: the first is sporulation, which stimulates the division of cells into two distinct cell types with differing developmental fates [8,9], and the other is competence, which allows cells to take up exogenous DNA [10,11]. In microbiological studies, *B. subtilis* has been widely used as a model organism, and systems for the genetic transformation of *B. subtilis* have been well established. Therefore, *B. subtilis* is often referred to as a model bacterium for genetic studies of Gram-positive bacteria in the same way that *Escherichia coli* serves as a model Gram-negative bacterium. Numerous molecular genetic studies examining the natural genetic competence of *B. subtilis* have been performed, and a quorum-sensing pheromone, called ComX, involved in the complex signal transduction cascade of genetic competence has been identified [12].

IV.   Primary Gene Cluster for ComX Pheromone Production

The gene cluster for initiation of genetic competence, *comQXPA*, was identified in *B. subtilis* and related bacilli, and encodes proteins results in the production of posttranslationally modified ComX pheromone (Figure 10.2) [13,14]. Within this cluster, ComP and ComA have homology to a membrane-bound histidine kinase and response regulator, respectively [13–16]. Therefore, these two proteins comprise a two-component system, which is a member of the large family of two-component regulatory systems found in bacteria. ComP, an eight-transmembrane domain receptor, is autophosphorylated in response to pheromonal signal, and ComA donates the phosphate for the phosphorylation reaction and subsequently transports the signal into cells.

The first gene in the *comQXPA* cluster, *comQ*, encodes a protein with a similarity to the isoprenyl diphosphate synthase of *Methanobacterium thermoautotrophicum*, an enzyme that catalyzes the condensation of an isopentenyl diphosphate and dimethylallyl diphosphate to form long-chain isoprenoids [17,18]. Notably, an aspartate-rich motif, DDXXD, is crucial for the interaction and transfer of isoprenoid and is highly conserved in ComQ. The aspartates in this putative isoprenoid-binding site of ComQ are required for the expression of pheromone activity [18]. Judging by these characteristics and the location of the *comQ* gene directly upstream of the *comX* gene, ComQ appears to modify and process ComX to produce mature pheromone containing an isoprenoidal group [18,19].

The active ComX pheromones of *B. subtilis* strains and related bacilli are derived from the *C*-terminus of ComX, and the phylogenetic analysis of ComX coding sequences shows that an invariant tryptophan residue is located near the C-terminus in the region of the mature pheromone (Figure 10.3). However, with the exception of this conserved residue, natural isolates of

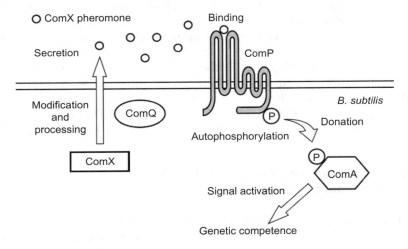

FIG. 10.2. Signal transduction cascade leading to genetic competence in *B. subtilis*.

Bacillus strain	Amino acid sequence of ComX
168	-MQDLINYFLNYPEALKKLKNKEACLIGFDVQETETIIKAYNDYYL--ADPITRQWGD-
RO-C-2	-MQDLINYFLSYPEVLKKLKNREACLIGFSSNETETIIKAYNDYHLSSPT--TREWDG-
RO-E-2	MKQDMIDYLMKNPQVLTKLENGEASLIGIPDKLIPSIVDIFNKKMTLSKKCKGIFWEQ-
RO-H-1	-MQEMVGYLIKYPNVLREVMEGNACLLGVDKDQSECIINGFKGLEI----YSMLDWKY-
RS-B-1	-MQEMVGYLIKYPNVLREVMEGNACLLGVDKDQSECIINGFKGLEI----YSMMDWHY-
RO-B-2	-MQEIVGYLVKNPEVLDEVMKGRASLLNIDKDQLKSIVDAFGGLQI----YTNGNWVPS

FIG. 10.3. Alignment of ComX amino acid sequences from several strains of *Bacillus* spp. Sequences of mature pheromones are underlined, and modified W residues are marked in boldface.

bacilli exhibit striking polymorphism in the amino acid sequence of ComX pheromones [19–22]. Molecular genetic analyses and molecular biological studies indicate that these tryptophan residues are similarly modified with isoprenoid, but the nature of the modification remains to be elucidated.

V. Posttranslational Modification of ComX Pheromone

Recent organic chemical studies have revealed that $ComX_{RO-E-2}$ phero-mone from *B. subtilis* strain RO-E-2 contains a tryptophan residue uniquely modified with a geranyl group at the 3-position of its indole ring, resulting in the formation of a tricyclic structure (Figure 10.4) [23–25]. It has also been confirmed that the $ComX_{RO-C-2}$ pheromone from *Bacillus mojavensis* strain RO-C-2 contains a tryptophan residue modified with a farnesyl group at the identical position as the $ComX_{RO-E-2}$ pheromone [26]. These find-ings, taken together with those of previous studies, suggest that posttrans-lational isoprenoidal modifications of ComX pheromones can be classified into two types, consisting of either geranyl or farnesyl modifications (Figure 10.5) [23–27]. For instance, the $ComX_{168}$ pheromone, produced by *B. subtilis* strain 168 used as a standard strain in genetic studies, is also modified with a farnesyl group, similar to the $ComX_{RO-C-2}$ pheromone. In living organisms, the modification of tryptophan residues with isopren-oid is unprecedented in peptide and protein. Additionally, the posttransla-tional isoprenoidal modification of ComX represents the first example of isoprenoidal peptide in prokaryote, although posttranslational farnesyla-tion or geranylgeranylation of cysteine residues is widely observed among many important proteins and peptides in eukaryotes, and also plays a crucial role for their functions [28–33]. Notably, with the exception of the plant kingdom, monoterpenes derived from geranyl pyrophosphate are rare, and only a few geranylated compounds have been identified in pro-karyotes [34]. Hence, ComX pheromone represents the only example of posttranslational geranylation in prokaryotes to date.

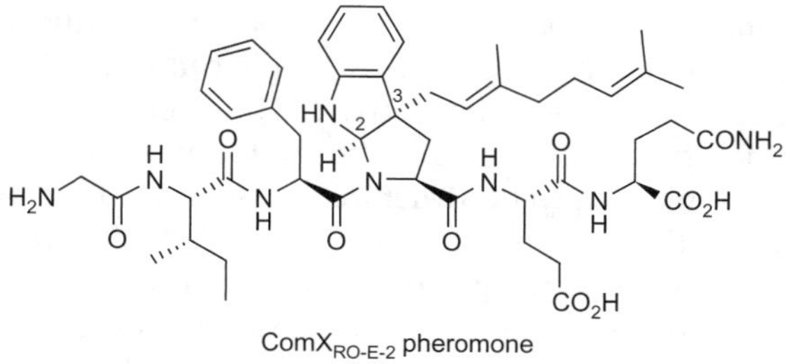

ComX$_{RO-E-2}$ pheromone

FIG. 10.4. Chemical structure of the ComX$_{RO-E-2}$ pheromone from *B. subtilis* strain RO-E-2.

Bacillus strain	Amino acid sequence	Chemical structure of w
168	ADPITRQWGD	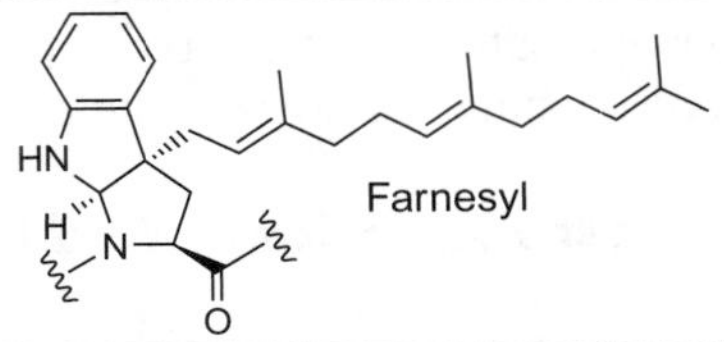
RO-C-2	TREWDG	Farnesyl
RO-E-2	GIFWEQ	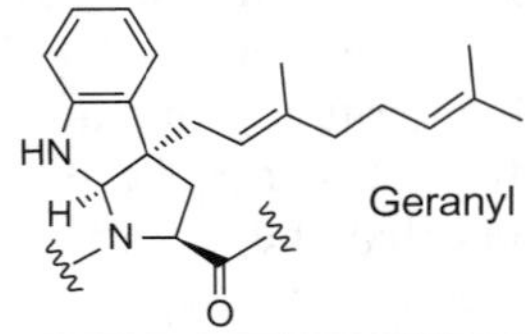
RO-H-1	MLDWKY	
RS-B-1	MMDWHY	Geranyl
RO-B-2	YTNGNWVPS	

FIG. 10.5. The two types of posttranslational modifications of ComX variants identified to date.

VI. Structure–Activity Relationships of ComX$_{RO-E-2}$ Pheromone

Structure–activity relationship studies on ComX$_{RO-E-2}$ pheromone were carried out by using synthetic alanine-substituted ComX$_{RO-E-2}$ derivatives. All amino acid residues of ComX$_{RO-E-2}$ pheromone, with the exception of the geranylated tryptophan residue, can be replaced by alanine without significant reduction of activity (Table 10.1). Addition of alanine residues on both the N- and C-terminal does not affect the biological activity and both the N- and C-terminal residues are dispensable. Presently, the smallest functional peptide that has been identified is a tripeptide, F-W(modified)-E (Table 10.1) [35]. These studies have demonstrated that the geranylated tryptophan residue is an essential role for expressing biological activity.

We synthesized six geranylated tryptophans, which geranyl group substitutes 1-, 2-, 4-, 5-, 6-, and 7-position of hydrogen on indole ring, and using these geranylated tryptophans, hexapeptides were synthesized with the same amino acid sequence as ComX$_{RO-E-2}$ pheromone [36]. But, all peptides synthesized did not show any activity. Thus, the geranyl side chain of peptide does not have only the role to addition of lipophilicity to the peptide, but the typical tricyclic structure is essential to biological activity. The newly formed five-membered ring in the modified tryptophan residue is resembled to proline. Since proline is well known as a key amino acid residue to define the three-dimensional structure of protein, it will be possible that the modified tryptophan residue strongly affects the conformation of ComX pheromone. When we determined the chemical structure

TABLE 10.1

STRUCTURE–ACTIVITY RELATIONSHIPS OF $COMX_{RO-E-2}$

Substance	Amino acid sequence	EC_{50}[a] (nM)	ED_{MAX}[b] (%)
$ComX_{RO-E-2}$ pheromone	Gly-Ile-Phe-Trp*(Ger)-Glu-Gln	1	100
Ala-$ComX_{RO-E-2}$	Ala-Gly-Ile-Phe-Trp*(Ger)-Glu-Gln		
[G1A] $ComX_{RO-E-2}$	Ala-Ile-Phe-Trp*(Ger)-Glu-Gln	1	100
[I2A] $ComX_{RO-E-2}$	Gly-Ala-Phe-Trp*(Ger)-Glu-Gln		
[F3A] $ComX_{RO-E-2}$	Gly-Ile-Ala-Trp*(Ger)-Glu-Gln	3	100
[E5A] $ComX_{RO-E-2}$	Gly-Ile-Phe-Trp*(Ger)-Ala-Gln	10	95
[Q6A] $ComX_{RO-E-2}$	Gly-Ile-Phe-Trp*(Ger)-Glu-Ala		
$ComX_{RO-E-2}$-Ala	Gly-Ile-Phe-Trp*(Ger)-Glu-Gln-Ala	20	90
[2–6] $ComX_{RO-E-2}$	Ile-Phe-Trp*(Ger)-Glu-Gln	6	100
[1–5] $ComX_{RO-E-2}$	Gly-Ile-Phe-Trp*(Ger)-Glu		
[3–5] $ComX_{RO-E-2}$	Phe-Trp*(Ger)-Glu	20	80

[a]Concentration of each peptide showing the same activity to observed activity at EC_{50} of $ComX_{RO-E-2}$ pheromone.
[b]Ratio of maximum activity compared with $ComX_{RO-E-2}$ pheromone.

of $ComX_{RO-E-2}$ pheromone, we determined the direction of geranyl side chain and the proton at 2-position of an original indole ring in the same side of rings with the NOE data of NMR [23,24]. The absolute stereochemistry was elucidated by computer simulation using coupling constant comparing with model compounds. To confirm the proposed structure, we synthesized four stereoisomers of geranylated tryptophan, which had stereochemistry of Lα, Lβ, Dα, and Dβ. Only the peptide having Lα geranyl tryptophan residue showed biological activity and other three isomers showed no activity [23]. These results indicate that the stereochemistry of geranylated tryptophan is also essential expressing the activity, and in other word, the receptor of $ComX_{RO-E-2}$ pheromone, $ComP_{RO-E-2}$, can recognize the exact stereochemistry of modified tryptophan residue.

In contrast, earlier work on the structure–activity relationships of tremerogen A-10, which is the sex pheromone of basidiomycetous yeast and contains an S-farnesyl-modified cysteine residue, revealed that removal of the N-terminal residues induced loss of biological activity [37]. Replacement of the farnesyl group with an alkyl chain had no significant effect on the biological activity of tremerogen A-10; moreover, replacement with a longer isoprenoid caused an increase in activity. The activity spectrum of tremerogen A-10 strongly contrasts with those of ComX pheromones. These results also indicate that the pattern of specific interaction of the ComX pheromone with its receptor is unique from that of the S-isoprenoidal peptide pheromone.

Modification pattern	Producer strain	Bioactivity of tester strain			
		168	RO-C-2	RO-E-2	RO-H-1
Farnesylation	168	+ +	+ +	–	–
	RO-C-2	+	+ +	–	–
Geranylation	RO-E-2	–	–	+ +	+ –
	RO-H-1	–	–	+	+ +

FIG. 10.6. The correlation between modification pattern and specific bioactivity. Symbols: + +, strong activity; +, moderate activity; + –, weak activity; –, no activity.

The two isoprenoidal groups may also influence the species (or group)-specific bioactivity exhibited by ComX pheromones. Experiments utilizing conditioned media of producer strains and partially purified ComX variants have revealed a correlation between specific activation pattern and type of modifying isoprenoid (Figure 10.6) [19–22]. In fact, synthetic ComX phero-mone modified with a farnesyl group showed significant activity in strains that produce farnesylated ComX, but had little activity in strains that produce geranylated pheromone, and vice versa (unpublished data). Since geranyl- and farnesyl-modified tryptophan residues have identical core structures that differ only in side-chain length, the length of the isoprenyl side chain appears to be a more influential determinant of group specificity than the amino acid sequence of the pheromone. The geranyl and farnesyl moieties in modified ComX pheromone are directly involved in ligand-receptor interactions. Inter-estingly, N-acyl-homoserine lactones, Gram-negative bacterial quorum-sens-ing pheromones, also have group-specific activity that is influenced by the length of the acyl side chain [2–4]. As quorum-sensing signal receptors have binding pockets precisely adapted to the length of the target acyl chain [38], bacilli may have evolved the isoprenoidal moiety in ComX pheromones to fit their own receptor, ComP, in a similar manner.

VII.   Is Posttranslational Isoprenylation of Tryptophan Universal?

Farnesylated peptides were also discovered to serve as mating factors of basidiomycetous yeasts as mentioned above [28–31]. Presently, numerous studies have reported posttranslational isoprenylation of cysteine residues, involving attachment of either a farnesyl or geranylgeranyl group at a C-terminal residue via a thioether linkage [32,33]. It has been elucidated that the modifying enzymes, farnesyltransferase and geranylgeranyltransferase 1, recognize a CaaX box in their substrates; "a" refers to an aliphatic amino acid, and X is typically a methionine or leucine for farnesyltransferase and geranylgeranyltransferase 1, respectively. Therefore, it is possible to search protein databases for the CaaX motif and identify potential proteins and peptides with an isoprenoidal cysteine. This approach has led to the discovery of several new isoprenylated proteins.

In contrast to cysteine residues, the posttranslational isoprenylation of tryptophan residues has not been identified in any proteins or peptides other than Com X. Several residues upstream of the cleavage site among ComX variants are conserved in contrast to the extreme polymorphism in C-terminus of ComX corresponding to a mature pheromone [19–22]; however, a consensus amino acid sequence recognized by ComQ has not been defined. To facilitate the identification of peptides and/or proteins with isoprenoidal modification, we have attempted to produce an antibody specific for geranylated tryptophan, but have yet to be successful. There is a strong possibility that other proteins possessing isoprenoidal tryptophan residues exist but have remained unnoticed due to the difficulty in detecting these proteins.

Since cysteine isoprenylation is observed only in eukaryotes, tryptophan isoprenylation in bacteria may have been one of the early steps in the evolution of posttranslational isoprenylation. If so, the posttranslational isoprenylation of tryptophan will be discovered in other prokaryotes and primitive eukaryotic microorganisms and may represent a universal posttranslational modification. Recently, we have obtained preliminary evidence of the consensus sequence recognized by ComQ with *in vitro* system (T. Tsuji *et al.*, unpublished results). Using this sequence, we will be able to determine the extent of the distribution of peptides and proteins geranylated on tryptophan residues.

VIII.   Summary and Future Prospects

B. subtilis and related bacilli produce a posttranslationally modified oligopeptide, termed ComX pheromone, which stimulates natural genetic competence controlled by quorum sensing. ComX pheromones are modified with either a geranyl or farnesyl group on a conserved tryptophan residue at the 3-position of its indole ring, resulting in the formation of a tricyclic structure, including a newly formed five-membered ring, similar to proline. Notably, ComX pheromone is the first example not only of isoprenoidal modification of tryptophan residues in living organisms but also of posttranslational isoprenylation in prokaryotes. The modification of ComX pheromones with isoprenoid plays a more essential role for establishing genetic competence than the amino acid sequence. It should be one of the most interesting problems in the future: how ComP recognizes its ligand, ComX pheromone. Based on the initial discovery of posttranslational isoprenylation of cysteine in pheromones secreted by eukaryotic microorganisms, which was later revealed to be a widespread phenomenon, we speculate that posttranslational isoprenylation of tryptophan will also be recognized as a universal modification and respond to important biological function in the near future.

ACKNOWLEDGMENT

We greatly thank Dr. David Dubnau at the Public Health Research Institute Center, NJ, USA, for collaborative studies of ComX pheromone.

REFERENCES

1. Fuqua, W.C., Winans, S.C., and Greenberg, E.P. (1994). Quorum sensing in bacteria: the LuxR-LuxI family of cell density-responsive transcriptional regulators. *J Bacteriol* 176:269–275.
2. Williams, P., Winzer, K., Chan, W., and Cámara, M. (2007). Look who's talking: communication and quorum sensing in bacterial world. *Philos Trans R Soc Lond B* 362:1119–1134.
3. Bassler, B.L., and Losick, R. (2006). Bacterially speaking. *Cell* 125:237–246.
4. Camilli, A., and Bassler, B.L. (2006). Bacterial small-molecule signaling pathways. *Science* 311:1113–1116.
5. Johnsborg, O., Eldholm, V., and Håvarstein, L.S. (2007). Natural genetic transformation: prevalence, mechanisms and function. *Res Microbiol* 158:767–778.
6 At present, these categories are not comprehensive. For details see in Refs. [2–4].
7. Recently, it was reported that quorum sensing was also mediated in fungi as follows: Hornby, J.M., Jensen, E.C., Lisec, A.D., Tasto, J.J., Jahnke, B., Shoemaker, R., Dussault, P., and Nickerson, K.W. (2001). Quorum sensing in the dimorphic fungus Candida albicans is mediated by farnesol. *Appl Environ Microbiol* 67:2982–2992. Nickerson, K.W., Atkin, A.L., and Hornby, J.M. (2006). Quorum sensing in dimorphic fungi: farnesol and beyond. *Appl Environ Microbiol* 72:3805–3813.
8. Lazazzera, B.A. (2000). Quorum sensing and starvation: signals for entry into stationary phase. *Curr Opin Microbiol* 3:177–182.
9. Piggot, P.J., and Hilbert, D.W. (2004). Sporulation of Bacillus subtilis. *Curr Opin Microbiol* 7:579–586.
10. Hamoen, L.W., Venema, G., and Kuipers, O.P. (2003). Controlling competence in Bacillus subtilis: shared use of regulators. *Microbiology* 149:9–17.
11. Tortosa, P., and Dubnau, D. (1999). Competence for transformation: a matter of taste. *Curr Opin Microbiol* 2:588–592.

12. Magnuson, R., Solomon, J., and Grossman, A.D. (1994). Biochemical and genetic characterization of a competence pheromone from B. subtilis. *Cell* 77:207–216.
13. Weinrauch, Y., Guillen, N., and Dubnau, D. (1989). Sequence and transcription mapping of Bacillus subtilis competence genes comB and comA, one of which is related to a family of bacterial regulatory determinants. *J Bacteriol* 171:5362–5375.
14. Weinrauch, Y., Penchev, R., Dubnau, E., Smith, I., and Dubnau, D. (1990). A Bacillus subtilis regulatory gene product for genetic competence and sporulation resembles sensor protein members of the bacterial two-component signal-transduction systems. *Genes Dev* 4:860–872.
15. Mueller, J.P., Bukusoglu, G., and Sonenshein, A.L. (1992). Transcriptional regulation of Bacillus subtilis glucose starvation-inducible genes: control of gsiA by the ComP-ComA signal transduction system. *J Bacteriol* 174:4361–4373.
16. Piazza, F., Tortosa, P., and Dubnau, D. (1999). Mutational analysis and membrane topology of ComP, a quorum-sensing histidine kinase of Bacillus subtilis controlling competence development. *J Bacteriol* 181:4540–4548.
17. Weinrauch, Y., Msadek, T., Kunst, F., and Dubnau, D. (1991). Sequence and properties of comQ, a new competence regulatory gene of Bacillus subtilis. *J Bacteriol* 173:5685–5693.
18. Schneider, K.B., Palmer, T.M., and Grossman, A.D. (2002). Characterization of comQ and comX, two genes required for production of ComX pheromone in Bacillus subtilis. *J Bacteriol* 184:410419.
19. Ansaldi, M., Marolt, D., Stebe, T., Mandic-Mulec, I., and Dubnau, D. (2002). Specific activation of the bacillus quorum-sensing systems by isoprenylated pheromone variants. *Mol Microbiol* 44:1561–1573.
20. Tortosa, P., Logsdon, L., Kraigher, B., Itoh, Y., Mandic-Mulec, I., and Dubnau, D. (2001). Specifity and genetic polymorphism of the Bacillus competence quorum-sensing system. *J Bacteriol* 183:451–460.
21. Ansaldi, M., and Dubnau, D. (2004). Diversifying selection at the Bacillus quorum-sensing locus and determinants of modification specificity during synthesis of the ComX pheromone. *J Bacteriol* 186:15–21.
22. Stefanic, P., and Mandic-Mulec, I. (2009). Social interactions and distribution of Bacillus subtilis pherotypes at microscale. *J Bacteriol* 191:1756–1764.
23. Okada, M., Sato, I., Cho, S.J., Iwata, H., Nishio, T., Dubnau, D., and Sakagami, Y. (2005). Structure of the Bacillus subtilis quorum-sensing peptide pheromone ComX. *Nat Chem Biol* 1:23–24.
24. Okada, M., Sato, I., Cho, S.J., Dubnau, D., and Sakagami, Y. (2006). Chemical synthesis of ComX pheromone and related peptides containing isoprenoidal tryptophan residues. *Tetrahedron* 62:8907–8918.
25. Okada, M., Qi, J., and Sakagami, Y. (2009). Chemistry of microbial signal compounds. *Mycotoxins* 59:55–66.
26. Okada, M., Yamaguchi, H., Sato, I., Tsuji, F., Dubnau, D., and Sakagami, Y. (2008). Chemical structure of posttranslational modification with a farnesyl group on tryptophan. *Biosci Biotechnol Biochem* 72:914–918.
27. Okada, M., Yamaguchi, H., Sato, I., Tsuji, F., Qi, J., Dubnau, D., and Sakagami, Y. (2007). Acid labile ComX pheromone from Bacillus mojavensis RO-H-1. *Biosci Biotechnol Biochem* 71:1807–1810.
28. Kamiya, Y., Sakurai, A., Tamura, S., Takahashi, N., Tsuchiya, T., Abe, K., and Fukui, S. (1979). Structure of rhodotoruicine A, a peptidyl factor inducing mating tube formation in Rhodosporidium toruloides. *Agric Biol Chem* 43:363–369.

29. Sakagami, Y., Isogai, A., Suzuki, A., Tamura, S., Kitada, C., and Fujino, M. (1979). Structure of tremerogen A-10, a peptidal hormone inducing conjugation tube formation in *Tremella mesenterica. Agric Biol Chem* 43:2643–2645.
30. Sakagami, Y., Yoshida, M., Isogai, A., and Suzuki, A. (1981). Peptidal sex hormones inducing conjugation tube formation in compatible mating-type cells of Tremella mesenterica. *Science* 212:1525–1527.
31. Ishibashi, Y., Sakagami, Y., Isogai, A., and Suzuki, A. (1984). Structure of tremerogen A-9291-I and A-9291-VIII; peptidal sex hormones of *Tremella brasiliensis. Biochemistry* 23:1399–1404.
32. Zhang, F.L., and Casey, P.J. (1996). Protein prenylation: molecular mechanisms and functional consequences. *Annu Rev Biochem* 65:241–269.
33. Clarke, S. (1992). Protein isoprenylation and methylation at carbonyl-terminal cysteine residues. *Annu Rev Biochem* 61:355386.
34. Kuzuyama, T., Noel, J.P., and Richard, S.B. (2005). Structural basis for the promiscuous biosynthetic prenylation of aromatic natural products. *Nature* 435:982–987.
35. Okada, M., Sato, I., Cho, S.J., Suzuki, Y., Ojika, M., Dubnau, D., and Sakagami, Y. (2004). Towards structural determination of the ComX pheromone: synthetic studies on peptides containing geranyltryptophan. *Biosci Biotechnol Biochem* 68:2374–2387.
36. Okada, M., Yamaguchi, H., Sato, I., Cho, S.J., Dubnau, D., and Sakagami, Y. (2007). Structure-activity relationship studies on quorum sensing $ComX_{RO-E-2}$ pheromone. *Bioorg Med Chem Lett* 17:1705–1707.
37. Fujino, M., Kitada, C., Sakagami, Y., Isogai, A., Tamura, S., and Suzuki, A. (1980). Biological activity of synthetic analogs of tremerogen A-10. *Naturwissenschaften* 67:406–408.
38. Zhang, R.G., Pappas, T., Brace, J.L., Miller, P.C., Oulmassov, T., Molyneaux, J.M., Anderson, J.C., Bashkin, J.K., Winans, S.C., and Joachimiak, A. (2002). Structure of a bacterial quorum-sensing transcription factor complexed with pheromone and DNA. *Nature* 417:971–974.

11

Global Analysis of Prenylated Proteins by the Use of a Tagging via Substrate Approach

LAI N. CHAN • FUYUHIKO TAMANOI

Department of Microbiology, Immunology, and Molecular Genetics
Molecular Biology Institute, Jonsson Comprehensive Cancer Center
University of California
Los Angeles, USA

I. Abstract

One of the major goals of the study on protein prenylation is to gain a comprehensive understanding of all prenylated proteins inside the cell. Such an analysis should not only provide a global view of prenylated proteins but also lead to the discovery of novel prenylated proteins. This knowledge should contribute to our understanding of the biological significance of protein prenylation. With this in mind, we have developed methods for global analysis of prenylated proteins. In our first attempt, we used FPP-azide or farnesol-azide to metabolically tag farnesylated proteins. Conjugation with biotinylated phosphine reagent enabled isolation and identification of a variety of farnesylated proteins. A similar method was developed to identify and characterize geranylgeranylated proteins. In this method, azido-geranylgeranyl alcohol was used to label geranylgeranylated proteins which were then linked to fluorescent tetramethylrhodamine-alkyne (TAMRA) by using the "click" chemistry. Geranylgeranylated proteins were separated by pH fractionation and 2D gel electrophoresis, and spots were cut out to

ISSN NO: 1874-6047
DOI: 10.1016/B978-0-12-381339-8.00011-1

identify geranylgeranylated proteins. Both methods provide a powerful means to identify prenylated proteins but further improvements are needed.

II. Introduction

Proteins such as the Ras superfamily G-proteins [1–3] are prenylated, and this is important for their function. For example, farnesylation of Ras proteins is important for their membrane association and transforming activity. Many of the Rho family proteins as well as the Rab family proteins are geranylgeranylated. In addition, proteins such as nuclear lamins and Hdj2 are farnesylated. Farnesylated proteins end with the so-called CAAX motif where C is cysteine, A is an aliphatic amino acid, and X is the C-terminal amino acid that is usually serine, methionine, glutamine, cysteine, and threonine. Geranylgeranylated proteins come in two flavors, one type ending with the CAAL motif (similar to the CAAX motif but the C-terminal amino acid is leucine or phenylalanine) and the other type ending with the CC (two cysteines) or CXC motif.

One of the recent excitements in the study of protein prenylation is the development of methods to determine the number of prenylated proteins inside the cell and gain insight into the ratio of farnesylated and geranylgeranylated proteins. It is also of major interest to examine intracellular localization of prenylated proteins. Having a complete list of all prenylated proteins will deepen our understanding of the biological significance of protein prenylation and provide insights into the functions of these proteins. An approach to apply bioinformatics identified gene products ending with the CAAX or CAAL motifs. Further, a prediction of the presence of a large number of farnesylated proteins has been made based on this type of approach [4,5].

Various attempts have been made to develop an experimental method to identify prenylated proteins. Initial methods used radioactive mevalonate ([^{3}H]mevalonate) to label mevalonate derivatives such as farnesyl pyrophosphate (FPP) and geranylgeranyl pyrophosphate [6–8]. However, this method is time consuming as it requires long exposure time (months) to detect radioactive bands on a gel. The second approach is to tag prenylated proteins. For example, we have made use of the azide chemistry to tag a prenyl moiety [9,10]. Similar methods using alkyne-modified isoprenoids have been reported [11–13]. Another approach to tag proteins with anilinogeraniol and use antibodies against anilinogeraniol has been developed [14]. Further, Nguyen et al. [15] used biotin-geranylgeranylpyrophosphate. In this chapter, we review how we used azide-modified isoprenoids to tag and identify prenylated proteins and discuss some of the observations obtained by the application of these methods.

III. General Approach: Tagging via Substrate Approach Utilizing Azide Chemistry

The use of azide modified precursors to tag posttranslationally modified proteins has been pioneered by Bertozzi and others [16,17]. The basic idea is to produce azide-linked proteins so that they can be coupled with phosphine reagents with functional moieties such as biotin using the Staudinger reaction. This phosphine catalysis is powerful and can be performed at room temperature in aqueous solution. Initial application of this method was on glycosylated proteins that can be tagged with azide-modified sugars [16]. We have applied this method to detect and characterize farnesylated and geranylgeranylated proteins.

The reason for using azide is severalfold. First, azide is nonpolar and has a small size. Thus, it is expected that the modification does not cause structural changes and does not interfere with the functions of the tagged proteins. A number of experiments we carried out support this point (discussed below). Second, azide is relatively inert and is not altered by a variety of conditions inside the cells or during the process of isolating tagged proteins. Third, azide is not present in any known molecules including proteins, nucleotides, and carbohydrates.

IV. Detection of Farnesylated Proteins

Farnesylation occurs by the addition of a farnesyl group to the cysteine in the CAAX motif, a reaction catalyzed by protein farnesyltransferase. The donor for the farnesyl moiety is FPP. Therefore, azido-farnesyl diphosphate (FPP-azide) was used as a means to tag farnesylated proteins. Because farnesol can be taken up into cells and converted to farnesyl-pyrophosphate, azido-farnesyl alcohol (F-azide-OH) was also used to tag farnesylated proteins. In order to prevent dilution of FPP-azide or F-azide-OH by the pool of FPP, mevalonate synthesis was blocked by treating cells with lovastatin, an HMG-CoA reductase inhibitor [18]. Cells thus treated were collected and then lysed. Prenylated proteins with azide modification were linked with biotinylated phosphine capture reagent (bPPCR) using the Staudinger reaction. Figure 11.1 describes the tagging method.

We have carried out a number of experiments to demonstrate that azide-modified proteins are functional [9]. First, we showed that FPP-azide can be used as a substrate for farnesyltransferase reaction *in vitro*. This point was further confirmed by *in vivo* studies. Lovastatin inhibits farnesylation of Ras and Hdj2 proteins, and this was detected by the shift of the mobility of

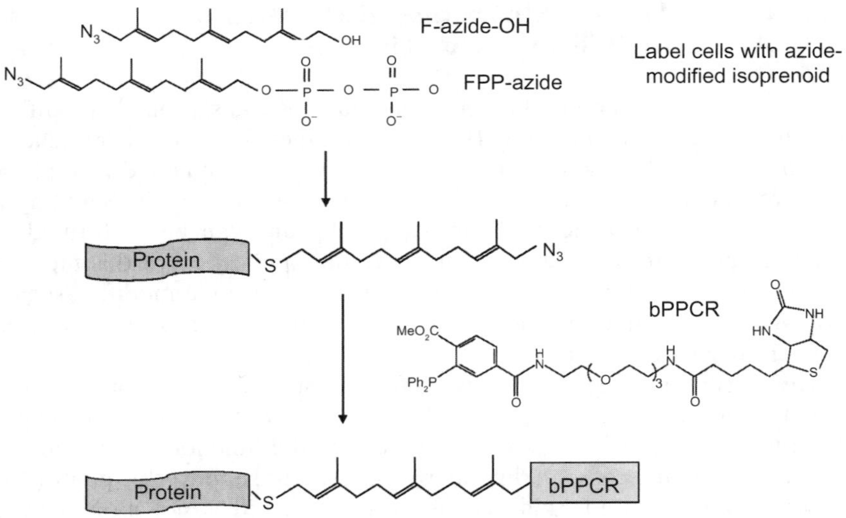

FIG. 11.1. Labeling of farnesylated proteins by using FPP-azide or F-azide-OH. FPP-azide and F-azide-OH are shown. Labeling cells with these compounds leads to the generation of azide-modified proteins that can be linked to phosphine capture reagent.

these proteins to a slow migrating form on a SDS polyacrylamide gel. This mobility shift was reversed by the addition of FPP or FPP-azide. In contrast, the mobility shift of a geranylgeranylated protein Rap1 was not reversed by FPP-azide. Second, FPP-azide and F-azide-OH were able to restore membrane association of Ras protein after the treatment with lovastatin. Cell fractionation experiments showed that lovastatin blocked membrane association of Ras proteins but the addition of FPP-azide or F-azide-OH restored the membrane association. Third, we showed that the function of Ras to activate the Raf/MEK/ERK pathway is unaffected by the azide modification. This was demonstrated by shutting down the Ras signaling by serum starvation followed by the activation by EGF. The activation of the Raf/MEK/ERK signaling as monitored by the phosphorylation of MEK was inhibited by the treatment with lovastatin. However, addition of FPP-azide restored the MEK activation. Finally, we showed that the azide modified compounds block apoptosis induced by lovastatin in H-ras transformed NIH3T3 cells. NIH3T3 cells transformed with activated H-ras were sensitive to apoptosis induction by the treatment with lovastatin, as detected by the activation of caspase-3. This apoptosis induction was blocked by the addition of farnesol or mevalonate. FPP-azide or F-azide-OH showed similar efficiency to block apoptosis induction.

Analysis of farnesylated proteins was first carried out by separating proteins on a SDS polyacrylamide gel. F-azide-modified proteins were conjugated with bPPCR and the resulting mixture was separated by SDS/PAGE. Prenylated proteins were then detected by using HRP-conjugated streptavidine. A profile of farnesylated proteins as detected by the use of the azide tagging method was somewhat reminiscent of that detected by the use of radioactive mevalonic acid with major bands in the 45–75 kDa range. Multiple bands in lower molecular weight range were also observed. The signal detected was confirmed as that of farnesylated proteins, as the treatment with farnesyltransferase inhibitor (FTI) resulted in the loss of the signal.

Finally, a mass spectrometry analysis was developed to identify farnesylated proteins. F-azide-modified proteins were conjugated to bPPCR. Streptavidin-agarose beads were used to isolate the biotinylated proteins. The affinity-purified proteins were digested with trypsin, and the resulting tryptic peptides were subjected to nano-HPLC/LCQ MS analysis. A protein sequence database search using the MS/MS data led to the identification of 21 proteins including 17 CAAX motif containing proteins. All three forms of Ras (K-Ras, N-Ras, and H-Ras) as well as Rheb, Rap2C, and Rab21 proteins were detected. In addition, a variety of DnaJ family member (Hdj2) proteins, nuclear lamins B1 and B2 as well as peroxisomal farnesylated proteins were identified. Interestingly, nucleosome assembly protein-1-like protein (NAP-1) which ends with the sequence CKQQ and is involved in nucleosome remodeling was identified in this analysis. We also identified annexin A2 that ends with the sequence CGGDD. Known geranylgeranylated proteins Cdc42 and TC21 were identified in this analysis. This might be due to a small amount of GGPP-azide arising from the conversion of FPP-azide.

V. Detection of Geranylgeranylated Proteins

We have developed a similar global method to identify geranylgeranylated proteins [10]. In our protocol shown in Figure 11.2, we used azido-geranylgeranyl analog to tag geranylgeranylated proteins. The tagged proteins were conjugated to tetramethylrhodamine-alkyne (TAMRA) by using the "click" chemistry. The resulting fluorescent proteins were separated by gel electrophoresis followed by mass spectrometry analysis. The azido-geranylgeranyl analog we used is azido-geranylgeranyl alcohol (N_3-GG-OH) which is taken up into cells and is converted to a diphosphate derivative that replaces geranylgeranyl diphosphate.

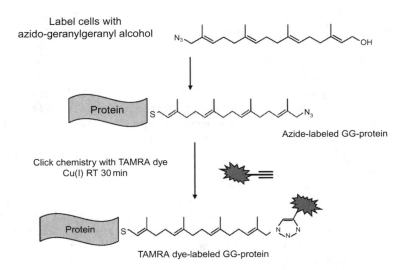

FIG. 11.2. Labeling of geranylgeranylated proteins by using azido-GG alcohol. Labeling cells with azido-geranylgeranyl alcohol leads to the generation of azide-modified proteins that can be linked to TAMRA.

Using Jurkat, COS-7, and MCF7 cells, we performed labeling of geranylgeranylated proteins and analyzed them on a SDS polyacrylamide gel. We found that the fluorescent bands appeared predominantly in a molecular weight range between 20 and 35 kDa. These bands disappeared when the tagging was carried out in the presence of excess geranylgeraniol, demonstrating that the bands represent those of geranylgeranylated proteins. For labeling prenylated proteins with radioactive mevalonic acid or for detecting farnesylated proteins using F-azide-OH, prior treatment with lovastatin was necessary to enhance labeling efficiency. This is because lovastatin inhibits HMG-CoA reductase, thereby blocking the mevalonate pathway and leading to the depletion of a pool of endogenous FPP. However, a prior treatment with lovastatin was not needed for detecting geranylgeranylated proteins using azido-geranylgeranyl alcohol.

Improved separation of geranylgeranylated proteins can be achieved by pH fractionation isoelectric focusing (IEF), a separation by isoelectric point. We used five different pH fractions (3–4.6. 4.6–5.4, 5.4–6.2, 6.2–7.0, and 7.0–10.0) and resolved each fraction by 1D gel electrophoresis. Each fraction contained different sets of proteins, thus demonstrating the ability to separate geranylgeranylated proteins. Further separation of geranylgeranylated proteins was accomplished by carrying out narrow pH range 2D SDS-PAGE. Multiple spots in the region around 21 kDa were separated by the 2D gel electrophoresis. Each spot was excised, digested with trypsin, and subjected to LC–MS/MS. Geranylgeranylated proteins were identified by peptide sequence using the search program MASCOT and X! Tandem. A variety of Rab proteins including Rab7, Rab11B, Rab14, Rab1A, Rab3D, Rab5, Rab6A, Rab2, and Rab18 were identified by this method. In addition, Rap2C was identified.

VI.   Overall Profiles of Farnesylated and Geranylgeranylated Proteins and Further Improvements

Overall profiles of farnesylated proteins and geranylgeranylated proteins on a SDS/PAGE differ significantly. With farnesylated proteins, we observed strong bands at positions corresponding to apparent molecular weights of 49, 60, and 80 kDa. Some of these are likely to represent nuclear lamins A and C which appear as a band of 72 and 62 kDa, respectively. In addition, minor bands were detected in the region corresponding to 20–30 kDa proteins. Some of these likely represent the Ras superfamily G-proteins. In contrast, a very different profile of geranylgeranylated proteins was observed. In our experiments, most of the bands of geranylgeranylated proteins were clustered in a region that corresponds to proteins of 20–30 kDa. This may suggest that the majority of geranylgeranylated proteins are the Ras superfamily G-proteins including the Rab and Rho family proteins.

Improving sensitivity of the detection of prenylated proteins is of major importance. In our analysis to identify farnesylated proteins, a majority of the proteins were known proteins. Thus, it is necessary to improve the efficiency of detection to identify novel prenylated proteins which are likely to be of low abundance. Ways to increase the efficiency, reduce background and increase sensitivity, need to be explored.

VII.   Applications of the TAS Methods

To explore possible applications of the tagging method, we have examined prelamin A that is normally farnesylated using the geranylgeranylated protein detection method. A genetic disorder called Hutchinson–Gilford progeria syndrome is associated with phenotypes that resemble premature aging [19]. The disease is caused by the synthesis of progerin, a mutant prelamin A that contains an internal 50 amino acid deletion. In a series of studies carried out by Young and his colleagues, mice expressing a geranylgeranylated version of progerin were generated [20]. Fibroblasts were obtained from these mice, and the expression of the geranylgeranylated progerin was confirmed by labeling cells with azido-geranylgeranyl alcohol.

Accumulation of unprenylated Rab proteins in lung and stomach tissues from *gunmetal* (*gm*) and *Chm3^lox* mice was detected by using 16-azido gernaylgeranyl pyrophosphate in the in-gel fluorescence assay [21]. The gunmetal mice have a mutation in a splice acceptor site within the α-subunit of RabGGTase decreasing its activity. The *Chm3^lox* mice have a conditional knockout of REP-1. Unprenylated Rab proteins accumulated in the cyto-plasm of cells were detected by using enzymatic transfer of a tagged GGPP analog to target Rab proteins.

VIII. Other Tagging Methods and Toward Constructing Prenylome

In addition to the azide tagging, different tags have been used and these are shown in Figure 11.3. One method is to use alkyne-modified isoprenoids such as C10-alkyne and C15-alkyne [11,13]. Alcohol forms of these alkyne

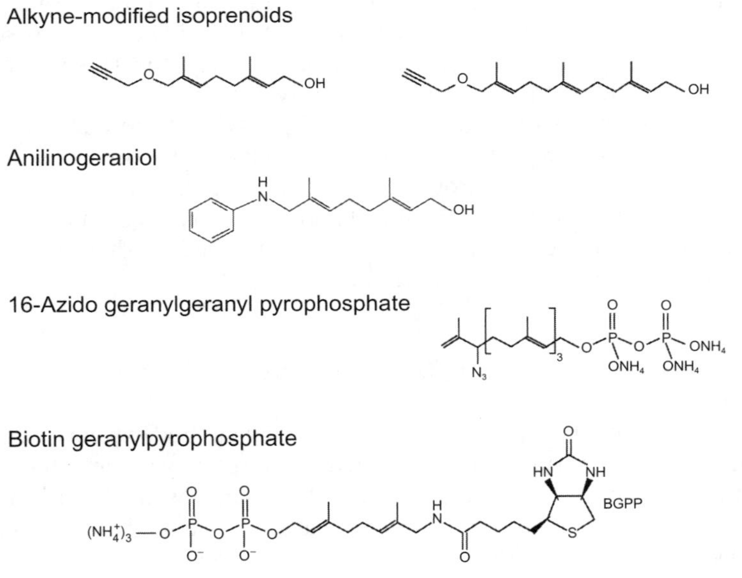

FIG. 11.3. A variety of reagents used to tag prenylated proteins. References for alkyne modified isoprenoids, anilinogeraniol, 16-azido-geranylgeranyl pyrophosphate, and biotin ger-anylpyrophosphate are [11], [12], [21], and [15], respectively.

isoprenoids have been added to cells. Cells were lysed and treated with azido- or propargyl-labeled tetramethylcarboxyrhodamine (TAMRA) fluorophore. This links prenylated proteins with the TAMRA dye via Cu (I)-catalyzed Click reaction. It appears that the C15-alkyne-modified isoprenoid is used by FTase and GGTase-I, while the C10-modified isoprenoid is used by FTase. The second approach used anilinogeraniol, the alcohol precursor to 8-anilinogeranyl diphosphate which is a FPP analog [14]. Anilinogeranyl-modified proteins were detected by using antibodies specific for the anilinogeranyl moiety. The third method is to use biotin-geranylpyrophosphate (BGPP). BGPP was efficiently transferred to substrate proteins by RabGGTase but not by FTase or GGTase-I [15]. Endogenous Rab proteins were shown to be modified by incubating cells with BGPP.

Prenylated proteins identified by the use of different tagging methods are listed in Table 11.1. Looking at these proteins, we begin to gain some insight into what kind of proteins are actually prenylated. As expected, a large number of the Ras superfamily proteins are detected. Various Ras proteins as well as Rheb proteins are identified by labeling with F-azide-OH as well as with anilino-geraniol. These methods detected other farnesylated proteins such as DnaJ proteins, peroxisomal proteins, and nuclear lamins. The anilinogeraniol tagging method identified Rho, Rac, and Rap family proteins that are geranylgeranylated. Azido-geranyl alcohol labeling identified a variety of Rab proteins.

Novel insights were obtained by these tagging methods. It is interesting that some proteins that have not been well characterized have been identified by multiple methods. One example is nucleosome assembly protein 1-like protein (NAP1L1). This protein ends with the CKQQ motif, and we have shown that a GST fusion protein containing C-terminal 11 amino acids of NAP1L1 can be farnesylated by FTase in vitro [9]. Interestingly, plant NAP1 (AtNAP1;1) is shown to be farnesylated [22]. Ectopic expression of nonfarnesylated AtNAP1;1 disrupts leaf development [22]. Another intriguing observation is Annexin proteins that have been detected by two different methods: Annexin A2 by the use of farnesol-azide and Annexin A3 by the use of C10-alkyne. These proteins end with a conserved motif CGGDD which is different from the CAAX motif. However, prenylation of Annexin A2 was not detected when anilinogeraniol was used [12]. Annexins are a family of calcium-dependent phospholipid binding proteins that are present in all eukaryotes [23]. Further studies are needed to investigate possible prenylation of these proteins.

Some proteins including RhoB protein are known to be both farnesylated and geranylgeranylated. Interestingly, we have identified Cdc42, a known geranylgeranylated protein, when conducting experiments with

TABLE 11.1

PRENYLATED PROTEINS IDENTIFIED BY TAGGING METHODS

	C-terminal	F-azide[a]	GG-azide[b]	Anilinogeraniol[c]	C10-alkyne[d]
H-Ras	CVLS	+		+	
K-Ras	CVIM	+		+	
N-Ras	CVVM	+		+	
Rheb	CSVM	+			
Pex19	CLIM	+		+	
Hdj2	CQTS	+		+	
DnaJ A2	CAHQ	+		+	
Dnj3	CAHQ	+			
NAP1L1	CKQQ	+		+	
Prelamin A	CSIM			+	+
Lamin B1	CAIM	+		+	
Lamin B2	CYVM	+		+	
Pex1pR633Ter	CKAL	+			
Annexin A2	CGGDD	+			+
Annexin A3	CGGDD			+	
LKB1	CKQQ			+	
INPP5A	CVVQ			+	
RhoA	CLVL			+	
RhoB	CKVL			+	
RhoC	CPIL			+	
Rac2	CSLL			+	
Rac1	CLLL			+	
Rap1A	CLLL			+	
Rap1B	CQLL			+	
Rap2A	CNIQ			+	
Rap2B	CVIL	+		+	
Rap2C	CVVQ	+	+		
Cdc42	CVLL	+		+	
TC21	CVIF				+
Rab7	SCSC		+		
Rab11B	CQNL		+		
Rab14	GCGC		+		
Rab1A	GGCC		+		
Rab3D	SCSC		+		
Rab5	CCSN		+		+
Rab6A	GCSC		+		+
Rab2	GGCC		+		
Rab18	CSVL	+			+
Rab1B	GGCC				+
GNBP	CGLY				

[a]F-azide (see Ref. [9]).
[b]GG-azide (see Ref. [10]).
[c]Anilinogeraniol (see Ref. [12]).
[d]C10-alkyine (see Ref. [11]).

F-azide-OH. In addition, Rap2C was identified by using F-azide-OH as well as by using geranylgeraniol-azide. It is possible that these proteins could be both farnesylated and geranylgeranylated. Alternatively, these results may reflect the conversion of FPP-azide to GGPP-azide inside the cell.

In conclusion, a variety of tagging methods developed over the years are beginning to provide an overview of prenylated proteins inside the cell. Further studies should enable us to group these prenylated proteins based on their functions and/or subcellular localizations, etc. It will be interesting to know how many different types of proteins are prenylated.

ACKNOWLEDGMENT

This work is supported by NIH Grant CA41996. We thank Gloria Lee for her help in the preparation of this chapter.

REFERENCES

1. Tamanoi, F. and Sigman, D.S. (eds.) (2001). The Enzymes Volume 21: Protein Lipidation Academic Press, San Diego, CA.
2. Zhang, F.L., and Casey, P.J. (1996). Protein prenylation: molecular mechanisms and functional consequences. *Annu Rev Biochem* 65:241–270.
3. Glomset, J.A., Gelb, M.H., and Farnsworth, C.C. (1990). Prenyl proteins in eukaryotic cells: a new type of membrane anchor. *Trends Biochem Sci* 15:139–142.
4. Maurer-Stroh, S., and Eisenhaber, F. (2005). Refinement and prediction of protein prenylation motifs. *Genome Biol* 6:R55.
5. Maurer-Stroh, S., Koranda, M., Benetka, W., Schneider, G., Sirota, F.L., and Eisenhaber, F. (2007). Towards complete sets of farnesylated and geranylgeranylated proteins. *PLoS Comput Biol* 3:e66.
6. Hancock, J.F. (1995). Reticulcyte lysate assay for in vitro translation and posttranslational modification of Ras proteins. *Methods Enzymol* 255:60–65.
7. Peter, M., Chavrier, P., Nigg, E.A., and Zerial, M. (1992). Isoprenylation of rab proteins on structurally distinct cysteine motifs. *J Cell Sci* 102:857–865.
8. Benetka, W., Koranda, W., Maurer-Stroh, S., Pittner, F., and Eisenhaber, F. (2006). Farnesylation or geranylgeranylation? Efficient assays for testing protein prenylation in vitro and in vivo *BMC Biochem* 7:6.
9. Kho, Y., Kim, S.C., Jiang, C., Barma, D., Kwon, S.W., Cheng, J., Jaunbergs, J., Weinbaum, C., Tamanoi, F., Falck, J., and Zhao, W. (2004). A tagging-via-substrate technology for detection and proteomics of farnesylated proteins. *Proc Natl Acad Sci USA* 101:12479–12484.
10. Chan, L.N., Hart, C., Guo, L., Nyberg, T., Davies, B.S., Fong, L.G., Young, S.G., Agnew, B.J., and Tamanoi, F. (2009). A novel approach to tag and identify geranylgeranylated proteins. *Electrophoresis* 30:3598–3606.
11. DeGraw, A.J., Palsuledesai, C., Ochocki, J.D., Dozier, J.K., Lenevich, S., Rashidian, M., and Distefano, M.D. (2010). Evaluation of alkyne-modified isoprenoids as chemical reporters of protein prenylation. *Chem Biol Drug Des* 76:460–471.

12. Onono, F.O., Morgan, M.A., Spielmann, H.P., Andres, D.A., Subramanian, T., Ganser, A., and Reuter, C.W. (2010). A tagging-via-substrate approach to detect the farnesylated proteome using two-dimensional electrophoresis coupled with Western blotting. *Mol Cell Proteomics* 9(4):742–751.
13. Charron, G., Tsou, L.K., Maguire, W., Yount, J.S., and Hang, H.C. (2011). Alkynyl-farnesyl reporters for detection of protein S-prenylation in cells. *Mol Biosyst* 7(1):67–73.
14. Troutman, J.M., Roberts, M.J., Andres, D.A., and Spielmann, H.P. (2005). Tools to analyze protein farnesylation in cells. *Bioconjug Chem* 16:1209–1217.
15. Nguyen, U.T.T., Guo, Z., Delon, C., Wu, Y., Deraeve, C., Franzel, B., Bon, R.S., Blankenfeldt, W., Goody, R.S., Waldmann, H., Wolters, D., and Alexandrov, K. (2009). Analysis of eukaryotic prenylome by isoprenoid affinity tagging. *Nat Chem Biol* 5(4):227–235.
16. Saxon, E., Luchansky, S.J., Hang, H.C., Yu, C., Lee, S.C., and Bertozzi, C.R. (2002). Investigating cellular metabolism of synthetic azidosugars with the Staudinger ligation. *J Am Chem Soc* 124:14893–14902.
17. Kiich, K.L., Saxon, E., Tirrell, D.A., and Bertozzi, C.R. (2002). Incorporation of azides into recombinant proteins for chemoselective modification by the Staudinger ligation. *Proc Natl Acad Sci USA* 99:19–24.
18. Sinensky, M., Beck, L.A., Leonard, S., and Evans, R. (1990). Differential inhibitory effects of lovastatin on protein isoprenylation and sterol synthesis. *J Biol Chem* 265:19937–19941.
19. Debusk, F.L. (1972). The Hutchinson–Gilford progeria syndrome. Report of 4 cases and review of the literature. *J Pediatr* 80:697–724.
20. Davies, B.S.J., Yang, S.H., Farber, E., Lee, R., Buck, S.B., Andres, D.A., Spielmann, H.P., et al. (2009). Increasing the length of isoprenyl anchor does not worsen bone disease or survival in mice with Hutchinson–Gilford progeria syndrome. *J Lipid Res* 50:126–134.
21. Berry, A.F.H., Heal, W.P., Tarafder, A.K., Tolmachova, T., Baron, R.A., Seabra, M.C., and Tate, E.W. (2010). Rapid multilabel detection of geranylgeranylated proteins by using bioorthogonal ligation chemistry. *Chembiochem* 11:771–773.
22. Galichet, A., and Gruissem, W. (2006). Developmentally controlled farnesylation modulates AtNAP1;1 function in cell proliferation and cell expansion during Arabidopsis leaf development. *Plant Physiol* 142:1412–1426.
23. Fatimathas, L., and Moss, S.E. (2010). Annexins as disease modifiers. *Histol Histopathol* 25:527–532.

12

Global Identification of Protein Prenyltransferase Substrates: Defining the Prenylated Proteome

CORISSA L. LAMPHEAR[a] • ELAINA A. ZVERINA[b] •
JAMES L. HOUGLAND[c,1] • CAROL A. FIERKE[a,b,c]

[a]*Department of Biological Chemistry*
University of Michigan
Ann Arbor, Michigan, USA

[b]*Chemical Biology Program*
University of Michigan
Ann Arbor, Michigan, USA

[c]*Department of Chemistry*
University of Michigan
Ann Arbor, Michigan, USA

I.   Abstract

The protein prenyltransferases, protein farnesyltransferase (FTase) and protein geranylgeranyltransferase-I (GGTase-I), catalyze the attachment of a 15-carbon farnesyl or 20-carbon geranylgeranyl moiety, respectively, to a cysteine near the C-terminus of a substrate protein targeting it to the membrane. Substrates of the prenyltransferases are involved in a myriad of signaling pathways and processes within the cell, therefore inhibitors targeting FTase and GGTase-I are being developed as therapeutics for treatment of diseases such as cancer, parasitic infection, asthma, and progeria.

---

[1] Present address: Department of Chemistry, Syracuse University, Syracuse, New York, USA

THE ENZYMES, Vol. XXIX   207   ISSN NO: 1874-6047
                    DOI: 10.1016/B978-0-12-381339-8.00012-3

FTase and GGTase-I were proposed to recognize a Ca_1a_2X motif, where C is the cysteine where the prenyl group is attached, a_1 and a_2 are small aliphatic amino acids, and X confers specificity between FTase and GGTase-I with X being methionine, serine, glutamine, and alanine for FTase and leucine or phenylalanine for GGTase-I. Recent studies indicate that the Ca_1a_2X paradigm should be expanded; therefore, further studies are needed to define the prenylated proteome, to understand normal cellular processes, and to determine the targets of prenyltransferase inhibitors. This review highlights the multiple approaches currently used to identify and define FTase and GGTase-I substrates. Direct identification approaches involve identifying FTase and GGTase-I substrates "one by one" or by using lipid donor analogs. A complementary approach to identify the prenylated proteome is to define the modes of FTase and GGTase-I substrate recognition using structure–function studies, peptide library studies, and computational methods.

II.   Introduction

Prenylation is an important posttranslational modification where an isoprenoid moiety is transferred to a cysteine near the C-terminus of a substrate protein [1,2]. This hydrophobic modification helps to localize proteins to cellular membranes to carry out their function as well as to facilitate protein–protein interactions [3,4]. FTase catalyzes the transfer of the 15-carbon farnesyl moiety from farnesyldiphosphate (FPP) to a cysteine residue near the C-terminus of the target protein; protein GGTase-I and GGTase-II (also called RabGGTase) catalyze the analogous attachment of a 20-carbon geranylgeranyl group from geranylgeranyldiphosphate (GGPP) to cysteine(s) near the C-terminus of the substrate protein (Figure 12.1) [2,5]. Additionally, *in vivo*, substrates can undergo further modification after prenylation. The last three amino acids of prenylated substrates can be proteolyzed by zinc metalloprotease Ste24 (ZMPSTE24) or Ras-converting enzyme 1 (RCE1) at the endoplasmic reticulum, followed by methylation of the carboxy terminus, catalyzed by isoprenylcysteine methyl transferase (ICMT) [6–8]. These additional modifications can aid in membrane localization, but it is not yet known whether these modifications occur on every prenylated substrate. This review will primarily focus on FTase and GGTase-I, the "CaaX prenyltransferases," which have similar modes of recognition (see Chapter 8 for a review on GGTase-II). Both of these enzymes are heterodimeric, containing identical α subunits and distinct but homologous β subunits [9], with active sites in both enzymes

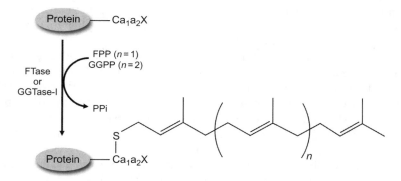

FIG. 12.1. FTase and GGTase-I catalyze the attachment of a farnesyl or geranylgeranyl group to the sulfur atom of a cysteine in the proposed C-terminal Ca_1a_2X sequence of a substrate protein using farnesyldiphosphate (FPP) or geranylgeranyldiphosphate (GGPP) as the lipid donor [2,5].

composed predominantly of β subunit residues near the α/β interface [10]. Currently, although many prenyltransferase substrates have been identified, the full extent of prenylation within the cell is still unclear [11–15].

Many proteins are modified by the prenyltransferases, including the small Ras and Rho GTPase superfamilies [1,16] and the nuclear lamins [17], and often the prenyl modification is essential for the biological function [1,18]. Prenyltransferases are being investigated as targets for inhibitors to treat a variety of diseases, including cancer [19], Hutchinson–Gilford progeria syndrome [20], and parasitic diseases such as malaria [21]. Prenyltransferase inhibitors were initially developed to target Ras protein signaling pathways implicated in cancer, but it was later determined that inhibitor efficacy is the result of modulating prenylation of non-Ras proteins [22]. Studies of the pleiotropic effects of statin treatment, which blocks the biosynthesis of FPP and GGPP and therefore may affect prenylation, suggest the involvement of prenylated proteins in diseases such as leukemia [23], asthma [24], and cardiovascular disease [25]. Therefore, understanding how prenyltransferases recognize their substrates, which substrates are prenylated *in vivo*, and finally, what substrates are responsible for inhibitor efficacy are important outstanding questions.

FTase and GGTase-I are proposed to recognize the "Ca_1a_2X" motif on substrate proteins [1,5,26]. This Ca_1a_2X motif is generally defined as follows: "C" is a cysteine four amino acids from the C-terminus where the prenyl group is attached forming a thioether bond; a_1 and a_2 are small aliphatic amino acids; and X is proposed to confer specificity of substrates for modification by FTase or GGTase-I, with X being methionine, serine, glutamine, and alanine for FTase and leucine or phenylalanine for GGTase-I [27–30]. Although many substrates are described by this paradigm, recent studies have indicated that the Ca_1a_2X model should be revised, as many substrates fall outside the traditional Ca_1a_2X definition [15]. Additionally, there is evidence for a large pool of dual substrates for FTase and GGTase-I, with these proteins potentially modified by both enzymes [31]. Understanding how FTase and GGTase-I recognize substrates would aid in defining the prenylated proteome and in understanding the biological signaling pathways involving prenylated proteins.

An emerging field is the study of prenyltransferases from pathogenic and parasitic organisms as potential therapeutic targets [32–35]. For instance, the pathogenic yeast strain *Candida albicans*, known to affect immunocompromised patients, utilizes both protein FTase and GGTase-I enzymes to catalyze prenylation of substrate proteins [36,37]. The specificity of these two enzymes is currently being studied to define the complement of prenylation of proteins in this organism, since it is possible that inhibitors of the prenyltransferases could be used as antifungal therapeutics [36,38]. Additionally, inhibitors are being developed as therapeutics for the treatment of parasitic diseases such as malaria, caused by *Plasmodium falciparum* [39]. These FTase inhibitors, based upon ethylenediamine [34] and tetrahydroquinoline [33] scaffolds, are thought to block *P. falciparum* FTase activity, which is essential for the organism. Based upon modeling and resistance mutation studies [40], these inhibitors may chelate the catalytic Zn^{2+} and bind to the enzyme in the lipid substrate grove. Recently, it has also been discovered that the Gram-negative bacterium, *Legionella pneumophila*, hijacks the host prenyltransferase machinery to catalyze prenylation of bacterial proteins, raising the possibility that current mammalian prenyltransferase inhibitors could also be used as antibiotics [41,42].

Defining the extent of prenylation within the proteome of an organism can be approached using two complementary tactics: (1) direct *in vivo* determination of proteins that are prenylated or (2) definition of the recognition elements used by FTase and GGTase-I to select peptide and protein substrates. This review covers both approaches. The summary of the direct *in vivo* identification modes includes a brief discussion of methods for detecting prenylation on a small or large scale (Sections III.A and III.B). Additionally, methods used to define molecular recognition in prenyltransferases are described, including structure–function studies (Section III.C), peptide library studies (Section III.D), and computational predictive methods (Section III.E). Finally, a brief discussion of pathogenic prenyltransferase enzyme structure and specificity is included in Section IV.

III. Methods for Discovering and Predicting Prenyltransferase Substrates

A. IDENTIFICATION OF SUBSTRATES ONE BY ONE

In the past, the methodology to determine prenyltransferase substrates has been challenging and, typically, was carried out by studying one protein at a time. The prenyl modification status of a particular protein can be assessed through incubation of cells or lysate with radioactive (^{3}H or ^{14}C) molecules, including FPP or GGPP [43–45]; a metabolic precursor of FPP or GGPP such as mevalonate [43,44,46]; or an alcohol precursor of GGPP and FPP, such as geranylgeraniol (GGOH) or farnesol (FOH) [45,47], that is phosphorylated *in vivo* [48]. Treatment of cells with these compounds allows the protein of interest to be radiolabeled upon prenylation. A significant limitation of this method is the low signal from the prenylated proteins arising at least partly from the low specific activity of the radioactive molecules typically used [2,45,49]. Control experiments to eliminate false positive results from radiolabeling studies include mutation of the cysteine of the Ca_1a_2X in the target protein to serine to block prenylation and/or incubation with prenyltransferase inhibitors [49,50]. These studies are difficult to perform on a large pool of protein targets since there is not a facile method to pull down proteins containing prenyl groups. Antibodies have been raised to the prenyl modifications for detection using immunoblotting [51,52]; however, these antibodies can exhibit problematic cross-reactivity with other lipid modifications [53] and are not able to distinguish between farnesyl and geranylgeranyl modifications [24,52,53].

B. FPP AND GGPP ANALOGS: AIDING IN PRENYL GROUP DETECTION

Using synthetic organic chemistry, multiple research groups have developed FPP and GGPP donor analogs with properties that enhance prenylated protein isolation and identification [11,54–56]. FTase and GGTase can recognize various prenyl donor analogs as substrates for incorporating recognition tags into prenylated proteins, both *in vitro* and *in vivo*, allowing for parallel identification of multiple substrates within the available pool of proteins as well as the monitoring of prenylation status changes upon inhibitor treatment. We will briefly focus on three recently reported classes of analogs: immunogenic analogs, analogs with functional groups allowing for chemoselective bioorthogonal labeling following prenylation, and analogs that contain an affinity tag such as biotin (Figure 12.2).

Immunogenic analogs incorporate a chemical moiety that can serve as the epitope for an analog-specific antibody. An example of such an analog

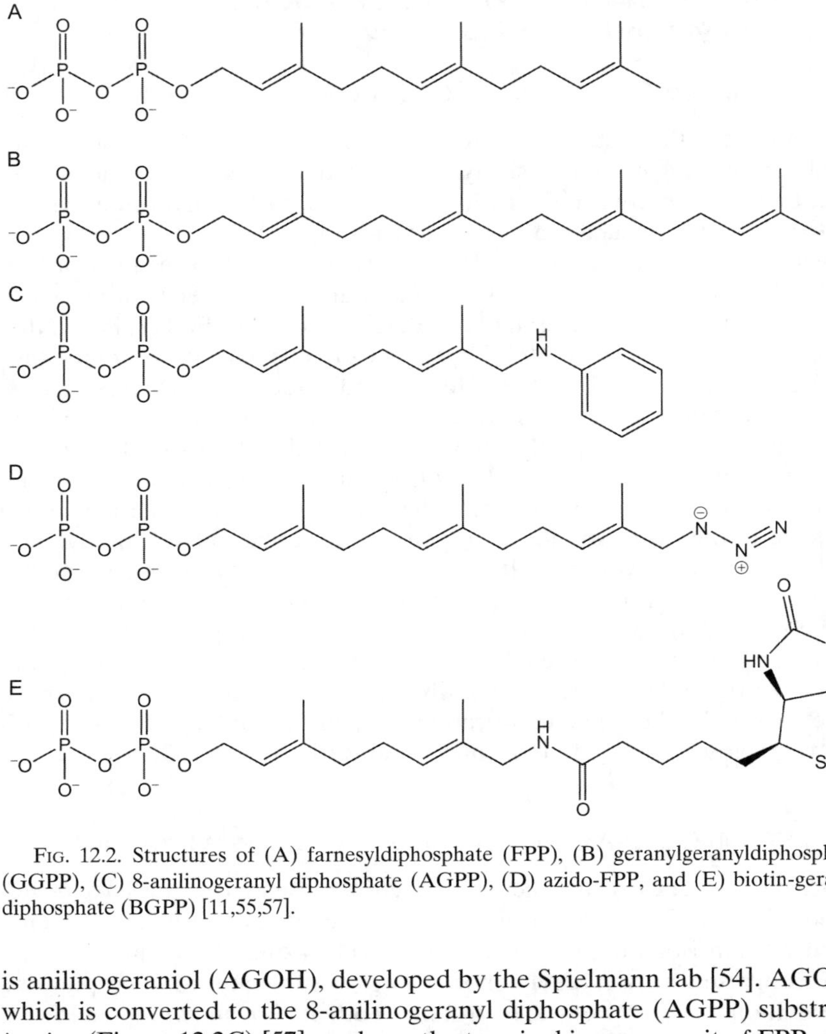

FIG. 12.2. Structures of (A) farnesyldiphosphate (FPP), (B) geranylgeranyldiphosphate (GGPP), (C) 8-anilinogeranyl diphosphate (AGPP), (D) azido-FPP, and (E) biotin-geranyl diphosphate (BGPP) [11,55,57].

is anilinogeraniol (AGOH), developed by the Spielmann lab [54]. AGOH, which is converted to the 8-anilinogeranyl diphosphate (AGPP) substrate *in vivo* (Figure 12.2C) [57], replaces the terminal isoprene unit of FPP with an aniline moiety that serves as an epitope. In studies employing AGOH as an immunogenic tag, cells are treated with the analog, lysed, and proteins are fractionated by 1- or 2-D gel electrophoresis. The farnesylated proteins are detected using Western blotting with an antibody raised against the AG-KLH protein [54,58]. This method detected several known FTase substrates using 1-D gel electrophoresis but failed to identify many small GTPases known to be FTase substrates; however, coupling the Western blot analysis with 2-D gel electrophoresis increases the efficiency of prenylated protein identification [58].

An emerging class of FPP and GGPP analogs employs bioorthogonal ligation methods, wherein analogs containing an azido or alkyne group on the terminal isoprenoid of FPP or GGPP are attached to substrate proteins. These functional groups are amenable to Staudinger ligation or Huisgen 1,3-dipolar cycloaddition ("click" chemistry) [59,60], which allows for chemoselective attachment of reporter molecules such as a fluorophore or biotin following protein prenylation. The first use of an azido analog for identification of prenylated proteins was reported by Kho et al. [11]. They treated COS-1 cells with azido-FPP (Figure 12.2D) or azido-FOH and demonstrated in vivo incorporation of this analog into proteins catalyzed by FTase. Following cell lysis, the azido-farnesylated proteins were labeled with a biotin-containing phosphine capture reagent using the Staudinger ligation, purified using streptavidin beads and identified using mass spectrometry [11]. The group used this method to detect 18 FTase substrates, including known farnesylated proteins such as H-Ras and Rheb. More recently, azido-GG alcohol has been used to identify geranylgeranylated substrates using click chemistry with TAMRA-alkyne, 2-D gels, and mass spectrometry [61] and to label substrates in prenylation-deficient mouse tissue [62]. Additionally, alkyne derivatives of FPP and GGPP are gaining popularity for identification of prenylated proteins in vitro and in vivo [63–65].

An alternative method, the "affinity tag" approach, uses as biotin-geranyl diphosphate analog (BGPP; Figure 12.2E) [55]. This analog serves as an efficient substrate for GGTase-II, but not FTase or GGTase-I. To expand the utility of this analog, Alexandrov and coworkers "engineered" FTase and GGTase-I enzymes with altered substrate selectivity by mutating residues in the active site (W102T/Y154T or W102T/Y154T/Y205T for FTase, and F53Y/Y126T or F52Y/F53Y/Y126T for GGTase-I), allowing the engineered enzymes to catalyze modification of substrate proteins using the BGPP analog. Compactin was used to block native FPP and GGPP synthesis, increasing the pool of unmodified FTase and GGTase-I substrates. The compactin-treated cell lysates were incubated in vitro with BGPP and mutant or wild-type prenyltransferases, pulled down using streptavidin beads, and modified proteins were identified using mass spectrometry. Many Rab proteins were detected and quantified as GGTase-II substrates, while various molecular weight substrates were identified from the engineered FTase- and GGTase-I-treated lysates.

Although there has been much success using analogs to identify prenylated substrates, there are caveats to these approaches. First of all, the FPP and GGPP analogs may alter the protein specificity of the enzyme [66–68] since the prenyl group forms a portion of the protein substrate-binding site

[14]. In particular, for use of the biotin analogs, the FTase and GGTase-I enzymes were engineered at positions in the active site that can alter the enzyme specificity [69]. Another concern is the "hit rate" for these analogs, as treatment with several of these analogs did not identify known substrates as prenylated proteins; these issues with false negatives may indicate that these analogs may not be incorporated into substrates at a high enough level to allow identification. One method to obtain higher incorporation of the analogs is to treat the cells with an inhibitor of FPP or GGPP synthesis, such as compactin [55,70]; however, this treatment may disrupt the homeostasis of the cell, not allowing for true identification of substrates under normal conditions. Additionally, the BGPP analog was incubated with cell lysates in an *in vitro* context, identifying potential substrates, but perhaps not identifying substrates that are prenylated under native *in vivo* conditions.

C.   IDENTIFYING PRENYLTRANSFERASE SUBSTRATES THROUGH STRUCTURAL
     AND STRUCTURE–FUNCTION BIOCHEMICAL STUDIES

In this section, we review the findings from crystallographic data and structure–activity studies that combine to provide a functional and structural picture of the interactions important for prenyltransferase selectivity. Such studies have yielded valuable insights that suggest the prenylated proteome may be much richer in size and diversity than has been previously proposed. Future studies, focusing on identifying additional interactions and quantifying the energetic contribution of each enzyme–substrate interaction, have the potential to provide the basis of a quantitative model for predicting the full complement of prenylated proteins within the cell.

1. *Structural Studies*

Over the past 15 years, a series of crystallographic structures of mammalian FTase and GGTase-I have provided insight into the active site interactions and structural context of the peptide substrate-binding site within prenyltransferases [14,71,72]. Using these structures as a reference, the selectivity for each amino acid within the Ca_1a_2X motif can be interpreted in light of potential contacts with active site residues. The binding site for the a_1 residue of the peptide substrate is exposed to solvent at the interface of the FTase α and β subunits [14], consistent with the relaxed specificity at the a_1 position observed in biochemical studies [13,15,29,30]. In contrast, the a_2 and X-residue binding sites lie within the solvent-excluded active site, suggesting that these two positions may be primarily responsible for prenyltransferase selectivity. The a_2 site in FTase is mainly composed of the side chains of W102β, W106β, and Y361β (Figure 12.3B), with analogous

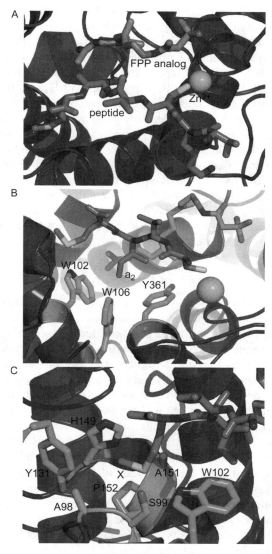

FIG. 12.3. The FTase active site is shown (PDB ID 1D8D) with FTase residues in purple or green, peptide substrate KKKSKTKCVIM in pink, FPP analog (FPT Inhibitor II) in orange, and the catalytic Zn^{2+} in light blue. For clarity, only a portion of the peptide substrate is shown: TKCVIM in (A) and CVIM in (B) and (C). (A) The structure of the rat FTase active site showing the position of the bound FPP analog and peptide; (B) structure of the contacts between the substrate a_2 residue (isoleucine) and FTase residues W102β, W106β, and Y361β (green) and the 3rd isoprene of FPP (orange); (C) structure of the contacts between the substrate X residue (methionine) and the X-residue binding pocket in FTase, including residues Y131α, A98β, S99β, W102β, H149β, A151β, and P152β (green) [14,106]. (See color plate section in the back of the book.)

residues T49β, F53β, and L321β in GGTase-I. In addition, the prenyl donor cosubstrate also contacts the a_2 residue in both enzymes [14]. Structures of FTase complexed with peptides containing different X groups suggest the possibility of two different X-residue binding pockets: S, Q, and M interact with Y131α, A98β, S99β, W102β, and H149β (Figure 12.3C), while F (and likely L, N, and H) interacts with L96β, S99β, W102β, W106β, and A151β [14]. These X-group binding sites in FTase generally confer a preference for prenylation of peptides with moderately polar amino acids such as Ser, Met, and Glu at the X position. In GGTase-I, amino acids T45β, T49β, and M124β replace the A98β, W102β, and P152β side chains in the X-group binding pocket in FTase. Unexpectedly, this less hydrophobic X-residue binding site in GGTase-I leads to a preference for catalyzing prenylation of peptides with more hydrophobic X residues, such as Leu and Phe. These contacts, acting in concert, suggest a set of preferences for FTase and GGTase-I substrates that can serve to guide studies for identifying novel prenyltransferase substrates.

2.   Structure–Function Studies of Peptide Reactivity

While structural studies provide insight into the interactions that engender FTase and GGTase-I peptide substrate specificity, the energetic benefit (or cost) of each interaction must be identified to ascertain the functional substrate preference at each position within the Ca_1a_2X sequence. Structure–function studies of peptide reactivity, wherein changes in peptide reactivity are correlated with changes in amino acid properties such as hydrophobicity or steric size, provide one method for identifying the properties that are used by FTase or GGTase-I to recognize substrates. At the X position, Hartman *et al.* assayed the reactivity of a series of peptide substrates derived from the C-terminal sequence of K-Ras-4B (-TKCVIM) wherein the terminal amino acid was substituted with 14 amino acids, including the natural Met [73]. This study illustrated that peptide selectivity for FTase versus GGTase-I arises from relative peptide reactivity, rather than relative binding affinity. Further, peptide reactivity correlates strongly with hydrophobicity, with FTase and GGTase-I displaying inverse reactivity; FTase reactivity decreases and GGTase-I reactivity increases with increasing hydrophobicity at the X position. This "reciprocal" reactivity pattern yielded three groups of peptides: (1) peptides terminating in moderately polar X residues (i.e., S and Q) that exhibit exclusive reactivity with FTase; (2) peptides terminating in nonpolar X residues (i.e., L and I) that exhibit exclusive GGTase-I reactivity; and (3) peptides terminating in a subset of X residues, such as Met or Phe, that react efficiently with both FTase and GGTase-I. Further, these studies identified a number of peptides that were rapidly prenylated under single turnover (STO) conditions but not under multiple turnover (MTO) conditions, presumably due to slow product dissociation. Further studies of peptide reactivity demonstrate that FTase is capable of catalyzing farnesylation of peptides with a variety of X-group residues, including leucine [74]. Additionally, the functional importance of positively charged residues upstream of Ca_1a_2X motif, often referred to as the polybasic region, has been explored by comparing the reactivity of FTase and GGTase-I with the C-terminal sequence of K-Ras4B (KKKSKTKCVIM vs. TKCVIM) [31]. This work demonstrated that the upstream sequence enhances dual prenylation of substrates by decreasing the efficiency of FTase-catalyzed farnesylation to a level comparable to that of geranylgeranylation catalyzed by GGTase-I. These structure–function studies of X-residue recognition indicate that both FTase and GGTase-I can recognize a much wider range of side chains at this position than had been previously proposed. In addition, the possibility for a class of substrates that can be prenylated by either enzyme underscores the potential role of "leaky prenylation" in affecting the makeup and biological role of the prenylated proteome. Taken together, these findings indicate that more than half of the 20 amino acids can serve as the X residue of a prenyltransferase-competent Ca_1a_2X substrate sequence.

A similar structure–activity profile at the a_2 position both underscored predictions from structural work regarding a_2 sequence preferences and uncovered a previously unknown example of context-dependent substrate recognition at the a_2 position of the Ca_1a_2X sequence [75]. Analysis of the relative reactivity of a series of peptide substrates varying at the a_2 position (-GCVa$_2$S and -GCVa$_2$A) indicated that FTase recognizes both the polarity and steric volume of the a_2 side chain simultaneously, discriminating against polar amino acids and both large and small amino acids at this position; maximal activity is observed for amino acids containing a steric volume near that of valine. These preferences match those predicted by structural studies [14], providing a functional picture of the energetic contribution of the structurally predicted interactions to substrate selectivity. Surprisingly, when the reactivity of FTase with analogous peptides with different X residues (-GCVa$_2$M and -GCVa$_2$Q) was analyzed, substrate recognition at the a_2 position was predominantly due to polarity. In these peptides, the steric volume of the amino acid at the a_2 position did not significantly affect reactivity as long as that amino acid was either weakly polar or nonpolar. This context-dependent a_2 selectivity, wherein a larger range of a_2 residues can be present in FTase substrates when the X residue is Met or Gln compared to when X is Ser or Ala, suggests that FTase can catalyze farnesylation of proteins with a wide range of a_2 residues as efficient substrates, provided that an appropriate X residue is present.

Structural studies of FTase and GGTase-I have provided an intricate picture of the interactions and active site microenvironment responsible for recognizing prenyltransferase substrates from among the milieu of all cellular proteins. Structure–function analysis of peptide reactivity indicates that both FTase and GGTase-I recognize a much wider range of protein sequences as substrates than was originally proposed. Further, the specific chemical properties recognized by prenyltransferases at positions within the Ca_1a_2X sequence have been characterized by correlation of peptide reactivity with amino acid properties such as size and polarity. These structural and functional insights will serve as essential foundations for development of models for comprehensive prediction of prenyltransferase substrates based on protein C-terminal sequence data.

D.   PEPTIDE LIBRARY STUDIES: A HIGH-THROUGHPUT METHOD OF
     IDENTIFYING POTENTIAL PRENYLTRANSFERASE SUBSTRATES AND
     DEFINING SUBSTRATE SPECIFICITY

An additional method of determining prenyltransferase substrates is to define the molecular recognition elements of FTase and GGTase-I using peptide library studies. Short peptides, as small as tetrapeptides, are efficient substrates for FTase and GGTase-I with comparable affinity and reactivity to full-length proteins [29,30,76–78]. Adding a dansyl group to the N-terminus of the peptide allows for continuous monitoring of the reaction using a fluorescence assay, as the dansyl group increases in fluorescence upon prenylation [77,79]. Conveniently, this assay can be carried out in a high-throughput manner using 96-well plates in a plate reader, and with large libraries, statistical analysis of peptide reactivity can be used to determine patterns of substrate recognition [15]. The use of these peptide libraries to study prenyltransferase specificity is advantageous, since potential substrates can be screened quickly and efficiently, using wild-type enzymes and the natural lipid substrates FPP and GGPP, although one limitation with this method is that, in some cases, the structure of protein substrates may alter recognition. Many research groups have tested small or nonhomogenous libraries of peptides as a means to identify prenyltransferase substrates [67,80,81]. For instance, groups have tested the Ca_1a_2X paradigm using GCxxS and GCxxL libraries [74,81], finding that their results generally correlate well with structural studies [14] and computational algorithms [12,13] (see Section III.E).

To more completely define the scope of the prenylated proteome, a large-scale peptide library study of the substrate selectivity of FTase was carried out, using statistical analysis to analyze FTase preference patterns [15]. The library peptides were screened for reactivity with FTase under both MTO, subsaturating peptide ($k_{cat}/K_M^{peptide}$) conditions and STO conditions, using the dansyl fluorescence assay or an *in vitro* radioactive assay with ^{3}H-FPP [15,77,79]. In experimental practice, the MTO reaction is performed under conditions with excess substrate such that $[E] \ll [S]$ while the STO reaction includes excess enzyme and peptide substrate with limiting concentrations of FPP ($[FPP] < [E]$).

Two kinetic parameters, $k_{cat}/K_M^{peptide}$ and $k_{farnesylation}$, can be measured to describe peptide reactivity with FTase. The value of $k_{cat}/K_M^{peptide}$, measured under MTO conditions, is also termed the "specificity constant" [82] and is most representative of the reactivity of a particular substrate in a biological context where all the protein substrates compete for prenylation catalyzed by FTase and GGTase-I [82]. *In vivo* the relative rate of prenylation of a given substrate (and hence the selectivity) depends on both the concentration of the protein substrate and the value of $k_{cat}/K_M^{peptide}$. For the prenyltransferase reactions, $k_{cat}/K_M^{peptide}$ represents all the reaction steps up to the first irreversible step in a reaction [82]. Previous kinetic studies of FTase suggest the basic kinetic pathway shown in Figure 12.4 [77,83–87]. Substrate binding is functionally ordered, with FPP binding before peptide, followed by a conformational rearrangement of the first two isoprene units of FPP required to position the C_1 of FPP near the sulfur of the peptide substrate for facile reaction [72,88–90]. After the chemical step, diphosphate dissociation is rapid [85]. For FTase, the k_{cat}/K_M parameter includes the rate constants for peptide binding to E · FPP (forming the ternary complex) through the formation of the prenylated peptide and pyrophosphate products, including the conformation change prior to chemistry and the chemical step ($k_{farnesylation}$, Figure 12.4) [72,85,86,90,91]. Under these conditions, dissociation of the diphosphate product is the first irreversible step [85,86]; therefore dissociation of the prenylated peptide product does not contribute to the observed value of $k_{cat}/K_M^{peptide}$. The MTO kinetic parameter measured at saturating concentrations of peptide and FPP, k_{cat}, includes all the rate constants describing the formation of dissociated products (prenylated peptide and diphosphate) from the ternary complex (E·FPP·peptide). Under these conditions, dissociation of the farnesylated peptide is frequently the rate-limiting step [85,87]. Further, dissociation of the prenylated product is enhanced by binding FPP, and possibly peptides, to the FTase·farnesylated-peptide complex (Figure 12.4B) [73,87,92].

In addition, it is possible to directly measure the rate constant ($k_{farnesylation}$) for the formation of the farnesylated product by detecting the formation of diphosphate using a coupled assay under STO conditions ($[E] > [S]$; Figure 12.4A) [85]. The $k_{farnesylation}$ parameter for FTase includes rate constants for binding the peptide substrate to FTase·FPP, the

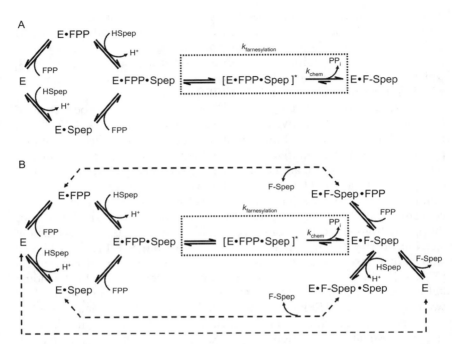

F‍IG. 12.4. The proposed catalytic cycle for FTase under (A) single turnover conditions (STO, [E] > S) and (B) multiple turnover conditions (MTO, [E] ≪ [S]). The STO-only substrates are proposed to undergo farnesylation, but not product release. FTase catalyzes turnover of MTO substrates where binding of either an additional peptide or FPP molecule to the E·farnesylated-peptide complex may facilitate product release. The rate constant $k_{\text{farnesylation}}$ includes both the conformational change and chemistry steps [15,83–87,89,107].

conformation change before the chemistry step, the farnesylation step, and the rapid dissociation of diphosphate, thereby including all the reaction steps up to but not including release of the prenylated protein [72,85,90,91]. At saturating concentrations of FTase, $k_{\text{farnesylation}}$ reflects solely the conformational change and farnesylation steps. Previous studies have demonstrated that, in some cases, peptide substrates are rapidly farnesylated by FTase, as measured under STO conditions, but that product dissociation is so slow that MTO activity is not observable by the current assays [73,93]. Measurement of both MTO and STO activity provides additional information to probe the molecular recognition modes of FTase.

The reactivity of FTase with two peptide libraries was measured: an "initial library" of 213 peptides was designed based on the Cxxx sequences taken from the human proteome with a mild bias toward sequences predicted by the Ca_1a_2X model, and a second library, the "targeted library" of

88 peptides, was similarly designed using human sequences but with a stronger bias toward sequences predicted to be FTase substrates based on the results from the initial library. The peptides in these libraries were of the form dansyl-TKCxxx, where x is any amino acid, and the upstream lysine was included to increase peptide solubility [89]. Overall, FTase catalyzed the farnesylation of a surprisingly large number of peptides [15]. FTase prenylated 77 peptides, or 36%, of the initial library under MTO conditions, and an additional 85 peptides (40%) under STO conditions. Together, this means that 76% of the initial 213 peptides were FTase substrates. Additionally, in the targeted library 29 (33%) and 45 (51%) of the peptides were farnesylated by FTase under MTO and STO conditions, respectively. In both libraries, FTase-catalyzed farnesylation of 78% of the peptides, including both MTO and STO conditions.

The values of $k_{cat}/K_M^{peptide}$ and $k_{farnesylation}$ were measured for a subset of FTase substrates [15]. The $k_{cat}/K_M^{peptide}$ parameters for the MTO substrate subset demonstrated a variation of approximately 100-fold. Further, the values for $k_{farnesylation}$ roughly correlate with $k_{cat}/K_M^{peptide}$, that is, more reactive substrates generally had higher $k_{farnesylation}$ values. For the STO substrates, the values for $k_{farnesylation}$ are comparable to those measured for the MTO substrates. Therefore, dissociation of the prenylated peptide must be sufficiently slow for the STO substrates such that MTO activity is not detectable. In this special case, product dissociation rates can alter substrate selectivity even in the presence of multiple competing substrates.

To further probe substrate recognition by FTase, the amino acid preferences at the a_1, a_2, and X positions of the MTO, STO, and nonsubstrate peptides from the initial library were compared using statistical analysis. For this analysis, the percentages of "canonical" and "noncanonical" Ca_1a_2X sequences at the a_2 and X positions, for the overall library, MTO, STO, and NON pools of peptides (Figure 12.5, top panel) were evaluated [15]. With "canonical" defined as V, I, L, M, and T for a_2, and A, S, M, Q, and F for X, the MTO peptide substrates were generally well described by the Ca_1a_2X paradigm; however, the STO substrates contain more varied sequences. A hypergeometric distribution model [15] was used to determine statistically significant enrichment (overrepresentation) or depletion (underrepresentation) of a specific amino acid as compared to the overall library. Using $p \leq 0.02$, sequence preferences for FTase could be divided into three classes of peptides (MTO, STO, and NON). MTO substrates are relatively well described by the original Ca_1a_2X paradigm: little sequence selectivity is observed at a_1; a nonpolar amino acid, like isoleucine and leucine, is preferred at a_2; and the X residue is preferentially a phenylalanine, methionine, or glutamine. Additionally, cysteine and lysine were depleted at the a_2 position in the reactive MTO substrates. Conversely,

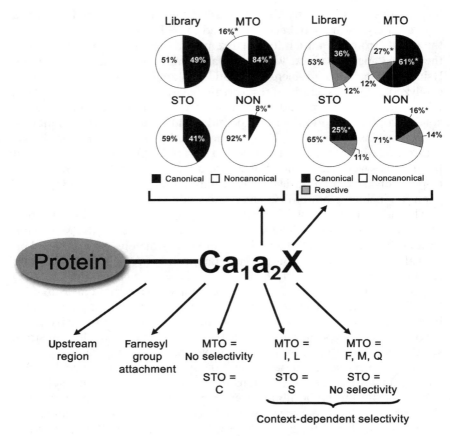

FIG. 12.5. Reactivity of FTase with a library of peptide substrates. *Top panel*: The percentages of "canonical," "noncanonical," and "reactive" pools of peptides in either the total library or the MTO, STO, or NON peptide pools catalyzed by FTase and sorted by the amino acid at either the a_2 or the X position. For a_2, canonical: V, I, L, M, T; noncanonical: all remaining amino acids; for X, canonical: A, S, M, Q, F; reactive: C, N, T; and noncanonical: I, L, R, Y, D, V, E, G, H, K, P, W. An asterisk (*) represents percentages which are statistically different ($p < 0.02$) as compared to the library. *Bottom panel*: FTase sequence preferences for MTO and STO substrates at the a_1, a_2, and X positions. Upstream region selectivity can include residues that form a flexible linker region. The a_2 and X position amino acids are recognized by FTase in a context-dependent manner [12,13,15,75].

STO substrates are not well described by the Ca_1a_2X paradigm or other more recently published substrate prediction models [12–14,73,75,89]. For the STO peptide substrates, cysteine is enriched, while leucine is depleted at the a_1 position; at the a_2 position, serine is enriched, while isoleucine and lysine are depleted; and at the X position, no amino acid is enriched, while methionine is depleted. The anticorrelation of methionine for the MTO and STO substrates at the X position suggests that this amino acid enhances product release [15]. For unreactive peptides, at the a_1 position, no sequence preferences were observed; at the a_2 position, D, K, and R were enriched, while V, I, L, and T were depleted; and at the X site, P and R were enriched. In this case, enrichment reflects amino acids that decrease reactivity. These statistically significant sequence preferences for FTase substrates derived from the human genome, along with other data [12,13,15,75], further broaden and define the Ca_1a_2X paradigm describing FTase selectivity. Figure 12.5 (bottom panel) shows a summary of FTase sequence preferences for substrates at the a_1, a_2, and X positions derived from these peptide library studies.

The observation of a large number of peptide substrates farnesylated under STO but not MTO conditions leads to the important question as to whether the STO substrates will likely be prenylated *in vivo*. Since slow dissociation of the farnesylated-peptide product is predicted to lead to STO behavior, it is possible that STO substrates could form long-lived product complexes with FTase in the cell, raising the possibility that these sequences function as inhibitors rather than biologically relevant substrates. However, at least two STO substrates, with C-terminal sequences corresponding to the peptides -CAVL and -CKAA [11,50], have been demonstrated to be farnesylated *in vivo*, suggesting that these proteins can turn over. One model consistent with these data is that an MTO substrate (or another "release factor") could facilitate the dissociation of the farnesylated-STO protein, similar to the FPP-catalyzed dissociation of product (see kinetic scheme in Figure 12.4B) [72,87] and this may have a regulatory function within the cell. This model enjoys some support from studies of small peptide reactivity [73,87,92]. Overall, these studies indicate that FTase may farnesylate a larger pool of proteins *in vivo* than would be predicted from the previously described Ca_1a_2X model.

E. Computational Work

In addition to the biochemical and structural methods used to identify FTase and GGTase-I substrates, which rely on various *in vitro* and *in vivo* approaches, *in silico* approaches have also been developed to help address this challenge of substrate identification. Various computational approaches have been introduced over the past several years to aid in large-scale prediction of prenylated proteins [12,13], with features that enable the user to rank the likelihood of a particular sequence being a

substrate for FTase, GGTase-I, both, or neither. The methods are itera-
tively improved by continuously incorporating new biochemical data as it
becomes available to further refine the predictive power of the computational analyses.

Several computer algorithms have been developed to help predict
protein substrates for mammalian prenyltransferases. Among prediction
software, one of the first tools built was Prosite protocol PS00294 which
used the consensus pattern C-{DENQ}-[LIVM]-x> (http://www.expasy.
org/prosite/) [94]. However, this tool is unable to distinguish between
FTase and GGTase-I substrates, nor does it predict prenylation by
GGTase-II enzyme. Crystallographic analysis of FTase and GGTase-I
complexed with eight cross-reactive substrates used interactions with the
binding pocket in the structures of the enzyme–substrate complexes to
draw inferences about FTase and GGTase-I substrate recognition elements [14]. The most significant drawback of this approach is that it only
identifies a subset of verified substrates, missing key substrate–protein
interactions that are not covered by the peptide diversity in the available
crystal structures.

PrePS is the most recent algorithm developed to predict prenyltransferase substrates [12]. To define this algorithm, the authors built a learning set
of known and homologous substrates (defined by specific rules) which
resulted in a set of 692 FTase and 486 GGTase-I substrates. One of the
difficulties in predicting prenylation substrates is the inherent complexity of
substrate recognition motifs, which may extend beyond the Ca_1a_2X box to
include the upstream region of the protein. To address this additional
complexity, the authors of PrePS included an 11-amino acid upstream
region of the Ca_1a_2X motif to refine their algorithm, as well as expanding
the list of acceptable amino acids for the Ca_1a_2X motif. This upstream
region typically consists of a flexible linker region that often has a compo-
sitional bias toward small or hydrophilic amino acids. Using this learning
set, the PrePS algorithm defines a set of rules that is used to predict the
likelihood of a 15-amino acid sequence being an FTase, GGTase-I, and/or
GGTase-II substrate. In cross-validation experiments, they were able to
establish 92.6% and 98.6% true positive rate for FTase and GGTase-I
substrates, respectively, with false positive rates of 0.11% and 0.02% for
FTase and GGTase-I, respectively. Consistent with this, analysis of the
substrates identified in the peptide library screen described above using
the PrePS algorithm yielded a very low number of false positive results [15].
However, the PrePS analysis led to a large number of false negative predictions, around 40%, indicating that the PrePS algorithm potentially
misses a large number of prenyltransferase substrates.

IV. Prenylation in Pathogenic Organisms: Structural and Biochemical Insights

A. C. ALBICANS PRENYLTRANSFERASES

Although most studies to date have focused on mammalian prenyltransferases, lower eukaryotic organisms, such as yeast, also contain FTase and GGTase-I and inhibition of these enzymes has been proposed as a possible treatment for pathogenic yeast infection [36–38,95–97]. C. albicans is a dimorphic yeast that is a major opportunistic human fungal pathogen which causes life-threatening infection in many immunocompromised patients. Although there are currently therapies on the market that combat C. albicans infections, resistance is always a concern, and thus novel therapies with unique mechanisms of action are highly sought after. Identifying C. albicans prenylation substrates and mining the differences between the substrate recognition of the human and C. albicans prenyltransferases could pave a way for novel drug candidates against this infection.

The first report of sequencing, cloning, and purification of C. albicans GGTase-I was published in 1999 by the Mazur group [37]. The α and β subunits of GGTase-I are 30% identical to their mammalian homologues, with conservation of the zinc ligands and substrate-binding residues of the β subunit. Based on sequence alignment and mammalian FTase crystal structure, His231, Arg296, and Lys299 were identified as prenyl donor contacts; and Arg160, Ala165, and Gly344 were implicated in protein substrate binding in C. albicans GGTase. In 2008, the Beese group solved the crystal structure of C. albicans GGTase-I complexed with GGPP at 1.58 Å resolution [98]. A noted difference between the mammalian and C. albicans GGTase-I structures is the size of the "exit groove." In the structure of rat GGTase complexed with prenylated peptide and GGPP, reflecting an intermediate in the dissociation of prenylated product during the catalytic cycle (Figure 12.4B), the prenyl group of the product binds in the exit groove while the new GGPP molecule binds in the active site pocket [71]. In the C. albicans crystal structure, the exit groove is narrowed, suggesting that the geranylgeranyl group of the product could not be accommodated in the same binding mode as observed for the mammalian GGTase-I. Although no peptide substrate was bound in the solved crystal structure of C. albicans GGTase-I, comparison of the putative Ca_1a_2X binding site of C. albicans GGTase-I to that of the mammalian enzyme revealed significant variation in the identities of amino acids, and most of these are nonconservative changes. This could suggest that the C. albicans GGTase-I has a different substrate recognition pattern than the mammalian enzyme.

Biochemical studies of a small number of peptide substrates have been carried out to investigate fungal prenyltransferase substrate specificity. The Mazur group initially measured the reactivity of partially purified *C. albicans* GGTase-I and FTase with seven peptides containing a Ca_1a_2X sequence (CVIL, CVVL, CTIL, CAIL, CVLM, CVLS, and CVIA) from prenylated proteins in either *C. albicans* or *Saccharomyces cerevisiae* with the upstream sequence from *S. cerevisiae* Ras [37]. Five of these substrates were geranylgeranylated upon incubation with *C. albicans* GGTase-I, although the K_M values were considerably higher than those measured for mammalian GGTase-I [37]. The two inactive sequences, CVLS and CVIA, are prototypical mammalian FTase substrates. Additionally, *C. albicans* FTase catalyzed farnesylation of all seven Ca_1a_2X sequences. Unexpectedly, the FTase-catalyzed prenylation of the -CaaL substrates was comparable to that of GGTase-I, suggesting both that *C. albicans* FTase has broad substrate specificity and that there is significant overlap in substrate specificity between the two enzymes. In a follow-up study in 2000, Mazur and coworkers explored the function of a polybasic sequence upstream of the Ca_1a_2X using biotinylated peptide substrates [35]. Overall, the upstream polybasic region had little impact on k_{cat} values, while the K_M values decreased; for example, addition of multiple lysine residues lowered the value of K_M ~15-fold for the CVIL and CAIL peptide sequences. In addition, the polybasic region broadened the substrate pool for GGTase, enhancing reactivity for peptides, like -CVIM, that contain an amino acid other than L as the terminal amino acid. In general, peptides ending in residues other than leucine are poor substrates for *C. albicans* GGTase-I unless they contain an upstream polybasic region.

Substrate specificity of the fungal prenyltransferases was also investigated *in vivo*. FTase and GGTase-I were shown to be essential for *C. albicans* growth as determined by knockout of the shared α subunit [36]. The β subunits of both FTase and GGTase-I are essential for growth of *S. cerevisiae* but not *C. albicans*, possibly due to cross-prenylation [99,100]. In *S. cerevisiae*, loss of FTase can be overcome by growth at lower temperatures, and lack of GGTase-I activity can be circumvented by overexpression of the GGTase-I substrates Rho1p and Cdc42p [99,101]. To address the issue of *in vivo* cross-prenylation in *C. albicans*, the prenylation status of the GGTase-I substrates Rho1p (-CVVL) and Cdc42p (-CTIL) was assessed in the absence of GGTase-I protein expression [38]. Under such conditions, Rho1p and Cdc42p were prenylated, as determined by localization of the proteins in the membrane fraction of the cells. This reaction was presumably catalyzed by FTase. These data suggest that a prenyltransferase inhibitor with dual specificity for FTase and GGTase-I would be necessary to obtain *in vivo* antifungal activity against *C. albicans*.

B. *L. PNEUMOPHILA*: UTILIZING HOST PRENYLTRANSFERASE MACHINERY

Recently, prenylation of proteins encoded by a Gram-negative pathogenic bacterium, *L. pneumophila*, has been discovered [41,42]. In this organism, the Dot/Icm secretion system of the bacterium injects into the host cell ~ 200 effectors that modulate and reprogram a number of host cellular processes to both allow bacterial proliferation as well as to induce host cell apoptosis [102]. One such effector, the F-box effector Ankyrin B (AnkB) protein, is essential for proliferation of *L. pneumophila* in mammalian cells and for manifestation of pulmonary disease in the mouse model of Legionnaires' disease [103]. AnkB functions as a scaffold for the docking of polyubiquitinated proteins to the *Legionella*-containing vacuole (LCV) membrane to enable bacterial intravacuolar proliferation in macrophages [104,105]. Recent work has shown that prenylation of the AnkB protein is essential for anchoring the protein to the LCV membrane. Further, prenylation is catalyzed by the host pathway [42], as evidenced by abolishment of membrane localization of AnkB in the presence of mammalian FTase and GGTase-I inhibitors and FTase-specific RNAi treatment. *L. pneumophila* does not encode any sequences homologous to the mammalian Ca_1a_2X-motif specific prenyltransferases in its genome and thus proteins that require lipidation for their biological function must be translocated into the host cytosol by the Dot/Icm type IV secretion system for modification. Addition of the Ca_1a_2X motif of *L. pneumophila* AnkB (CLVC) to a cytosolic protein is sufficient to localize the protein to the membrane, presumably due to prenylation. Recently, 10 proteins containing a C-terminal Ca_1a_2X motif have been identified in several *L. pneumophila* strains, including eight unique Ca_1a_2X sequences (such as CLVC for AnkB) [41]. Using GFP-Ca_1a_2X protein fusions, the membrane localization of these fusion proteins were verified, presumably due to prenylation of the Ca_1a_2X sequence. Consistent with this, most of these Ca_1a_2X sequences had previously been identified as a substrate for either mammalian FTase (CLVC, CVIS) or both FTase and GGTase-I (CSIL, CNLL, CVLM, CTIM, CSIL). Overall, these studies suggest that prenylation of *L. pneumophila* effector proteins is carried out by host farnesylation machinery. Blocking the membrane localization of these effector proteins by prenyltransferase inhibitors, and thus potentially disrupting effector protein function, may possibly lead to a novel approach for antibacterial treatment.

V. Conclusions

Multiple approaches have been employed to identify proteins that are prenylated *in vivo*, catalyzed by FTase and GGTase-I. Although prenyltransferase substrates can be identified and studied one at time, FPP and

GGPP analogs are being developed to allow detection of farnesylated and geranylgeranylated proteins in a "high-throughput" fashion. Further, structural and structure–function studies have provided insight into the interactions between substrates and the FTase and GGTase enzymes that are important for molecular recognition. Peptide library studies have supplied data regarding the amino acid composition of substrates and nonsubstrate pools, allowing the use of statistical analysis to determine patterns of FTase recognition and indicating that the Ca_1a_2X paradigm does not sufficiently describe the recognition of substrates by FTase. Further, these studies have uncovered the puzzle of the reactivity of STO substrates. Computational studies have allowed prediction of substrates *in vivo*, based upon both the C-terminus of the protein as well as the upstream region. All these studies in concert have helped to define the current pool of proteins known to be prenyltransferase substrates. Table S1 in reference [15] is a list of known prenylated substrates, identified by a variety of methods, highlighting the rapid identification of many proteins in the "prenylome." Through application and refinement of the approaches detailed in this chapter, identification of the entire prenylated proteome appears to a realistic goal in the next decade, if not sooner. Definition of the prenylated proteome will be essential to better understand the roles of prenylation in cellular signaling, disease processes, and the complex array of posttranslational modifications within the cell.

ACKNOWLEDGMENTS

We thank members of the Fierke laboratory for helpful comments and suggestions on the chapter. We especially thank Katherine A. Hicks, Heather L. Hartman, Rebekah A. Kelly, and Terry J. Watt for their work on the peptide library studies. This work was supported by National Institutes of Health (NIH) grant GM40602 (C. A. F.), PSTP training grant GM07767 (E. A. Z.), and NIH postdoctoral fellowship GM78894 (J. L. H.).

REFERENCES

1. Zhang, F.L., and Casey, P.J. (1996). Protein prenylation: molecular mechanisms and functional consequences. *Annu Rev Biochem* 65:241–269.
2. Benetka, W., Koranda, M., and Eisenhaber, F. (2006). Protein prenylation: an (almost) comprehensive overview on discovery history, enzymology, and significance in physiology and disease. *Monatsh Chem Chem Mon* 137:1241–1281.
3. Marshall, C.J. (1993). Protein prenylation: a mediator of protein–protein interactions. *Science* 259:1865–1866.
4. Casey, P.J. (1994). Lipid modifications of G proteins. *Curr Opin Cell Biol* 6:219–225.
5. Casey, P.J., and Seabra, M.C. (1996). Protein prenyltransferases. *J Biol Chem* 271:5289–5292.

6. Winter-Vann, A.M., and Casey, P.J. (2005). Post-prenylation-processing enzymes as new targets in oncogenesis. *Nat Rev Cancer* 5:405–412.
7. Barrowman, J., and Michaelis, S. (2009). ZMPSTE24, an integral membrane zinc metalloprotease with a connection to progeroid disorders. *Biol Chem* 390:761–773.
8. Trueblood, C.E., Boyartchuk, V.L., Picologlou, E.A., Rozema, D., Poulter, C.D., and Rine, J. (2000). The CaaX proteases, Afc1p and Rce1p, have overlapping but distinct substrate specificities. *Mol Cell Biol* 20:4381–4392.
9. Seabra, M.C., Reiss, Y., Casey, P.J., Brown, M.S., and Goldstein, J.L. (1991). Protein farnesyltransferase and geranylgeranyltransferase share a common alpha subunit. *Cell* 65:429–434.
10. Lane, K.T., and Beese, L.S. (2006). Thematic review series: lipid posttranslational modifications. Structural biology of protein farnesyltransferase and geranylgeranyltransferase type I. *J Lipid Res* 47:681–699.
11. Kho, Y., *et al.* (2004). A tagging-via-substrate technology for detection and proteomics of farnesylated proteins. *Proc Natl Acad Sci USA* 101:12479–12484.
12. Maurer-Stroh, S., and Eisenhaber, F. (2005). Refinement and prediction of protein prenylation motifs. *Genome Biol* 6:R55.
13. Maurer-Stroh, S., Koranda, M., Benetka, W., Schneider, G., Sirota, F.L., and Eisenhaber, F. (2007). Towards complete sets of farnesylated and geranylgeranylated proteins. *PLoS Comput Biol* 3:e66.
14. Reid, T.S., Terry, K.L., Casey, P.J., and Beese, L.S. (2004). Crystallographic analysis of CaaX prenyltransferases complexed with substrates defines rules of protein substrate selectivity. *J Mol Biol* 343:417–433.
15. Hougland, J.L., Hicks, K.A., Hartman, H.L., Kelly, R.A., Watt, T.J., and Fierke, C.A. (2010). Identification of novel peptide substrates for protein farnesyltransferase reveals two substrate classes with distinct sequence selectivities. *J Mol Biol* 395:176–190.
16. Samuel, F., and Hynds, D.L. (2010). RHO GTPase signaling for axon extension: is prenylation important? *Mol Neurobiol* 42:133–142.
17. Rusinol, A.E., and Sinensky, M.S. (2006). Farnesylated lamins, progeroid syndromes and farnesyl transferase inhibitors. *J Cell Sci* 119:3265–3272.
18. Basso, A.D., Kirschmeier, P., and Bishop, W.R. (2006). Lipid posttranslational modifications. Farnesyl transferase inhibitors. *J Lipid Res* 47:15–31.
19. Sousa, S.F., Fernandes, P.A., and Ramos, M.J. (2008). Farnesyltransferase inhibitors: a detailed chemical view on an elusive biological problem. *Curr Med Chem* 15:1478–1492.
20. Young, S.G., Meta, M., Yang, S.H., and Fong, L.G. (2006). Prelamin A farnesylation and progeroid syndromes. *J Biol Chem* 281:39741–39745.
21. Nallan, L., *et al.* (2005). Protein farnesyltransferase inhibitors exhibit potent antimalarial activity. *J Med Chem* 48:3704–3713.
22. Sepp-Lorenzino, L., *et al.* (1995). A peptidomimetic inhibitor of farnesyl:protein transferase blocks the anchorage-dependent and -independent growth of human tumor cell lines. *Cancer Res* 55:5302–5309.
23. Nonaka, M., *et al.* (2009). Role for protein geranylgeranylation in adult T-cell leukemia cell survival. *Exp Cell Res* 315:141–150.
24. Chiba, Y., Sato, S., Hanazaki, M., Sakai, H., and Misawa, M. (2009). Inhibition of geranylgeranyltransferase inhibits bronchial smooth muscle hyperresponsiveness in mice. *Am J Physiol Lung Cell Mol Physiol* 297:L984–L991.
25. Ridker, P.M., *et al.* (2008). Rosuvastatin to prevent vascular events in men and women with elevated c-reactive protein. *N Engl J Med* 359(21):2195–2207.
26. Casey, P.J. (1995). Protein lipidation in cell signaling. *Science* 268:221–225.

27. Caplin, B.E., Hettich, L.A., and Marshall, M.S. (1994). Substrate characterization of the *Saccharomyces cerevisiae* protein farnesyltransferase and type-I protein geranylgeranyl-transferase. *Biochim Biophys Acta* 1205:39–48.

28. Omer, C.A., *et al.* (1993). Characterization of recombinant human farnesyl-protein trans-ferase: cloning, expression, farnesyl diphosphate binding, and functional homology with yeast prenyl-protein transferases. *Biochemistry* 32:5167–5176.

29. Reiss, Y., Stradley, S.J., Gierasch, L.M., Brown, M.S., and Goldstein, J.L. (1991). Sequence requirement for peptide recognition by rat brain p21ras protein farnesyltrans-ferase. *Proc Natl Acad Sci USA* 88:732–736.

30. Moores, S.L., *et al.* (1991). Sequence dependence of protein isoprenylation. *J Biol Chem* 266:14603–14610.

31. Hicks, K.A., Hartman, H.L., and Fierke, C.A. (2005). Upstream polybasic region in peptides enhances dual specificity for prenylation by both farnesyltransferase and ger-anylgeranyltransferase type I. *Biochemistry* 44:15325–15333.

32. Gelb, M.H., *et al.* (2003). Protein farnesyl and N-myristoyl transferases: piggy-back medicinal chemistry targets for the development of antitrypanosomatid and antimalarial therapeutics. *Mol Biochem Parasitol* 126:155–163.

33. Bulbule, V.J., Rivas, K., Verlinde, C.L., Van Voorhis, W.C., and Gelb, M.H. (2008). 2-Oxotetrahydroquinoline-based antimalarials with high potency and metabolic stability. *J Med Chem* 51:384–387.

34. Fletcher, S., *et al.* (2008). Potent, plasmodium-selective farnesyltransferase inhibitors that arrest the growth of malaria parasites: structure–activity relationships of ethylenedia-mine-analogue scaffolds and homology model validation. *J Med Chem* 51:5176–5197.

35. Smalera, I., Williamson, J.M., Baginsky, W., Leiting, B., and Mazur, P. (2000). Expression and characterization of protein geranylgeranyltransferase type I from the pathogenic yeast *Candida albicans* and identification of yeast selective enzyme inhibitors. *Biochim Biophys Acta* 1480:132–144.

36. Song, J.L., and White, T.C. (2003). RAM2: an essential gene in the prenylation pathway of *Candida albicans*. *Microbiology* 149:249–259.

37. Mazur, P., *et al.* (1999). Purification of geranylgeranyltransferase I from *Candida albicans* and cloning of the CaRAM2 and CaCDC43 genes encoding its subunits. *Microbiology* 145(Pt 5):1123–1135.

38. Kelly, R., *et al.* (2000). Geranylgeranyltransferase I of *Candida albicans*: null mutants or enzyme inhibitors produce unexpected phenotypes. *J Bacteriol* 182:704–713.

39. Eastman, R.T., Buckner, F.S., Yokoyama, K., Gelb, M.H., and Van Voorhis, W.C. (2006). Thematic review series: lipid posttranslational modifications. Fighting parasitic disease by blocking protein farnesylation. *J Lipid Res* 47:233–240.

40. Eastman, R.T., *et al.* (2007). Resistance mutations at the lipid substrate binding site of *Plasmodium falciparum* protein farnesyltransferase. *Mol Biochem Parasitol* 152:66–71.

41. Ivanov, S.S., Charron, G., Hang, H.C., and Roy, C.R. (2010). Lipidation by the host prenyltransferase machinery facilitates membrane localization of *Legionella pneumo-phila* effector proteins. *J Biol Chem* 285:34686–34698.

42. Price, C.T., Al-Quadan, T., Santic, M., Jones, S.C., and Abu Kwaik, Y. (2010). Exploita-tion of conserved eukaryotic host cell farnesylation machinery by an F-box effector of *Legionella pneumophila*. *J Exp Med* 207:1713–1726.

43. Hancock, J.F. (1995). Reticulocyte lysate assay for in vitro translation and posttransla-tional modification of Ras proteins. *Methods Enzymol* 255:60–65.

44. Wilson, A.L., and Maltese, W.A. (1995). Coupled Translation/Prenylation of Rab Proteins in-Vitro. Lipid Modifications of Proteins. Academic Press Inc., San Diego, Vol. 250, pp. 79–91.

45. Gibbs, B.S., Zahn, T.J., Mu, Y., Sebolt-Leopold, J.S., and Gibbs, R.A. (1999). Novel farnesol and geranylgeraniol analogues: a potential new class of anticancer agents directed against protein prenylation. *J Med Chem* 42:3800–3808.
46. Peter, M., Chavrier, P., Nigg, E.A., and Zerial, M. (1992). Isoprenylation of Rab proteins on structurally distinct cysteine motifs. *J Cell Sci* 102:857–865.
47. Corsini, A., Farnsworth, C.C., McGeady, P., Gelb, M.H., and Glomset, J.A. (1999). Incorporation of radiolabeled prenyl alcohols and their analogs into mammalian cell proteins. A useful tool for studying protein prenylation. *Methods Mol Biol* 116:125–144.
48. Andres, D.A., Crick, D.C., Finlin, B.S., and Waechter, C.J. (1999). Rapid identification of cysteine-linked isoprenyl groups by metabolic labeling with [3H]farnesol and [3H]geranylgeraniol. *Methods Mol Biol* 116:107–123.
49. Benetka, W., Koranda, M., Maurer-Stroh, S., Pittner, F., and Eisenhaber, F. (2006). Farnesylation or geranylgeranylation? Efficient assays for testing protein prenylation in vitro and in vivo *BMC Biochem* 7:6.
50. Liu, Z.H., *et al.* (2009). Membrane-associated farnesylated UCH-L1 promotes alpha-synuclein neurotoxicity and is a therapeutic target for Parkinson's disease. *Proc Natl Acad Sci USA* 106:4635–4640.
51. Baron, R., *et al.* (2000). RhoB prenylation is driven by the three carboxyl-terminal amino acids of the protein: evidenced in vivo by an anti-farnesyl cysteine antibody. *Proc Natl Acad Sci USA* 97:11626–11631.
52. Lin, H.P., Hsu, S.C., Wu, J.C., Sheen, I.J., Yan, B.S., and Syu, W.J. (1999). Localization of isoprenylated antigen of hepatitis delta virus by anti-farnesyl antibodies. *J Gen Virol* 80 (Pt 1):91–96.
53. Liu, X.H., Suh, D.Y., Call, J., and Prestwich, G.D. (2004). Antigenic prenylated peptide conjugates and polyclonal antibodies to detect protein prenylation. *Bioconjug Chem* 15:270–277.
54. Troutman, J.M., Roberts, M.J., Andres, D.A., and Spielmann, H.P. (2005). Tools to analyze protein farnesylation in cells. *Bioconjug Chem* 16:1209–1217.
55. Nguyen, U.T., *et al.* (2009). Analysis of the eukaryotic prenylome by isoprenoid affinity tagging. *Nat Chem Biol* 5:227–235.
56. Nguyen, U.T.T., Goody, R.S., and Alexandrov, K. (2010). Understanding and exploiting protein prenyltransferases. *Chembiochem* 11:1194–1201.
57. Chehade, K.A., Andres, D.A., Morimoto, H., and Spielmann, H.P. (2000). Design and synthesis of a transferable farnesyl pyrophosphate analogue to Ras by protein farnesyltransferase. *J Org Chem* 65:3027–3033.
58. Onono, F.O., *et al.* (2010). A tagging-via-substrate approach to detect the farnesylated proteome using two-dimensional electrophoresis coupled with Western blotting. *Mol Cell Proteomics* 9:742–751.
59. Saxon, E., and Bertozzi, C.R. (2000). Cell surface engineering by a modified Staudinger reaction. *Science* 287:2007–2010.
60. Moses, J.E., and Moorhouse, A.D. (2007). The growing applications of click chemistry. *Chem Soc Rev* 36:1249–1262.
61. Chan, L.N., *et al.* (2009). A novel approach to tag and identify geranylgeranylated proteins. *Electrophoresis* 30:3598–3606.
62. Berry, A.F., *et al.* (2010). Rapid multilabel detection of geranylgeranylated proteins by using bioorthogonal ligation chemistry. *Chembiochem* 11:771–773.
63. Charron, G., Tsou, L.K., Maguire, W., Yount, J.S., and Hang, H.C. (2011). Alkynylfarnesol reporters for detection of protein S-prenylation in cells. *Mol Biosyst* 7:67–73.
64. DeGraw, A.J., *et al.* (2010). Evaluation of alkyne-modified isoprenoids as chemical reporters of protein prenylation. *Chem Biol Drug Des* 76:460–471.

65. Labadie, G.R., Viswanathan, R., and Poulter, C.D. (2007). Farnesyl diphosphate analogues with omega-bioorthogonal azide and alkyne functional groups for protein farnesyl transferase-catalyzed ligation reactions. *J Org Chem* 72:9291–9297.

66. Reigard, S.A., Zahn, T.J., Haworth, K.B., Hicks, K.A., Fierke, C.A., and Gibbs, R.A. (2005). Interplay of isoprenoid and peptide substrate specificity in protein farnesyltransferase. *Biochemistry* 44:11214–11223.

67. Krzysiak, A.J., *et al.* (2007). Combinatorial modulation of protein prenylation. *ACS Chem Biol* 2:385–389.

68. Troutman, J.M., Subramanian, T., Andres, D.A., and Spielmann, H.P. (2007). Selective modification of CaaX peptides with ortho-substituted anilinogeranyl lipids by protein farnesyl transferase: competitive substrates and potent inhibitors from a library of farnesyl diphosphate analogues. *Biochemistry* 46:11310–11321.

69. Terry, K.L., Casey, P.J., and Beese, L.S. (2006). Conversion of protein farnesyltransferase to a geranylgeranyltransferase. *Biochemistry* 45:9746–9755.

70. Luckman, S.P., Hughes, D.E., Coxon, F.P., Russell, R.G.G., and Rogers, M.J. (1998). Nitrogen-containing bisphosphonates inhibit the mevalonate pathway and prevent posttranslational prenylation of GTP-binding proteins, including Ras. *J Bone Miner Res* 13:581–589.

71. Taylor, J.S., Reid, T.S., Terry, K.L., Casey, P.J., and Beese, L.S. (2003). Structure of mammalian protein geranylgeranyltransferase type-I. *EMBO J* 22:5963–5974.

72. Long, S.B., Casey, P.J., and Beese, L.S. (2002). Reaction path of protein farnesyltransferase at atomic resolution. *Nature* 419:645–650.

73. Hartman, H.L., Hicks, K.A., and Fierke, C.A. (2005). Peptide specificity of protein prenyltransferases is determined mainly by reactivity rather than binding affinity. *Biochemistry* 44:15314–15324.

74. Krzysiak, A.J., Aditya, A.V., Hougland, J.L., Fierke, C.A., and Gibbs, R.A. (2009). Synthesis and screening of a CaaL peptide library versus FTase reveals a surprising number of substrates. *Bioorg Med Chem Lett* 20:767–770.

75. Hougland, J.L., Lamphear, C.L., Scott, S.A., Gibbs, R.A., and Fierke, C.A. (2009). Context-dependent substrate recognition by protein farnesyltransferase. *Biochemistry* 48:1691–1701.

76. Goldstein, J.L., Brown, M.S., Stradley, S.J., Reiss, Y., and Gierasch, L.M. (1991). Non-farnesylated tetrapeptide inhibitors of protein farnesyltransferase. *J Biol Chem* 266:15575–15578.

77. Pompliano, D.L., Gomez, R.P., and Anthony, N.J. (1992). Intramolecular fluorescence enhancement—a continuous assay of ras farnesyl—protein transferase. *J Am Chem Soc* 114:7945–7946.

78. Hightower, K.E., Huang, C.C., Casey, P.J., and Fierke, C.A. (1998). H-Ras peptide and protein substrates bind protein farnesyltransferase as an ionized thiolate. *Biochemistry* 37:15555–15562.

79. Cassidy, P.B., Dolence, J.M., and Poulter, C.D. (1995). Continuous fluorescence assay for protein prenyltransferases. *Methods Enzymol* 250:30–43.

80. Boutin, J.A., *et al.* (1999). Investigation of S-farnesyl transferase substrate specificity with combinatorial tetrapeptide libraries. *Cell Signal* 11:59–69.

81. Krzysiak, A.J., Scott, S.A., Hicks, K.A., Fierke, C.A., and Gibbs, R.A. (2007). Evaluation of protein farnesyltransferase substrate specificity using synthetic peptide libraries. *Bioorg Med Chem Lett* 17:5548–5551.

82. Fersht, A. (1999). Structure and Mechanism in Protein Science. W.H. Freeman and Company, New York, NY.

83. Pompliano, D.L., Schaber, M.D., Mosser, S.D., Omer, C.A., Shafer, J.A., and Gibbs, J.B. (1993). Isoprenoid diphosphate utilization by recombinant human farnesyl:protein transferase: interactive binding between substrates and a preferred kinetic pathway. *Biochemistry* 32:8341–8347.

84. Zimmerman, K.K., Scholten, J.D., Huang, C.C., Fierke, C.A., and Hupe, D.J. (1998). High-level expression of rat farnesyl:protein transferase in *Escherichia coli* as a translationally coupled heterodimer. *Protein Expr Purif* 14:395–402.

85. Pais, J.E., Bowers, K.E., Stoddard, A.K., and Fierke, C.A. (2005). A continuous fluorescent assay for protein prenyltransferases measuring diphosphate release. *Anal Biochem* 345:302–311.

86. Furfine, E.S., Leban, J.J., Landavazo, A., Moomaw, J.F., and Casey, P.J. (1995). Protein farnesyltransferase: kinetics of farnesyl pyrophosphate binding and product release. *Biochemistry* 34:6857–6862.

87. Tschantz, W.R., Furfine, E.S., and Casey, P.J. (1997). Substrate binding is required for release of product from mammalian protein farnesyltransferase. *J Biol Chem* 272:9989–9993.

88. Bowers, K.E., and Fierke, C.A. (2004). Positively charged side chains in protein farnesyltransferase enhance catalysis by stabilizing the formation of the diphosphate leaving group. *Biochemistry* 43:5256–5265.

89. Long, S.B., Hancock, P.J., Kral, A.M., Hellinga, H.W., and Beese, L.S. (2001). The crystal structure of human protein farnesyltransferase reveals the basis for inhibition by CaaX tetrapeptides and their mimetics. *Proc Natl Acad Sci USA* 98:12948–12953.

90. Pickett, J.S., *et al.* (2003). Kinetic studies of protein farnesyltransferase mutants establish active substrate conformation. *Biochemistry* 42:9741–9748.

91. Pais, J.E., Bowers, K.E., and Fierke, C.A. (2006). Measurement of the alpha-secondary kinetic isotope effect for the reaction catalyzed by mammalian protein farnesyltransferase. *J Am Chem Soc* 128:15086–15087.

92. Troutman, J.M., Andres, D.A., and Spielmann, H.P. (2007). Protein farnesyl transferase target selectivity is dependent upon peptide stimulated product release. *Biochemistry* 46:11299–11309.

93. Spence, R.A., Hightower, K.E., Terry, K.L., Beese, L.S., Fierke, C.A., and Casey, P.J. (2000). Conversion of Tyr361 beta to Leu in mammalian protein farnesyltransferase impairs product release but not substrate recognition. *Biochemistry* 39:13651–13659.

94. Falquet, L., *et al.* (2002). The PROSITE database, its status in 2002. *Nucleic Acids Res* 30:235–238.

95. McGeady, P., Logan, D.A., and Wansley, D.L. (2002). A protein-farnesyl transferase inhibitor interferes with the serum-induced conversion of *Candida albicans* from a cellular yeast form to a filamentous form. *FEMS Microbiol Lett* 213:41–44.

96. Murthi, K.K., Smith, S.E., Kluge, A.F., Bergnes, G., Bureau, P., and Berlin, V. (2003). Antifungal activity of a *Candida albicans* GGTase I inhibitor-alanine conjugate. Inhibition of Rho1p prenylation in *C. albicans*. *Bioorg Med Chem Lett* 13:1935–1937.

97. Nishimura, S., *et al.* (2003). Massadine, a novel geranylgeranyltransferase type I inhibitor from the marine sponge Stylissa aff. massa. *Org Lett* 5:2255–2257.

98. Hast, M.A., and Beese, L.S. (2008). Structure of protein geranylgeranyltransferase-I from the human pathogen *Candida albicans* complexed with a lipid substrate. *J Biol Chem* 283:31933–31940.

99. Trueblood, C.E., Ohya, Y., and Rine, J. (1993). Genetic evidence for in vivo cross-specificity of the CaaX-box protein prenyltransferases farnesyltransferase and geranylgeranyltransferase-I in *Saccharomyces cerevisiae*. *Mol Cell Biol* 13:4260–4275.

100. He, B., Chen, P., Chen, S.Y., Vancura, K.L., Michaelis, S., and Powers, S. (1991). RAM2, an essential gene of yeast, and RAM1 encode the two polypeptide components of the farnesyltransferase that prenylates a-factor and Ras proteins. *Proc Natl Acad Sci USA* 88:11373–11377.

101. Ohya, Y., Qadota, H., Anraku, Y., Pringle, J.R., and Botstein, D. (1993). Suppression of yeast geranylgeranyl transferase I defect by alternative prenylation of two target GTPases, Rho1p and Cdc42p. *Mol Biol Cell* 4:1017–1025.

102. Zink, S.D., Pedersen, L., Cianciotto, N.P., and Abu-Kwaik, Y. (2002). The Dot/Icm type IV secretion system of *Legionella pneumophila* is essential for the induction of apoptosis in human macrophages. *Infect Immun* 70:1657–1663.

103. Price, C.T., Al-Khodor, S., Al-Quadan, T., and Abu Kwaik, Y. (2010). Indispensable role for the eukaryotic-like ankyrin domains of the ankyrin B effector of *Legionella pneumophila* within macrophages and amoebae. *Infect Immun* 78:2079–2088.

104. Price, C.T., *et al.* (2009). Molecular mimicry by an F-box effector of *Legionella pneumophila* hijacks a conserved polyubiquitination machinery within macrophages and protozoa. *PLoS Pathog* 5:e1000704.

105. Al-Khodor, S., Price, C.T., Habyarimana, F., Kalia, A., and Abu Kwaik, Y. (2008). A Dot/Icm-translocated ankyrin protein of *Legionella pneumophila* is required for intracellular proliferation within human macrophages and protozoa. *Mol Microbiol* 70:908–923.

106. Long, S.B., Casey, P.J., and Beese, L.S. (2000). The basis for K-Ras4B binding specificity to protein farnesyltransferase revealed by 2 A resolution ternary complex structures. *Structure* 8:209–222.

107. Pompliano, D.L., Rands, E., Schaber, M.D., Mosser, S.D., Anthony, N.J., and Gibbs, J.B. (1992). Steady-state kinetic mechanism of Ras farnesyl:protein transferase. *Biochemistry* 31:3800–3807.

13

Structural Biochemistry of CaaX Protein Prenyltransferases

MICHAEL A. HAST • LORENA S. BEESE

Department of Biochemistry
Duke University Medical Center, Durham
North Carolina, USA

I.   Abstract

Protein prenylation is a posttranslational lipid modification required for proper function by over 100 proteins in the eukaryotic cell. A family of structurally related protein prenyltransferase enzymes carry out this reaction: protein farnesyltransferase (FTase), protein geranylgeranyltransferase-I (GGTase-I), and Rab geranylgeranyltransferase (GGTase-II or Rab GGTase). This chapter concerns the structural biology of Ca_1a_2X protein prenyltransferases (FTase and GGTase-I). These enzymes recognize a well-defined C-terminal motif on substrate proteins: cysteine (C), followed by two generally aliphatic amino acids (aa) and a variable (X) residue. FTase and GGTase-I catalyze the addition of a 15-carbon or 20-carbon isoprenoid lipid, respectively. FTase and GGTase-I have been shown to be important targets for the development of cancer chemotherapeutics because prenylated signal transduction proteins play significant roles in oncogenesis. More recently, it has been demonstrated that protein prenyltransferases also show promise as drug targets for treating a variety of infectious diseases caused by fungi and protozoans. We review the structural features of the enzymatic reaction cycle, and how these relate to the mechanisms of inhibition by small molecules. We also review recent structural studies of human pathogen protein prenyltransferases, and how this

THE ENZYMES, Vol. XXIX   235   ISSN NO: 1874-6047
DOI: 10.1016/B978-0-12-381339-8.00013-5

information can be used for developing species-specific prenyltransferase inhibitors to treat infectious diseases.

II. Protein Farnesyltransferase Structure and Reaction Cycle

The first crystal structure of mammalian protein farnesyltransferase (FTase) was reported in 1997 [1], revealing a number of structural features that have been shown to be critical for understanding the similarities and differences in the reaction mechanisms and substrate recognition of all protein prenyltransferases, thereby forming a framework for the development of specific inhibitors [2–5]. FTase is a zinc metalloenzyme α/β heterodimer with α and β subunits of 48 and 46 kDa, respectively. The α subunit comprises 15 α-helices, 14 of which are arranged in antiparallel pairs (Figure 13.1). These paired helices form a right-handed superhelical,

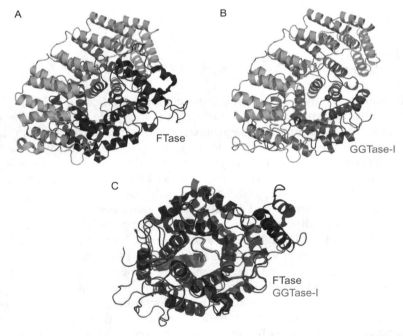

FIG. 13.1. (A) Protein farnesyltransferase heterodimer: red, α subunit; blue, β subunit; catalytic Zn^{2+}, pink. (B) Protein geranylgeranyltransferase-I heterodimer. The α subunit is shared with FTase. (C) Superposition of the β subunits of FTase (blue) and GGTase-I (purple) shows the high degree of structural homology. (See color plate section in the back of the book.)

crescent-shaped molecule that envelops the β subunit. The β subunit is also predominantly α-helical, with its 14 major α-helices arranged into an α–α barrel fold. About 3000 Å^2 are buried within the extensive interface between the α and β subunits. The heterodimer surrounds a large, funnel-shaped cavity, formed predominantly by hydrophobic residues contributed by the β subunit. This barrel is closed at one end and contains the active site of the enzyme near its opening [1,6,7].

A Zn^{2+} ion is bound by the β subunit at the rim of the barrel, marking the site at which chemistry is carried out by the enzyme (Figure 13.1). This Zn^{2+} is coordinated in a distorted pentacoordinate geometry by three residues: H362β, C299β, and D297β [1]. The interaction with the aspartic acid is bidentate, with 2.1 and 2.7 Å bond lengths between the Zn^{2+} and carboxylate oxygens. Although this coordination arrangement has been a subject of debate in the field [8], recent high-resolution crystal structures (1.5 Å resolution) of FTase support this model with little room for ambiguity [9]. The fifth vertex of the Zn^{2+} coordination sphere is open, and at the various stages of the reaction cycle (see below), is occupied alternately by the γ-sulfur of the cysteine residue of the Ca_1a_2X motif peptide substrate, the thioether of the prenylated peptide product, or a water molecule [6].

The active site cavity contains distinct clefts. Two clefts run roughly parallel to each other from the top of the barrel near the catalytic zinc into the barrel, and accommodate the isoprenoid and Ca_1a_2X substrates (Figure 13.2). The third cleft is formed by helices 2β, the loop linking helices 2β and 3β, and helix 13β and is termed the "exit groove," which binds the prenylated product [6].

The isoprenoid binding cleft is composed of predominantly aromatic residues that are highly conserved across species. The diphosphate binding pocket at the top of the cleft (adjacent to the catalytic Zn^{2+} ion) is lined with positively charged residues, a tyrosine, and a histidine. These residues make either direct or water-mediated hydrogen bonds to the diphosphate moiety of FPP [6,7,11]. In FTase, the FPP binding cleft is only large enough to accommodate a 15-carbon lipid chain, being restricted at the bottom of the pocket by a tryptophan residue (W102β). In GGTase-I, a much smaller threonine residue (T49β) replaces this tryptophan, thereby lengthening the pocket and enabling the longer 20-carbon GGPP to bind [12] (Figure 13.3). Mutation of W102β to threonine in FTase broadens the lipid substrate specificity, enabling the enzyme to function as a GGTase-I, while retaining its original Ca_1a_2X substrate preferences [9].

The Ca_1a_2X peptide binding site is located adjacent to the isoprenoid binding site. The substrate peptide is anchored at the cysteine of the Ca_1a_2X motif by the Zn^{2+} at the top of this cleft, and at its carboxy terminus by a

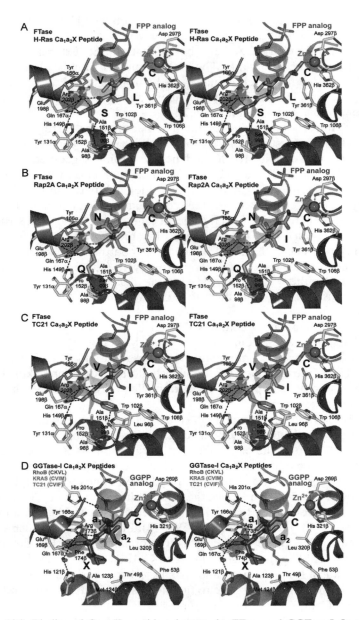

FIG. 13.2. Binding of Ca$_1$a$_2$X peptide substrates by FTase and GGTase-I. In all cases, Ca$_1$a$_2$X peptide adopts an extended conformation and is anchored at its cysteine by the catalytic zinc ion (pink), and by Q167α at its carboxy terminus. Peptide substrate makes

glutamine residue ($Q167\alpha$) at the bottom of the cleft (Figure 13.2). Between these two anchors points, a predominantly hydrophobic surface interacts with the side chains of the a_2 residue of the Ca_1a_2X peptide as well as the X residue [7,10]. The a_1 residue is oriented toward solvent in the crystal structures and does not contribute significantly to recognition by the enzyme (Figure 13.2).

Solution kinetic studies showed that substrate binding is strictly ordered [13]. The crystal structures of FTase with bound substrates suggest a molecular basis for this observation. A ternary complex between FTase, a nonreactive, high-affinity analog of FPP (FPT-II), and a CaaX substrate peptide revealed that these two substrates bind next to each other in the cavity, forming extensive van der Waals contacts between them [6,7]. The Ca_1a_2X substrate restricts the accessibility of the lipid substrate into the cavity. Further, the lipid forms part of the peptide binding site, consistent with the lipid binding first, followed by the peptide (Figure 13.2).

Crystal structures of mammalian FTase define the major steps along the reaction coordinate (Figure 13.3) [6]. In these studies, binary complexes (with lipid only) and ternary complexes (with a nonreactive lipid analog and Ca_1a_2X peptide) define the two major substrate binding grooves and the structural basis for Zn^{2+} ion function in orienting the Ca_1a_2X peptide and mediating the alkylation reaction. Solution studies have shown that the product release step of the FTase reaction cycle is the rate-limiting step (over 300-fold slower than the rate of chemical bond formation) [13]. The slow kinetic step of product release enabled the prenylated product complex to be captured in the crystal (Figure 13.3) [6]. In this product complex, the peptide binding conformation remains largely unchanged from the ternary

---

extensive van der Waals contact with lipid substrate (FPP analog or GGPP analog). The residue at the a_1 position is solvent exposed. Recognition of the Ca_1a_2X motif is controlled by the a_2- and X-residue binding pockets. Examples of Ca_1a_2X recognition by FTase and GGTase-I are shown in (A)–(D). (A) CVLS peptide by FTase. The leucine residue at a_2 position interacts with aromatic residues in this binding pocket. The serine at the X position makes a hydrogen bond to a water molecule bound to other residues in the binding pocket. (B) CNIQ peptide bound to FTase. Like leucine, isoleucine interacts with aromatic residues in the a_2-binding pocket. Glutamine at the X position makes a hydrogen bond with the side chain of $W102\beta$. (C) CVIF peptide bound to FTase. The phenylalanine side chain interacts with a secondary aromatic recognition pocket of X-residue-binding site that is distinct from the primary recognition pocket occupied by Ser and Gln (A and B). Phenylalanine is predicted to be excluded by the primary recognition pocket based on its size. (D) GGTase-I recognition of CaaX peptides. GGTase-I substrates CKVL, CVIM, and CVIF are superimposed. Unlike FTase, there is no secondary binding pocket for X-residues in GGTase-I. The GGTase-I a_2 residue binding site exhibits similar residue preferences to FTase. The X-residue binding site is aliphatic in character and preferentially binds leucine or isoleucine residues. Figure reproduced from Ref. [10] with permission. (See color plate section in the back of the book.)

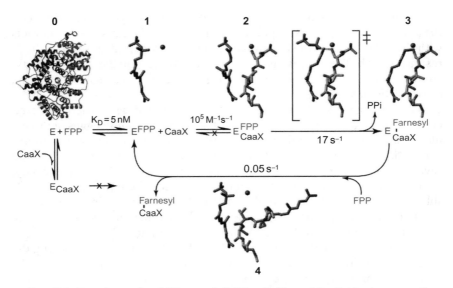

FIG. 13.3. Reaction cycle of FTase and GGTase-I. Blue sticks, lipid substrate; yellow, peptide; pink, catalytic Zn^{2+}. FTase and GGTase-I reaction chemistries are identical; only the cycle for FTase is shown. The stages of the reaction cycle captured in X-ray crystal structures are numbered 0–4. Ordered substrate binding starts with the lipid (1), followed by the Ca_1a_2X substrate (2). A proposed conformational change of the lipid in the transition state is modeled as an intermediate conformation between ternary substrate complex and isoprenylated product (3). The isoprenylated product is partially displaced to the exit groove (4) by the incoming lipid substrate for the next turn of the reaction cycle. Subsequently, the product is fully displaced, binding a new Ca_1a_2X substrate, and initiating the next cycle. Figure reproduced from Ref. [6] with permission. (See color plate section in the back of the book.)

complexes, although the coordination distance between the Zn^{2+} ion and the thioether product (2.7 Å) is longer than the substrate (2.3 Å). However, the first two isoprenes of the lipid make a conformational change compared to their substrate structures (Figure 13.3). In the ternary substrate complex, the C1 carbon of FPP is located ~ 7.5 Å away from the Zn^{2+} ion and does not support covalent bond formation. To bring the C1 carbon sufficiently close to the Zn^{2+} ion to react, the first two isoprenes must rotate and translate toward the Zn^{2+}-bound cysteine thiolate. Spectroscopic studies and kinetic isotope effect analysis of the chemical step also support a mechanism in which the isoprenoid substrate undergoes a conformational change, while the Ca_1a_2X peptide remains stationary during the reaction [14,15].

When crystals containing a farnesylated peptide product are soaked in excess isoprenoid substrate, an unusual, partially displaced product intermediate can be captured (Figure 13.3) [6]. In this complex, a fresh isoprenoid lipid substrate is bound in the same conformation and position as the lipid observed in the binary complex. The a_2 and X residues of the prenylated peptide remain in the conformation observed in either the lipid, Ca_1a_2X ternary substrate, or the prenylated peptide complexes. However, the lipid moiety of the farnesylated product is flipped into the third major cleft of the active site: the exit groove (Figure 13.3). This shallow, hydrophobic cavity is oriented perpendicular to the clefts that bind the two substrates. To accommodate the flipped farnesyl moiety, the cysteine and a_1 residues of the Ca_1a_2X peptide undergo a conformational change, thereby disrupting the interaction with the catalytic Zn^{2+} ion such that the thioether bond is nearly 9 Å displaced from this ion. These observations suggest that the product release mechanism is unusual, through displacement by substrate binding of the next reaction cycle [13]. Further, this mechanism may account for the slow product release kinetics and has been suggested to play a role in ordered handover between successive processing steps [6].

III.   Protein Geranylgeranyltransferase-I Structure and Reaction Cycle

The first crystal structures of mammalian GGTase-I, and reaction intermediates captured in the crystal, were determined in 2003 [12]. These observations enabled the determination structural features that are common to all Ca_1a_2X protein prenyltransferases, and those that determine the differences in peptide and lipid selection by GGTase-I.

The overall fold observed for GGTase-I is homologous to FTase, as was predicted by sequence alignments (Figure 13.1). The crescent of the α subunit of FTase and GGTase-I exhibits slightly different curvature in the two enzymes. This observation is attributed to the size differences of the FTase and GGTase-I β subunit (45 vs. 43 kDa, respectively), which is enveloped by the α subunit. The GGTase-I β subunit adopts an α–α barrel fold. The central, funnel-shaped active site is similarly proportioned to FTase [12]. The three major substrate binding clefts in GGTase-I (for the lipid, Ca_1a_2X, and displaced product) are readily identifiable, with the same relative spatial orientation observed in FTase (Figure 13.2). The catalytic Zn^{2+} ion also is coordinated with pentacoordinate geometry by conserved histidine (321β), aspartic acid (D269β), and cysteine (271β) residues.

GGPP is bound by the enzyme in a cleft similar to that observed in FTase, lined with highly conserved residues. The lipid diphosphate group also forms hydrogen bonds with basic residues, histidine and tyrosine.

The fourth isoprene unit of GGPP (absent in FPP) adopts a conformation that introduces a right angle relative to the first three isoprene units in the lipid chain (Figure 13.4). Although the GGPP binding site exhibits high conservation with the FPP binding site of FTase, a critical residue substitution at the bottom of the lipid binding pocket (W102β in FTase to T49β in GGTase-I) forms a pocket that is large enough to accommodate a 20-carbon isoprenoid substrate (Figure 13.4). This observation forms the basis of the "molecular ruler" hypothesis of isoprenoid discrimination by FTase and GGTase-I [9,12]. GGTase-I can accommodate the shorter FPP ligand; however, it transfers FPP with markedly reduced efficiency compared with its cognate substrate [9].

The reaction cycle of GGTase-I is largely indistinguishable from that of FTase [12]. In both cases, the lipid substrate initially binds in a conformation that is unproductive for chemistry, necessitating a conformational change for the reaction to take place. The conformation of the prenylated

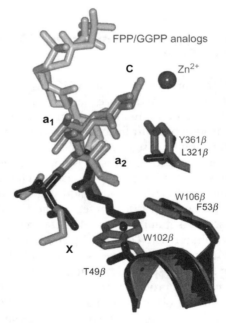

FIG. 13.4. Lipid substrate selection by FTase and GGTase-I. In FTase (red sticks and ribbon), the length of the lipid binding pocket is restricted by W102β. In GGTase-I (blue sticks and ribbon), the residue at the equivalent position is T49β. Substitution of the smaller residue in GGTase-I permits binding of the longer, 20-carbon GGPP substrate. Figure reproduced from Ref. [10] with permission. (See color plate section in the back of the book.)

peptide product is similar in GGTase-I and FTase (except for the differences in length between the two lipid moieties). A displaced product complex also can be captured in the exit groove when GGTase-I crystals containing prenylated product are soaked with excess isoprenoid substrate.

Unlike FTase, mammalian GGTase-I exhibits no dependence on Mg^{2+}. In the crystal structure of GGTase-I, it was noted that lysine (K311β) replaces D352β which binds Mg^{2+} [12]. Solution kinetics studies confirmed that the positively charged amine of the K311β side chain accelerates the chemical step GGTase-I (Figure 13.5) [16]. GGTase-I catalyzes prenylation at a rate nearly 40-fold slower than FTase, consistent with the decreased ability of a lysine residue to stabilize the transition state relative to a divalent metal ion. Mutagenesis studies (K311βA or K311βD) indicated

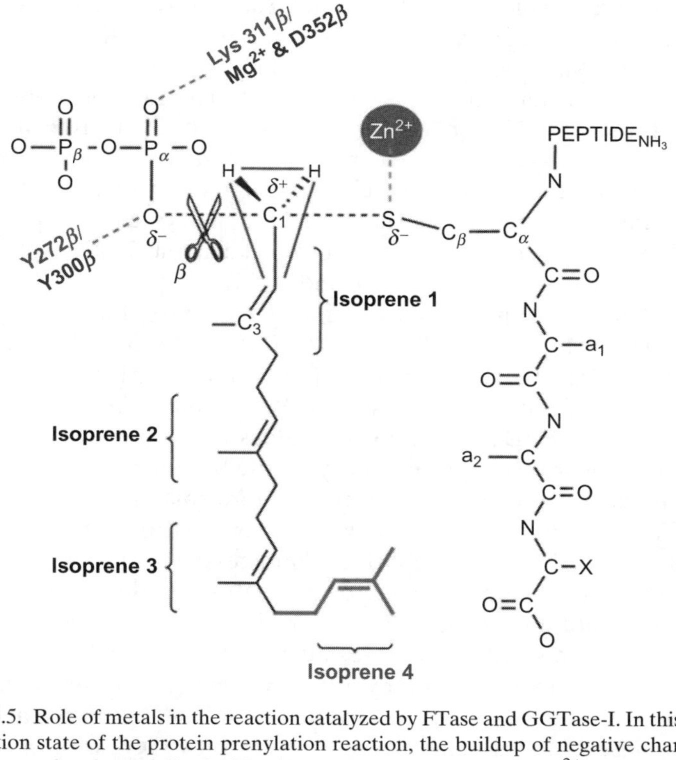

FIG. 13.5. Role of metals in the reaction catalyzed by FTase and GGTase-I. In this model of the transition state of the protein prenylation reaction, the buildup of negative charge on the diphosphate moiety is partially stabilized by a transiently associated Mg^{2+} ion in FTase, or the positively charged side chain of K311β in GGTase-I. The Zn^{2+} ion activates the cysteine thiolate for attack on the C1 carbon of the lipid. Figure reproduced from Ref. [12] with permission.

that deletion of the positive charge conferred Mg^{2+}-dependent rate acceleration to GGTase-I [16]. In the presence of Mg^{2+}, the K311βD mutant catalyzed the prenylation reaction faster than wild-type GGTase-I [16].

IV.   Determinants of Ca_1a_2X Substrate Selection in FTase and GGTase-I

Over 100 proteins in the eukaryotic cell possess a Ca_1a_2X motif that is prenylated by either FTase or GGTase-I [17]. FTase and GGTase-I exhibit both distinct and overlapping preferences for Ca_1a_2X peptide substrates [18–20]. The a_2 and X residues in the Ca_1a_2X sequence together comprise the recognition element for FTase and GGTase-I, with the X residue playing the largest role in specificity [10]. The a_1 residue is not specifically recognized by the enzymes, as it is oriented toward solvent.

In mammals, substrates with M, Q, S, T, and A at the X position are most efficiently farnesylated by FTase [5,10], whereas in most substrates of GGTase-I, the Ca_1a_2X sequence terminates in L or I [21]. Recent solution kinetics studies suggested that there is an interplay between the a_2 and X-position residues that can also affect the degree to which a given Ca_1a_2X sequence can be prenylated by the noncognate enzyme [22–24]. Structural studies have provided insight into the molecular mechanisms underlying Ca_1a_2X substrate selection by FTase and GGTase-I (Figure 13.2) [7,10,12].

Stereochemical complementarity of the FTase X-residue binding pocket (comprising Y131α, A98β, S99β, W102β, H149β, A151β, and P152β) determines the set of residues that can be easily accommodated (M, Q, S, T, and A; Figure 13.2) [10]. Bulky residues such as W, Y, and R are excluded, as are charged ones (K, E, D). Small polar residues (S, T) form water-mediated hydrogen bonds with the protein, burying the bridging water molecule. In general, FTase X-residue binding pocket is more polar in character, whereas the GGTase-I X-residue binding pocket is more hydrophobic (Figure 13.2) [10]. As in FTase, the shape and surface character of the GGTase-I X-residue pocket determines the preference for X-position residues. The shape and hydrophobic nature of the GGTase-I X-residue binding pocket accommodates aliphatic and small aromatic residues (L, I, F; Figure 13.2).

FTase and GGTase-I can prenylate noncognate substrates both *in vitro* and *in vivo*. K-Ras is normally prenylated by FTase, but in the absence of functional FTase, K-Ras can also be prenylated by GGTase-I *in vivo* [25]. This cross-reactivity has been postulated to be one mechanism by which some K-Ras-driven tumors are resistant to treatment by highly selective FTase inhibitors [25,26].

Interactions between the protein and the a_2 residue also encode substrate recognition features by FTase and GGTase-I. In FTase, the a_2 residue interacts with W102β, W106β, Y361β, and the third isoprene of FPP [7,10] (Figure 13.2). Although this pocket can accommodate amino acids of varying volumes and shapes, its aromatic character dictates a preference for L, F, I, and V at this position, with W and Y being too large to fit [10]. The a_2 pocket of GGTase also is composed of largely hydrophobic groups (isoprenes 3 and 4 of GGPP, F53β, L321β, and T49β) and exhibits similar preferences to FTase for I and L at this position [10,12] (Figure 13.2). Structural studies have led to the development of a predictive model for assigning Ca_1a_2X substrates to either GGTase-I or FTase based upon sequence [10]. Many solution studies of peptide libraries have largely affirmed the strength of this predictive model [23,24].

Recent studies have further subdivided substrates of FTase into two categories: multiple turnover (MTO) substrates and single turnover (STO) substrates [22]. The MTO substrates were farnesylated under standard stead-state conditions (where $[E] \ll [S]$), whereas STO substrates are farnesylated only if $[E] > [S]$, leading to a single prenylation event. The CaaX sequences of most MTO substrates conform quite well to the predictive models outlined by structural studies. However, the STO substrates included sequences that were not predicted to be substrates for FTase based upon structural predictive models. The differences between STO and MTO behavior may reflect product release steps. As mentioned previously, product release by FTase and GGTase-I is accelerated by the addition of a fresh substrate molecule. STO products may interact with the enzyme with much higher affinity than observed in MTO reactions and therefore cannot be efficiently displaced by the next STO substrate, or the STO substrates could have weaker affinity.

V.   Structure of *Candida albicans* Protein Geranylgeranyltransferase-I

The opportunistic fungus *C. albicans* is the most common fungal infection found in the U.S. hospitals, and the incidence of its drug-resistant infections is rapidly rising [27]. Deletion of the FTase and GGTase-I α subunit is lethal, confirming an essential role for prenylated proteins in *C. albicans* [28]. This protein can be produced by heterologous expression in *Escherichia coli*, enabling determination of its kinetic properties and testing of inhibitors [29–32]. The crystal structure of *C. albicans* GGTase-I (CaGGTase-I) has been determined recently [33].

CaGGTase-I is the first nonmammalian protein prenyltransferase to be studied structurally. The overall fold of the heterodimer and locations of

binding grooves for substrates are conserved compared to higher eukaryotes (Figure 13.6). Many residues in the Ca_1a_2X substrate binding pocket
are nonconservatively substituted in CaGGTase-I, compared with mammalian GGTase-I (Figure 13.6). The residues in the Ca_1a_2X X-residue binding
pocket are nearly all different from the corresponding residues in

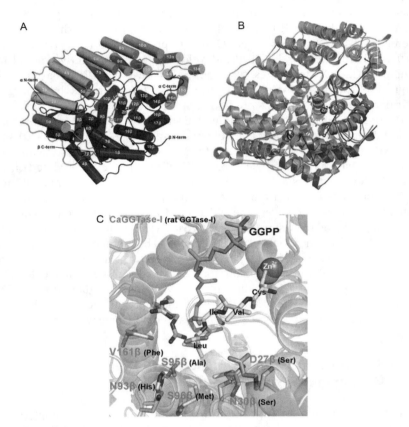

Fig. 13.6. Structure of *C. albicans* GGTase-I. (A) Structural overview of CaGGTase-I.
Red, α subunit; blue, β subunit; green sticks, lipid; pink, catalytic Zn^{2+}. (B) Superposition of
CaGGTase-I (blue, red) with the mammalian GGTase-I (gray, PDB ID 1N4P) illustrates the
structural homology of the fold. (C) Model of ternary substrate complex of GGPP (green
sticks) and Ca_1a_2X peptide (yellow sticks) in the active site of CaGGTase-I (green ribbon,
orange sticks). For comparison, mammalian GGTase-I is superimposed (gray ribbon, gray
sticks). The composition of the Ca_1a_2X X-residue-recognition site is completely different
between CaGGTase-I and mammalian GGTase-I, indicating significant opportunities for the
design of inhibitors that are selective for the fungal enzyme. Figure reproduced from Ref. [33]
with permission. (See color plate section in the back of the book.)

mammalian GGTase-I. Nevertheless, the substrate preference of CaGG-Tase-I, as assayed by limited steady-state kinetics studies, appears to be similar to that of mammalian GGTase-I, suggesting a level of degeneracy in recognition of Ca_1a_2X sequences. Further structural studies will permit elucidation of the structural basis of Ca_1a_2X selection in CaGGTase-I.

Unlike mammalian GGTase-I, CaGGTase-I activity is dependent on Mg^{2+} [30]. Although both enzymes possess positively charged side chains at equivalent sequence positions near the lipid diphosphate, R339β of CaGGTase-I is not structurally equivalent to K311β in mammalian GGTase-I, which partially replaces Mg^{2+}. The R339β in CaGGTase-I does not point toward the diphosphate moiety of GGPP; the lack of electron density for this residue in the structure suggests multiple conformations [33]. The neighboring residue in CaGGTase-I is an aspartate (D340β), which may interact with Mg^{2+}, but which is not observed in the structure.

The prenylated product exit groove in CaGGTase-I is narrower than that of mammalian GGTase-I and diverges in sequence. Although the rates of product release of CaGGTase-I have not yet been experimentally tested, the structure predicts that kinetic parameters may be altered compared with the mammalian ortholog. Taken together, the structural variations observed in the active site of CaGGTase-I are sufficiently distinct from the human enzyme that it may be possible to develop species-specific inhibitors of this enzyme, which may lead to new classes of antifungal therapeutics.

VI.   Inhibitors of Protein Prenyltransferases as Cancer Chemotherapeutics

The discovery and development of protein prenyltransferase inhibitors for the treatment of cancer have been extensively reviewed [5,34–37]. Understanding the structural determinants of inhibitor binding in mammalian FTase and GGTase-I provides a framework for recent studies of the structure–activity relationships of new series of farnesyltransferase inhibitors (FTIs) targeting both the human enzyme and FTase enzymes from human pathogens.

A.   CA₁A₂X PEPTIDOMIMETIC INHIBITORS

Early studies characterizing the activity of protein prenyltransferases showed that several CaaX tetrapeptides inhibited the enzymes. Although peptides are poor drug candidates, scaffolds that mimic a tetrapeptide motif were designed to inhibit protein prenyltransferases [38,39]. Specific FTase and GGTase-I inhibitors can be designed by changing the identity of the substituents at the positions that correspond to the a_2 and X residues in the Ca_1a_2X mimetic. Two interactions with the enzyme that mimic those of native Ca_1a_2X substrates are common to all Ca_1a_2X mimetics: a thiol or a imidazole moiety coordinates the catalytic Zn^{2+}, and a carboxyl group binds $Q167\alpha$. The inhibitors adopt an extended conformation in the active site of the enzymes and take advantage of extensive interactions in the a_2- and X-residue binding pockets. The Merck peptidomimetic inhibitor L-739,750, the active metabolite of the prodrug compound L-744,832 (Figure 13.7) [40], was among the first to show potent tumor regression in mice, with few observable biological toxicities [41]. L-744,832 possesses an ester-linked group on the carboxy terminus that assists in cell permeability. Following uptake, this group is later cleaved by cellular esterases to yield an active inhibitor. L-739,750 is a mimetic of the Ca_1a_2X sequence CIFM. The benzyl substituent of the inhibitor binds in the largely aromatic a_2 residue pocket analogous to a phenylalanine side chain in the a_2 position of a Ca_1a_2X substrate (Figure 13.7) [40]. At the equivalent of the X-residue position in the L-739,750 peptidomimetic, a sulfonyl-containing moiety mimics a methionine side chain. However, the sulfonyl forms stronger hydrogen bonds with the enzyme than the thioether of a methionine side chain. The pharmacokinetic properties of peptidomimetic compounds such as L-744,832 *in vivo* have limited further development [38,42].

B.   OTHER SMALL MOLECULE PRENYLTRANSFERASE INHIBITORS

By far, the most clinically successful and widely studied inhibitors of protein prenyltransferases are small molecules that are not substrate mimetics. Here, we describe the structural features of four such FTIs that have advanced into the clinic for the treatment of cancer (L-778,123, R115777, SCH66336, and BMS-214662). Comparison of crystal structures of all four FTIs bound to FTase reveals common themes of molecular recognition [2–4]. In principle, inhibitors of protein prenyltransferases can bind to the Zn^{2+}, the lipid binding groove, the CaaX substrate binding site, or the prenylated product exit groove. The inhibitors described here explore various aspects of these binding modes.

Originally developed by Merck, L-778,123 is one of the few inhibitors of FTase that also potently inhibits GGTase-I [2,43,44]. L-778,123 can be divided into three functional parts: an imidazole ring, a chlorobenzyl ring, and a benzonitrile ring. In both FTase and GGTase-I, the imidazole binds to the catalytic Zn^{2+} ion (Figure 13.8). However, the other two functional groups bind in distinct conformations, which can be best described as a 30° pivot relative to each other around the Zn^{2+} (Figure 13.8) [2]. In FTase, this

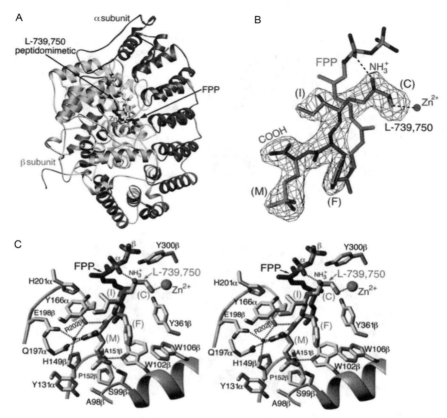

FIG. 13.7. Binding of Ca_1a_2X peptidomimetic inhibitor L-739,750 in the active site of FTase. L-739,750 is a mimetic of the Ca_1a_2X sequence CIFM. (A) Structural overview of FTase, with the α subunit in red and β subunit in blue. Inhibitor binds along with lipid substrate FPP in the active site in center of the β subunit. (B) Omit electron density enables unambiguous modeling of peptidomimetic inhibitor binding mode. (C) Stereo image of L-739,750 binding in FTase active site. Benzyl moiety of inhibitor binds in the Ca_1a_2X a_2 residue binding site, stabilized by aromatic interactions with tryptophan residues. The moiety mimicking the methionine adopts a similar conformation to the native substrate in the X-residue binding site and makes a hydrogen bond with the side chain of S99β. Figure reproduced from Ref. [40] with permission. (See color plate section in the back of the book.)

places the chlorophenyl group into the a_2 residue binding pocket, thereby blocking peptide binding (Figure 13.8). The benzonitrile group makes extensive interactions with the lipid chain of a bound FPP substrate. In GGTase-I, neither lipid nor peptide can bind in the presence of this inhibitor. In this enzyme, both the benzonitrile and chlorophenyl groups

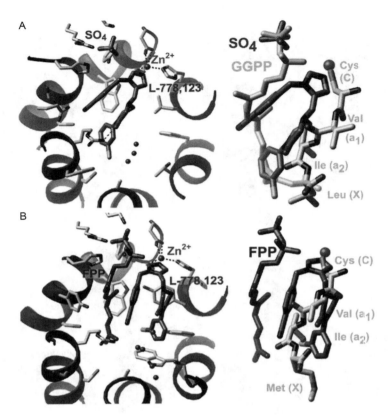

F<small>IG</small>. 13.8. Comparison of L-778,123 binding to FTase and GGTase-I. (A) Left, GGTase-I complex. Right, superposition of L-778,123 and substrate shows that binding of the inhibitor is competitive with respect to GGPP. The L-778,123 imidazole group coordinates the catalytic Zn^{2+} ion; the benzonitrile and chlorophenyl moieties bind in sites where isoprene units 2 and 3 of GGPP interact. (B) Left, FTase complex with inhibitor and a lipid substrate molecule. Right, superposition of L-778,123 and a substrate shows that the inhibitor is competitive with respect to Ca_1a_2X peptide. The L-778,123 imidazole group coordinates the catalytic Zn^{2+} ion; the chlorophenyl and benzonitrile moieties interact with the a_2 residue binding pocket and lipid substrate, respectively. (See color plate section in the back of the book.)

bind in regions normally occupied by isoprenes 2–4 of the GGPP lipid chain (Figure 13.8).

Two FTase-specific inhibitors, BMS-214663 (developed by Bristol-Myers Squibb) and R11577 (Tipifarnib/Zarnestra, developed by Johnson & Johnson), bind in a manner similar to L-778,123 (Figure 13.9) [3]. In both cases, an imidazole moiety coordinates the catalytic Zn^{2+}, a second aromatic group binds in the a_2 residue binding site, and FPP is observed to

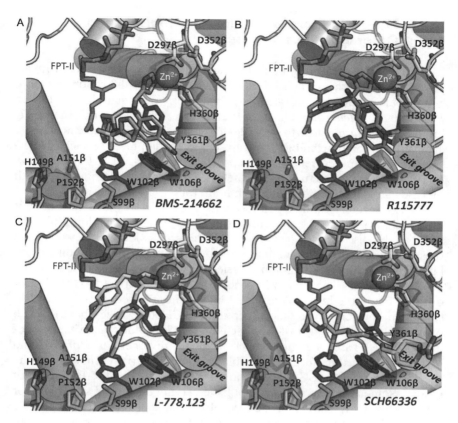

FIG. 13.9. Interactions of representative FTase inhibitors evaluated in clinical trials. (A) BMS-214662, (B) R115777, (C) L-778,123, and (D) SCH66336. All inhibitors shown are competitive with respect to Ca₁a₂X peptide and bind alongside FPP in the active site. All inhibitors interact in the aromatic a₂ residue binding pocket. Only SCH66336 does not coordinate the catalytic zinc ion, and makes significant interactions in the product exit groove. (See color plate section in the back of the book.)

bind, making interactions with a third substituent of the inhibitor (Figure 13.9). By contrast, the tricyclic scaffold SCH66336 (Lonafarnib, developed by Schering-Plough) interacts in a very different mode (Figure 13.9) [4]. It does not coordinate the catalytic Zn^{2+}, makes few interactions with the bound FPP substrate, but retains binding in the a_2 site, and, unlike the other inhibitors, makes extensive interactions in the exit groove (Figure 13.9). These crystal structures have provided insight for the design and development of new inhibitors, such as ethylenediamine derivatives, which will be discussed in the next section.

VII.   FTase Inhibitors for Treatment of Malaria and Other Infectious Diseases

Prenylated proteins play important signal transduction roles in all eukaryotes. Protein prenylation is critical for growth and proliferation of pathogenic eukaryotic microorganisms, including the malaria protozoan *Plasmodium falciparum* [45–47]. Preliminary screens reveal that FTIs inhibit *P. falciparum* growth [45]. Ideally, antimalarial FTIs have no activity against the human ortholog. Structure-based development of such species-specific inhibitors currently is hampered by the lack of a *P. falciparum* FTase (PfFTase) structure. Nevertheless, analysis of FTI complexes with hFTase can guide the design of such inhibitors.

PfFTase prefers Ca_1a_2X peptides with a Q or M as the X residue, but little is known about the preferences at the a_2 residue position [46]. One route to the discovery of new drugs is to repurpose known ones or to use these scaffolds as starting points for the development of new modifications [48]. Most known FTIs exhibit little selectivity between the mammalian and PfFTase enzymes [45,49,50], which perhaps can be attributed to the binding of these inhibitors in the a_2 residue binding site, the sequence of which is similar in both enzymes. One exception is SCH66336 (see above), which exhibits little activity against PfFTase [45]. This FTI binds the exit groove of mammalian FTase, a region which is highly divergent in PfFTase and hFTase. It may therefore be possible to identify variants that invert the specificity profile for SCH66336 and bind specifically to the PfFTase exit groove. Some antimalarial FTIs that have varying degrees of selectivity for PfFTase over hFTase have been developed [45,50–54]. Crystallographic analysis of a new ethylenediamine-scaffold series of inhibitors bound to mammalian FTase reveals the dominant determinants of inhibitor selectivity (Figure 13.10) [51]. In all compounds, an *N*-methylimidazole group coordinates the catalytic Zn^{2+}; most compounds contain aromatic substituents that bind in the a_2 residue site (Figures 13.10). Interestingly, selectivity of this series is controlled by substituents that bind in the exit groove and the X position of the peptide binding site (Figure 13.10). These two binding sites in FTase are predicted by sequence analysis to be the most divergent between hFTase and PfFTase.

Protein prenyltransferases have been shown to be attractive targets for the development of a number of therapeutics, including cancer and infectious diseases caused by fungal and trypanosomatid pathogens [32,33,55–57]. The modularity of the scaffolds used to synthesize FTIs is enabling development of variants that exploit structural diversities in the substrate and product binding regions of the active site to obtain species-specific inhibitors of

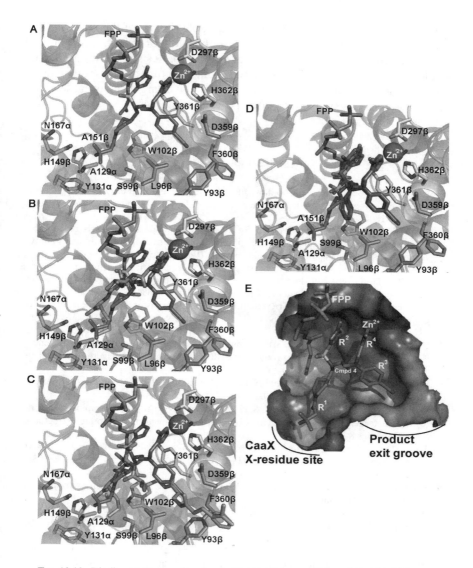

FIG. 13.10. Binding of ethylenediamine-scaffold inhibitors to FTase. (A) All inhibitors coordinate Zn^{2+} with an N-methylimidazole group and extend the p-benzonitrile moiety into the product exit groove (orange residues). This representative compound also interacts with both the Ca_1a_2X a_2 residue binding site and the X-residue binding site (green residues). (B) Superposition of inhibitor (blue) with Ca_1a_2X peptide (tan). (C) Superposition of inhibitor with displaced prenylated product in the exit groove (tan). (D) Superposition of multiple ethylenediamine-scaffold inhibitors illustrates common binding mode. (E) Surface representation of active site. Residues that are structurally divergent between human and *P. falciparum* FTase enzymes are highlighted: exit groove (orange); Ca_1a_2X X-residue binding site (green). The most selective inhibitor for malarial FTase (blue sticks) contacts both of the divergent regions. Figure reproduced from Ref. [51] with permission. (See color plate section in the back of the book.)

members in this enzyme family. The remarkable rigidity of these enzymes in which conformational changes that are important for catalysis are limited mainly to the substrates or products greatly simplifies structure-based drug development efforts. Together with the essential nature of FTases to cellular function, these features make FTases an attractive target for drug development efforts that leverage key contemporary technologies.

REFERENCES

1. Park, H.W., Boduluri, S.R., Moomaw, J.F., Casey, P.J., and Beese, L.S. (1997). Crystal structure of protein farnesyltransferase at 2.25 angstrom resolution. *Science* 275:1800–1804.
2. Reid, T.S., Long, S.B., and Beese, L.S. (2004). Crystallographic analysis reveals that anticancer clinical candidate L-778,123 inhibits protein farnesyltransferase and geranylgeranyltransferase-I by different binding modes. *Biochemistry* 43:9000–9008.
3. Reid, T.S., and Beese, L.S. (2004). Crystal structures of the anticancer clinical candidates R115777 (Tipifarnib) and BMS-214662 complexed with protein farnesyltransferase suggest a mechanism of FTI selectivity. *Biochemistry* 43:6877–6884.
4. Strickland, C.L., *et al.* (1999). Tricyclic farnesyl protein transferase inhibitors: crystallographic and calorimetric studies of structure–activity relationships. *J Med Chem* 42:2125–2135.
5. Lane, K.T., and Beese, L.S. (2006). Thematic review series: lipid posttranslational modifications. Structural biology of protein farnesyltransferase and geranylgeranyltransferase type I. *J Lipid Res* 47:681–699.
6. Long, S.B., Casey, P.J., and Beese, L.S. (2002). Reaction path of protein farnesyltransferase at atomic resolution. *Nature* 419:645–650.
7. Long, S.B., Casey, P.J., and Beese, L.S. (2000). The basis for K-Ras4B binding specificity to protein farnesyltransferase revealed by 2 A resolution ternary complex structures. *Struct Fold Des* 8:209–222.
8. Tobin, D.A., Pickett, J.S., Hartman, H.L., Fierke, C.A., and Penner-Hahn, J.E. (2003). Structural characterization of the zinc site in protein farnesyltransferase. *J Am Chem Soc* 125:9962–9969.
9. Terry, K.L., Casey, P.J., and Beese, L.S. (2006). Conversion of protein farnesyltransferase to a geranylgeranyltransferase. *Biochemistry* 45:9746–9755.
10. Reid, T.S., Terry, K.L., Casey, P.J., and Beese, L.S. (2004). Crystallographic analysis of CaaX prenyltransferases complexed with substrates defines rules of protein substrate selectivity. *J Mol Biol* 343:417–433.
11. Long, S.B., Casey, P.J., and Beese, L.S. (1998). Cocrystal structure of protein farnesyltransferase complexed with a farnesyl diphosphate substrate. *Biochemistry* 37:9612–9618.
12. Taylor, J.S., Reid, T.S., Terry, K.L., Casey, P.J., and Beese, L.S. (2003). Structure of mammalian protein geranylgeranyltransferase type-I. *EMBO J* 22:5963–5974.
13. Furfine, E.S., Leban, J.J., Landavazo, A., Moomaw, J.F., and Casey, P.J. (1995). Protein farnesyltransferase: kinetics of farnesyl pyrophosphate binding and product release. *Biochemistry* 34:6857–6862.
14. Pickett, J.S., *et al.* (2003). Kinetic studies of protein farnesyltransferase mutants establish active substrate conformation. *Biochemistry* 42:9741–9748.

15. Pais, J.E., Bowers, K.E., and Fierke, C.A. (2006). Measurement of the alpha-secondary kinetic isotope effect for the reaction catalyzed by mammalian protein farnesyltransferase. *J Am Chem Soc* 128:15086–15087.

16. Hartman, H.L., Bowers, K.E., and Fierke, C.A. (2004). Lysine beta311 of protein geranylgeranyltransferase type I partially replaces magnesium. *J Biol Chem* 279:30546–30553.

17. Casey, P.J., and Seabra, M.C. (1996). Protein prenyltransferases. *J Biol Chem* 271: 5289–5292.

18. Pompliano, D.L., Rands, E., Schaber, M.D., Mosser, S.D., Anthony, N.J., and Gibbs, J.B. (1992). Steady-state kinetic mechanism of ras farnesyl:protein transferase. *Biochemistry* 31:3800–3807.

19. Moores, S.L., *et al.* (1991). Sequence dependence of protein isoprenylation. *J Biol Chem* 266:14603–14610.

20. Yokoyama, K., and Gelb, M.H. (1993). Purification of a mammalian protein geranylgeranyltransferase. Formation and catalytic properties of an enzyme–geranylgeranyl pyrophosphate complex. *J Biol Chem* 268:4055–4060.

21. Yokoyama, K., McGeady, P., and Gelb, M.H. (1995). Mammalian protein geranylgeranyltransferase-I: substrate specificity, kinetic mechanism, metal requirements, and affinity labeling. *Biochemistry* 34:1344–1354.

22. Krzysiak, A.J., Aditya, A.V., Hougland, J.L., Fierke, C.A., and Gibbs, R.A. (2010). Synthesis and screening of a CaaL peptide library versus FTase reveals a surprising number of substrates. *Bioorg Med Chem Lett* 20:767–770.

23. Hougland, J.L., Hicks, K.A., Hartman, H.L., Kelly, R.A., Watt, T.J., and Fierke, C.A. (2010). Identification of novel peptide substrates for protein farnesyltransferase reveals two substrate classes with distinct sequence selectivities. *J Mol Biol* 395:176–190.

24. Krzysiak, A.J., Scott, S.A., Hicks, K.A., Fierke, C.A., and Gibbs, R.A. (2007). Evaluation of protein farnesyltransferase substrate specificity using synthetic peptide libraries. *Bioorg Med Chem Lett* 17:5548–5551.

25. Lerner, E.C., Zhang, T.T., Knowles, D.B., Qian, Y., Hamilton, A.D., and Sebti, S.M. (1997). Inhibition of the prenylation of K-Ras, but not H- or N-Ras, is highly resistant to CAAX peptidomimetics and requires both a farnesyltransferase and a geranylgeranyltransferase I inhibitor in human tumor cell lines. *Oncogene* 15:1283–1288.

26. Fiordalisi, J.J., Johnson, R.L., 2nd, Weinbaum, C.A., Sakabe, K., Casey, P.J., and Cox, A.D. (2003). High affinity for farnesyl transferase and alternative prenylation contribute individually to K-Ras4B resistance to farnesyl transferase inhibitors. *J Biol Chem* 278(43):41718–41727.

27. Mukherjee, P.K., Chandra, J., Kuhn, D.M., and Ghannoum, M.A. (2003). Differential expression of *Candida albicans* phospholipase B (PLB1) under various environmental and physiological conditions. *Microbiology* 149:261–267.

28. Song, J.L., and White, T.C. (2003). RAM2: an essential gene in the prenylation pathway of *Candida albicans*. *Microbiology* 149:249–259.

29. Smalera, I., Williamson, J.M., Baginsky, W., Leiting, B., and Mazur, P. (2000). Expression and characterization of protein geranylgeranyltransferase type I from the pathogenic yeast *Candida albicans* and identification of yeast selective enzyme inhibitors. *Biochim Biophys Acta* 1480:132–144.

30. Mazur, P., *et al.* (1999). Purification of geranylgeranyltransferase I from *Candida albicans* and cloning of the CaRAM2 and CaCDC43 genes encoding its subunits. *Microbiology* 145(Pt 5):1123–1135.

31. Mayer, M.L., Caplin, B.E., and Marshall, M.S. (1992). CDC43 and RAM2 encode the polypeptide subunits of a yeast type I protein geranylgeranyltransferase. *J Biol Chem* 267:20589–20593.

32. Murthi, K.K., Smith, S.E., Kluge, A.F., Bergnes, G., Bureau, P., and Berlin, V. (2003). Antifungal activity of a *Candida albicans* GGTase I inhibitor-alanine conjugate. inhibition of Rho1p prenylation in *C. albicans*. *Bioorg Med Chem Lett* 13:1935–1937.

33. Hast, M.A., and Beese, L.S. (2008). Structure of protein geranylgeranyltransferase-I from the human pathogen *Candida albicans* complexed with a lipid substrate. *J Biol Chem* 283(46):31933–31940.

34. Sousa, S.F., Fernandes, P.A., and Ramos, M.J. (2008). Farnesyltransferase inhibitors: a detailed chemical view on an elusive biological problem. *Curr Med Chem* 15:1478–1492.

35. Takemoto, Y., and Imoto, M. (2007). Development of farnesyltransferase inhibitor for anti-cancer drugs. *Tanpakushitsu Kakusan Koso* 52:1713–1718.

36. Puntambekar, D.S., Giridhar, R., and Yadav, M.R. (2007). Inhibition of farnesyltransferase: a rational approach to treat cancer? *J Enzyme Inhib Med Chem* 22:127–140.

37. Basso, A.D., Kirschmeier, P., and Bishop, W.R. (2006). Lipid posttranslational modifications. Farnesyl transferase inhibitors. *J Lipid Res* 47:15–31.

38. Dinsmore, C.J., and Bell, I.M. (2003). Inhibitors of farnesyltransferase and geranylgeranyltransferase-I for antitumor therapy: substrate-based design, conformational constraint and biological activity. *Curr Top Med Chem* 3:1075–1093.

39. Ohkanda, J. (2002). CAAX peptidomimetics: their farnesyltransferase inhibition activity and antitumor effect. *Seikagaku* 74:46–50.

40. Long, S.B., Hancock, P.J., Kral, A.M., Hellinga, H.W., and Beese, L.S. (2001). The crystal structure of human protein farnesyltransferase reveals the basis for inhibition by CaaX tetrapeptides and their mimetics. *Proc Natl Acad Sci USA* 98:12948–12953.

41. Kohl, N.E., *et al.* (1995). Inhibition of farnesyltransferase induces regression of mammary and salivary carcinomas in ras transgenic mice. *Nat Med* 1:792–797.

42. Avendano, C., and Menendez, J.C. (2007). Peptidomimetics in cancer chemotherapy. *Clin Transl Oncol* 9:563–570.

43. Martin, N.E., *et al.* (2004). A phase I trial of the dual farnesyltransferase and geranylgeranyltransferase inhibitor L-778,123 and radiotherapy for locally advanced pancreatic cancer. *Clin Cancer Res* 10:5447–5454.

44. Njoroge, F.G., *et al.* (1997). Discovery of novel nonpeptide tricyclic inhibitors of Ras farnesyl protein transferase. *Bioorg Med Chem* 5:101–113.

45. Nallan, L., *et al.* (2005). Protein farnesyltransferase inhibitors exhibit potent antimalarial activity. *J Med Chem* 48:3704–3713.

46. Chakrabarti, D., *et al.* (2002). Protein farnesyltransferase and protein prenylation in *Plasmodium falciparum*. *J Biol Chem* 277:42066–42073.

47. Chakrabarti, D., Azam, T., DelVecchio, C., Qiu, L., Park, Y.I., and Allen, C.M. (1998). Protein prenyl transferase activities of *Plasmodium falciparum*. *Mol Biochem Parasitol* 94:175–184.

48. Fischbach, M.A., and Walsh, C.T. (2009). Antibiotics for emerging pathogens. *Science* 325:1089–1093.

49. Gelb, M.H., *et al.* (2003). Protein farnesyl and N-myristoyl transferases: piggy-back medicinal chemistry targets for the development of antitrypanosomatid and antimalarial therapeutics. *Mol Biochem Parasitol* 126:155–163.

50. Ohkanda, J., *et al.* (2001). Peptidomimetic inhibitors of protein farnesyltransferase show potent antimalarial activity. *Bioorg Med Chem Lett* 11:761–764.

51. Hast, M.A., *et al.* (2009). Structural basis for binding and selectivity of antimalarial and anticancer ethylenediamine inhibitors to protein farnesyltransferase. *Chem Biol* 16:181–192.

52. Glenn, M.P., *et al.* (2005). Structurally simple farnesyltransferase inhibitors arrest the growth of malaria parasites. *Angew Chem Int Ed Engl* 44:4903–4906.

53. Kettler, K., Sakowski, J., Wiesner, J., Ortmann, R., Jomaa, H., and Schlitzer, M. (2005). Novel lead structures for antimalarial farnesyltransferase inhibitors. *Pharmazie* 60:323–327.

54. Wiesner, J., *et al.* (2004). Farnesyltransferase inhibitors inhibit the growth of malaria parasites in vitro and in vivo. *Angew Chem Int Ed Engl* 43:251–254.

55. Buckner, F.S., *et al.* (2002). Cloning, heterologous expression, and substrate specificities of protein farnesyltransferases from Trypanosoma cruzi and Leishmania major. *Mol Biochem Parasitol* 122:181–188.

56. Yokoyama, K., *et al.* (1998). The effects of protein farnesyltransferase inhibitors on trypanosomatids: inhibition of protein farnesylation and cell growth. *Mol Biochem Parasitol* 94:87–97.

57. Vallim, M.A., Fernandes, L., and Alspaugh, J.A. (2004). The RAM1 gene encoding a protein-farnesyltransferase beta-subunit homologue is essential in Cryptococcus neoformans. *Microbiology* 150:1925–1935.

14

Genetic Analyses of the CAAX Protein Prenyltransferases in Mice

MOHAMED X. IBRAHIM • OMAR M. KHAN •
MARTIN O. BERGO

Cancer Center Sahlgrenska
Department of Molecular and Clinical Medicine
Institute of Medicine, University of Gothenburg
Gothenburg, Sweden

I. Abstract

The RAS and RHO family proteins, the nuclear lamins, and other so-called *CAAX* proteins undergo three sequential posttranslational processing steps: prenylation, endoproteolysis, and carboxyl methylation. These steps are mediated by four different enzymes: farnesyltransferase (FTase) or geranylgeranyltransferase type I (GGTase-I), RAS converting enzyme 1 (RCE1), and isoprenylcysteine carboxyl methyltransferase (ICMT). Over the past 20 years, there has been an intense interest in understanding the biological significance of *CAAX* protein processing and its impact on *CAAX* protein stability, membrane targeting, protein–protein interactions, signaling, and function. Perhaps, the greatest interest has stemmed from the efforts of targeting the *CAAX* protein prenyltransferases FTase and GGTase-I as a strategy to interfere with the activity of disease-causing mutants of *CAAX* proteins such as RAS and prelamin A in the treatment of cancer and progeroid disorders. Mice with targeted conditional knockout alleles for FTase and GGTase-I have been used to shed light on those topics. Here, we review the genetic studies that have provided new information on the biochemical and medical importance of *CAAX* protein prenylation.

THE ENZYMES, Vol. XXIX   259   ISSN NO: 1874-6047
DOI: 10.1016/B978-0-12-381339-8.00014-7

II.   Introduction

Several hundred proteins in the mammalian genome have a cysteine amino acid four residues from the carboxyl terminus (i.e., C-X-X-X). A presently unknown subset of these proteins is classified as $CAAX$ proteins and includes small GTPases (e.g., RAS, RHO, RAC, RAL, RHEB), heterotrimeric G protein γ subunits, and the nuclear lamins (i.e., prelamin A and lamin B) [1]. The $CAAX$ motif triggers three sequential posttranslational processing steps: First, a 15-carbon farnesyl or 20-carbon geranylgeranyl lipid is attached to the cysteine residue by FTase and GGTase-I, respectively; second the last three amino acids (the -AAX) is clipped off by RCE1; and third, the newly exposed cysteine residue is methylated by ICMT. These modifications are constitutive and render the carboxyl terminus more hydrophobic, which facilitates interactions with membranes, stimulates protein–protein interactions, and can affect protein stability [2].

$CAAX$ protein processing is strongly conserved throughout evolution and is interesting from a purely biochemical perspective. However, most of the interest stems from the fact that several of the $CAAX$ proteins are involved in the pathogenesis of human disease. Hyperactive RAS signaling is implicated in the pathogenesis of many human cancers, including lung, colon, pancreatic, and hematopoietic cancer [3–8]. Increased RAS signaling is commonly caused by somatic mutations that lead to constitutive activation of RAS. However, increased RAS signaling can also be elicited by mutations in genes that encode proteins that interact with the RAS proteins, such as BCR/ABL, FLT3, and the tumor suppressor neurofibromin (NF1). Thus, the RAS proteins and the RAS signaling pathways are attractive anticancer targets.

One strategy to block RAS signaling is to inhibit the enzymes that process the $CAAX$ motif and thereby prevent RAS from reaching its site of action at the inner surface of the plasma membrane. FTase inhibitors (FTIs) prevent RAS from reaching the plasma membrane, and they have shown promise against a wide variety of malignancies in preclinical cell culture and mouse models [2,9]. However, clinical trials have been disappointing. One potential explanation is that several $CAAX$ proteins, including K-RAS (the isoform most commonly mutated in cancer), can be alternately prenylated by GGTase-I in the setting of FTI therapy, thereby retaining a mechanism for proper membrane attachment [1,10,11]. This alternate prenylation pathway fueled the interest in developing GGTase-I inhibitors (GGTIs) and also inhibitors of proteolysis and methylation— posttranslational modifications that are shared by farnesylated and geranylgeranylated $CAAX$ proteins. Thus, GGTIs could potentially be used in combination with FTIs to block the prenylation of K-RAS, and inhibitors of

RCE1 and ICMT might block K-RAS activity even if the protein was alternately prenylated. The rationale for inhibiting FTase, GGTase-I, RCE1, and ICMT in cancer therapy is bolstered by the fact that other *CAAX* proteins, including RHOA, RAC1, RALA, and RHEB, participate in tumor growth and metastasis. Another rationale for targeting FTase stems from the fact that mutations in prelamin A are involved in the pathogenesis of Hutchinson–Gilford Progeria syndrome. Blocking the far-nesylation of mutant prelamin A with an FTI reduces its toxicity and improves disease phenotypes in mouse models of progeria [12].

Although several FTIs and GGTIs have been developed and tested and compounds that inhibit RCE1 and ICMT are under development, compound-specific and off-target effects have made it difficult to tease out their potential utility. In this chapter, we review the genetic approaches in mice that have been used to define the impact of inhibiting the *CAAX* protein prenyltransferases FTase and GGTase-I on normal cell and tissue biology and on the development of RAS-induced malignancies. Genetic studies on RCE1 and ICMT are described elsewhere [13,14].

III. The Prenyltransferases FTase and GGTase-I

FTase and GGTase-I have been cloned from a number of nonmamma-lian species. Studies in yeast, fungi, flies, and plants yielded diverging results on the impact of disrupting FTase and GGTase-I activity. *RAM2*, a homo-log of *FNTA*, is essential in the yeast *Saccharomyces cerevisiae* [15] and in *Candida albicans* [16]. In contrast, disruption of *RAM1*, the homolog of *FNTB*, was not lethal but resulted in growth defects [15,17]. Null mutations in *CDC43*, a homolog of *PGGT1B*, however, were lethal in yeast [18]. Interestingly, *C. albicans* null *CDC43* mutants were viable, despite the lack of detectable GGTase-I activity, but were morphologically abnormal [19]. Further, null mutations in the β subunit of GGTase-I were lethal in *Drosophila melanogaster* [20] but not in *Arabidopsis thaliana* [21]. Based on these different results, it was impossible to predict the impact of FTase and GGTase-I deficiency in mammalian cells.

Mammalian FTase and GGTase-I were first identified and isolated from rat brain cytosol in the early 1990s [22–25]. FTase and GGTase-I share a common α-subunit but have unique β subunits. In humans, the common α subunit is encoded by *FNTA* and the β subunits are encoded by *FNTB* and *PGGT1B*, respectively. The β subunits dictate substrate specificity: in general, if the "*X*" residue of the *CAAX* motif is leucine, the protein is geranylgeranylated; otherwise it is farnesylated [1,26]. However, there are several exceptions to this rule and it is clear that the "*X*" residue only partially explains the substrate specificity. It is possible to mutate an exclusively geranylgeranylated substrate such as RHOA so that it becomes farnesylated, by replacing the wild-type *CAAX* sequence (CVLL) with the *CAAX* sequence of H-RAS (CVLS).

IV.   FTase

A few years ago, Mijimolle *et al.* developed mice with a conditional knockout allele for *Fntb* [27]. In their study, the inactivation of *Fntb* resulted in embryonic lethality, but the effects in adult tissues were very modest. Some findings were clearly inconsistent with previous studies. Inactivation of *Fntb* appeared to inhibit the farnesylation of HDJ2 and H-RAS, but only partially, and most remarkably, H-RAS remained in the membrane fraction of cells. Moreover, they found that *Fntb*-deficient fibroblasts proliferated in culture and that the development of K-RAS-induced tumors was unaffected by *Fntb* deficiency.

These results from this study were surprising for several reasons. First, several studies had established that membrane association of H-RAS is dependent on farnesylation [28,29]. Second, treating cells with FTIs usually results in cell cycle arrest and, in mouse models, FTIs are effective against many tumors, including tumors without RAS mutations [2,9]. A potential explanation for the differences between the genetic and pharmacologic studies is that FTIs might affect other proteins, aside from FTase. But the likeliest explanation is that the *Fntb* knockout allele generated by Mijimolle *et al.* [27] yielded a transcript with an in-frame deletion [30], and it is possible that this mutant transcript yielded a protein with some residual enzymatic activity.

We generated a new conditional *Fntb* knockout allele (*Fntb*^fl) where *loxP* sites flank exon 1 and upstream promoter sequences [31]. *Fntb*^fl/fl mice were healthy and fertile and FTase activity in tissues was normal. We isolated fibroblasts from *Fntb*^fl/fl mouse embryos fibroblasts (MEFs) and incubated them with a *Cre*-adenovirus to assess the impact of *Fntb* deficiency on the isoprenylation of FTase substrates. Western blots of lysates from *Fntb*^Δ/Δ (Δ, deleted, recombined) MEFs revealed that 100% of HDJ2 and H-RAS exhibited a reduced electrophoretic mobility indicating the FTase activity was eliminated. Moreover, H-RAS was cytosolic with very little or no traces of the protein in the membrane fraction. The levels of H-RAS protein increased two- to fourfold suggesting that farnesylation negatively affects the stability of the protein. As expected, the mobility of K-RAS and N-RAS were unchanged unless the cells were incubated with a GGTI, indicating that those proteins are geranylgeranylated in the *Fntb*-deficient MEFs. The inactivation of *Fntb* in primary, spontaneously immortalized, and K-RAS-transformed cells induced a G_2M cell cycle arrest and was associated with increased levels of $p21^{CIP1}$. We have never been able to culture *Fntb*-deficient cells, which suggests that FTase is essential for cell proliferation. Analyses of RAS downstream effectors revealed that the serum-induced phosphorylation of MEK and ERK was unaffected by *Fntb* deficiency. There was, however, a delayed serum-induced phosphorylation of AKT in *Fntb*-deficient MEFs.

To define the impact of FTase deficiency in tumor growth *in vivo*, we bred $Fntb^{fl/\Delta}$ and littermate control $Fntb^{fl/+}$ mice on a background of the lox-STOP-lox oncogenic $Kras2^{LSL}$ allele (K) and the lysozyme M-*Cre* allele (L) [32]. Earlier studies had shown that KL mice express oncogenic K-RAS (G12D) in type II pneumocytes and develop lung cancer that is fatal around 3 weeks of age (median, 22 days; maximum 24 days; Figure 14.1). The inactivation of *Fntb* in $Fntb^{fl/\Delta}KL$ mice reduced lung tumor burden and significantly increased survival (median, 50 days, maximum 142 days; Figure 14.1). A substantial proportion of H-RAS migrated slowly and accumulated in the cytosolic fraction as judged by Western blots of tumor lysates. Moreover, nonfarnesylated prelamin A accumulated in lungs of $Fntb^{fl/\Delta}KL$ mice as judged by Western blots and immunohistochemistry using a prelamin A-specific antibody [31]. Consistent with the finding that K-RAS was entirely membrane-associated in lungs of $Fntb^{fl/\Delta}KL$ mice, there was no impact of *Fntb* deficiency on the levels of phosphorylated ERK.

In an effort to broaden the understanding of protein farnesylation in different cell types, Yang and coworkers bred the $Fntb^{fl/fl}$ mice with keratin 14-Cre transgenic mice ($KCre$) with the goal of defining the impact of *Fntb* deficiency in skin keratinocytes [33]. The $KCre$ transgene is expressed during development and throughout adult life. The absence of *Fntb* in cultured keratinocytes reduced the farnesylation of HDJ2, mislocalized H-RAS to the cytosolic fraction, resulted in accumulation of nonfarnesylated prelamin A, and blocked proliferation. Thus, the consequences of inactivating *Fntb* are essentially identical in MEFs and skin keratinocytes. Immunohistochemical analyses revealed that prelamin A accumulated in the skin of $Fntb^{fl/fl}KCre$ mice. Further histological analyses revealed stunted hair follicles and abnormal hair shafts. Consequently, the $Fntb^{fl/fl}KCre$ mice developed essentially complete alopecia that persisted throughout life [33].

We conclude that FTase is essential for fibroblast and keratinocyte proliferation and that it is the only enzyme in mammalian cells capable of prenylating the exclusively farnesylated substrates HDJ2 and H-RAS. Targeting FTase significantly reduces K-RAS-induced tumor growth *in vivo*.

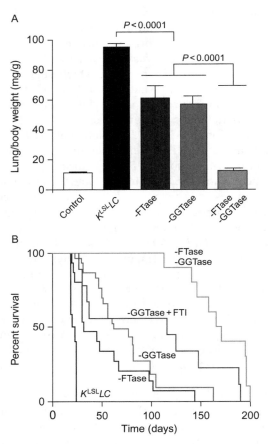

FIG. 14.1. Impact of FTase and GGTase-I inactivation on K-RAS-induced lung tumor development and survival. (A) Lung weight in 3-week-old healthy control (*Ctr*, $n = 16$), *KLC* ($n = 27$), *Fntb*$^{fl/\Delta}$*KLC* (-FTase, $n = 7$), *Pggt1b*$^{fl/\Delta}$*KLC* (-GGTase-I, $n = 13$), and *Fntb*$^{fl/\Delta}$*Pggt1b*$^{fl/\Delta}$*KLC* (-FTase, -GGTase-I, $n = 6$) mice. (B) Kaplan–Meier curve showing survival of *KLC* ($n = 12$), *Fntb*$^{fl/\Delta}$*KLC* (-FTase, $n = 16$), *Pggt1b*$^{fl/fl}$*KLC* (-GGTase-I, $n = 22$), *Pggt1b*$^{fl/fl}$*KLC* mice treated with an FTI in the diet (SCH66336; -GGTase-I + FTI, $n = 9$), and *Fntb*$^{fl/\Delta}$*Pggt1b*$^{fl/\Delta}$*KLC* (-FTase, -GGTase-I, $n = 10$) mice.

Because K-RAS was geranylgeranylated in the *Fntb*-deficient tumors—and likely fully functional—the antitumor activity of inactivating FTase must stem from inhibiting the farnesylation of *CAAX* proteins other than K-RAS.

V. GGTase-I

We generated mice with a conditional knockout allele for *Pggt1b*, the gene encoding the β subunit of GGTase-I [32]. In this allele, *loxP* sites flank exon 7 which contains sequences required for activity [34]. To validate the conditional allele, we isolated *Pggt1b*[fl/fl] MEFs and incubated them with the *Cre*-adenovirus. The inactivation of *Pggt1b* eliminated GGTase-I activity and blocked proliferation of primary fibroblasts, but did not affect cell viability. The *Pggt1b*[Δ/Δ] fibroblasts were arrested in the G_1 phase of the cell cycle and accumulated p21[CIP1]. Moreover, *Pggt1b*-deficient fibroblasts were small and spindled shaped, contained reduced amounts of polymerized actin and migrated poorly. In K-RAS[G12D]-expressing fibroblasts, inactivation of *Pggt1b* caused cell rounding and proliferation arrest, but not apoptosis. *Pggt1b*[Δ/Δ]K-RAS[G12D] cells also accumulated in the G_1 phase of the cell cycle, but expressed normal levels of p21[CIP1]. However, *Pggt1b* inactivation prevented the K-RAS[G12D]-induced increase in cyclin D1. The nonprenylated form of RAP1A accumulated in *Pggt1b*-deficient cells as judged by Western blots of cell lysates using the sc-1482 antibody from Santa Cruz—a widely used marker for absent geranylgeranylation. Most of the nonprenylated RAP1A was detected in the cytosolic fraction of cells. Moreover, the absence of *Pggt1b* increased the steady-state levels of RHOA. Thus, prenylation can reduce the stability of both farnesylated and geranylgeranylated *CAAX* proteins (e.g., H-RAS, RHOA).

To determine if farnesylation of well-known GGTase-I substrates could overcome phenotypes of *Pggt1b*-deficient cells, we generated cross-prenylation mutants by switching the *CAAX*-box of RHOA and CDC42 to C-V-L-S, the *CAAX* sequence of H-RAS. Coexpression of farnesyl-RHOA and farnesyl-CDC42 in *Pggt1b*[Δ/Δ]K-RAS[G12D] fibroblasts prevented cell rounding and partially restored proliferation. Expression of either construct alone had no effect [32]. Interestingly, expression of farnesyl-RHOA and farnesyl-CDC42 in *S. cerevisiae* was previously shown to overcome the lethality induced by a deficiency in CAL1, the yeast ortholog of *Pggt1b* [35]. Thus, although GGTase-I processes many different *CAAX* proteins in yeast and mammalian cells, only a few of them appear to be involved in critical phenotypes triggered by GGTase-I deficiency.

Knockout of *Pggt1b* in keratinocytes resulted in accumulation of nonprenylated RAP1A and proliferation arrest [33]. *Pggt1b*[fl/fl]*KCre* mice survived embryonic development but died shortly after birth. No abnormalities were identified in the interfollicular epidermis and the skin barrier was intact. At this point, the precise cause of death of *Pggt1b*[fl/fl]*KCre* mice is unknown.

To define the impact of GGTase-I deficiency on tumor development *in vivo*, we bred $Pggt1b^{fl/fl}KL$ and littermate control $Pggt1b^{fl/+}KL$ mice. The control $Pggt1b^{fl/+}KL$ mice developed a rapidly fatal lung cancer, consistent with the findings in $Fntb^{fl/+}KL$ mice. In this study, we also found that the $Pggt1b^{fl/+}KL$ and $Pggt1b^{+/+}KL$ mice develop myeloproliferative phenotypes including an increased proportion of myeloid cells in peripheral blood, infiltration of myeloid cells in the liver, and increased colony growth of bone marrow and spleen cells *in vitro*. The myeloproliferative phenotypes were expected because the lysozyme M-*Cre* allele is known to be expressed in myeloid cells [32]. Consequently, K-RAS^{G12D} was expressed in myeloid cells of the *KL* mice. The inactivation of *Pggt1b* in $Pggt1b^{fl/fl}KL$ mice reduced tumor development and improved survival to a similar extent as the inactivation of *Fntb* (Figure 14.1). Moreover, the myeloproliferative phenotypes were essentially eliminated. In all these studies, fibroblasts, keratinocytes, and hematopoietic cells were viable in the absence of GGTase-I, and myelopoiesis appeared to function normally. Thus, mammalian cells are viable in the absence of GGTase-I, and K-RAS-induced tumor development is significantly impaired.

The absence of myeloproliferative phenotypes in the $Pggt1b^{fl/fl}KL$ mice prompted us to further evaluate the impact of GGTase-I deficiency on hematopoiesis and the development of hematologic malignancies. The hematologic phenotypes of $Pggt1b^{fl/+}KL$ and $Pggt1b^{+/+}KL$ mice were mild and the mice succumbed to lung cancer before the development of overt myeloproliferative disease (MPD). We, and others, have previously bred the *K* mice on a background of the interferon-inducible Mx1-*Cre* (*M*) transgene [36–39]. Injection of interferon or pI-pC, an interferon stimulant, in *KM* mice switches on the expression of K-RAS^{G12D} in hematopoietic progenitors and results in a rapidly progressing and lethal MPD with leukocytosis, infiltration of myeloid cells in the liver and spleen, splenomegaly, and an ability of bone marrow and splenocytes to form colonies *in vitro* in the absence of exogenous growth factors. The disease in *KM* mice is classified as MPD [40] and many of the phenotypes are found in humans with RAS-induced MPDs. A relatively small proportion of *KM* mice (10–30%) also develop T-cell acute lymphoblastic leukemia (T-ALL) [36–39]. The drawback with the K-RAS-induced MPD mouse model is that the Mx1-*Cre* transgene triggers Cre and K-RAS^{G12D} expression in many other tissues and produces large tumors in the lung and skin and along the entire gastrointestinal tract. To overcome this issue, we harvested fetal liver cells from $Pggt1b^{fl/\Delta}KM$ and littermate control $Pggt1b^{fl/+}KM$ embryos and transplanted them into lethally irradiated wild-type recipient mice. The recipient mice were allowed to recover for 3 weeks and were then injected with pI-pC [38]. This strategy allowed us to switch on K-RAS^{G12D}

expression—and induce *Pggt1b* inactivation—exclusively in hematopoietic cells. As expected, the pI-pC-injected recipients of *Pggt1b*^{fl/+}*KL* fetal liver cells never developed tumors in lung, skin, and gastrointestinal tract. Although they developed the characteristic MPD phenotypes, those phenotypes were much less pronounced than in the primary pI-pC-injected *KM* mice. However, 100% of the mice developed aggressive T-ALL with enlarged thymuses and striking infiltration of lymphoblasts in peripheral tissues. Thus, the phenotypic consequences of expressing K-RAS^{G12D} in hematopoietic cells were different in pI-pC-injected primary *KM* mice compared to secondary recipients. One potential explanation is that K-RAS^{G12D} expression in the bone marrow microenvironment might stimulate myeloid or suppress lymphoid expansion in primary *KM* mice. The inactivation of *Pggt1b* in K-RAS^{G12D}-expressing hematopoietic cells significantly reduced MPD phenotypes but had no impact on the development of aggressive T-ALL. Consequently, the *Pggt1b* inactivation had no impact on survival. There are two potential explanations for why the absence of *Pggt1b* affected MPD but not T-ALL phenotypes. First, *Pggt1b* inactivation might have been incomplete in lymphoid cells. However, we had minimized the risk for this potential problem by using mice with one conditional allele and one recombined allele (i.e., *Pggt1b*^{fl/Δ}). Indeed, we observed robust staining for nonprenylated RAP1A in enlarged thymuses of mice transplanted with *Pggt1b*^{fl/Δ}*KM* cells. The second potential explanation is that geranylgeranylation is dispensable for the function and proliferation of lymphoid cells. More work is required to address this issue. It will also be important to understand the importance of GGTase-I in other cell types, not least because we need to understand the consequences of inhibiting the enzyme on the whole organism.

As discussed earlier, the RHO family proteins RAC1, RHOA, and CDC42, which are geranylgeranylated by GGTase-I, are important for proper function of inflammatory cells. For example, these proteins regulate the actin cytoskeleton during cell migration and phagocytosis and participate in intracellular signaling pathways [41,42]. It is widely believed that geranylgeranylation is required for membrane anchoring of RHO family proteins and is considered essential for their subcellular targeting and activation. Consequently, inhibiting GGTase-I has been proposed as a potential strategy to inhibit the proinflammatory activities of RHO family proteins and to treat inflammatory disorders including rheumatoid arthritis [43,44]. Inhibition of the geranylgeranylation of RHO family proteins has also been proposed to explain the anti-inflammatory properties of statins [45]. To determine if inhibiting GGTase-I might be a strategy to treat inflammatory disorders, we knocked out *Pggt1b* in macrophages by breeding *Pggt1b*^{fl/fl}*L* mice [46]. Contrary to our expectations, GGTase-I deficiency in macrophages did not impair migration or phagocytosis and resulted in accumulation of GTP-bound RAC1, increased production of ROS and proinflammatory cytokines, and a chronic progressive erosive arthritis.

The *Pggt1b*-deficient macrophages were reduced in number and were small and rounded (Figure 14.2). This result was not particularly surprising, given our findings with *Pggt1b*-deficient fibroblasts and the well-known role of RHO family proteins in regulating the cytoskeleton. The cell spreading defect was rescued by expressing farnesyl-RAC1 alone or farnesyl-RHOA alone, suggesting that these proteins play essential and overlapping roles in the spreading of macrophages in culture and that prenylation is required for this function. Thus, reduced cell spreading of *Pggt1b*-deficient macrophages is a loss-of-function phenotype. However, despite the cell spreading defect, there was no effect of *Pggt1b* deficiency on macrophage migration and phagocytosis. Similarly, the ability of macrophages to migrate into tissues and the peritoneal cavity *in vivo* was unaffected by *Pggt1b* deficiency. These findings suggest that geranylgeranylation of RHO family proteins is dispensable for macrophage migration and phagocytosis. RAC1, RHOA, and CDC42 accumulated in the active GTP-bound form. Increased GTP loading of RHO family proteins has been observed in previous studies where isoprenoid synthesis was reduced indirectly by mevalonate kinase deficiency or by statin treatment [47,48]. Perhaps, the most striking finding was that RAC1 remained associated with the plasma

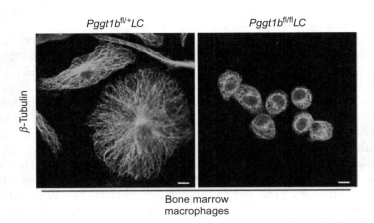

Bone marrow
macrophages

FIG. 14.2. Reduced cell adhesive area/cell rounding in GGTase-I-deficient macrophages. Confocal micrographs of bone marrow-derived *Pggt1b*$^{fl/fl}$*LC* and heterozygous control *Pggt1b*$^{fl/+}$*LC* macrophages stained with antibodies to β-tubulin (scale bars, 10 μm). (See color plate section in the back of the book.)

membrane in *Pggt1b*-deficient macrophages. RAC1 remained plasma membrane associated even if the cells were incubated with an FTI and a GGTI, suggesting that RAC1 was not prenylated by FTase or residual GGTase-I activity in the *Pggt1b*-deficient macrophages. Although, we cannot rule out the possibility that RAC1 was brought to the membrane through an as-yet unknown compensatory mechanism, the likeliest explanation is that the polybasic sequence upstream of the *CAAX* motif is sufficient for plasma membrane targeting.

The *Pggt1b*-deficient macrophages exhibited enhanced ROS production and increased p38 activation, NFκB activation, and expression and secretion of proinflammatory cytokines including TNFα, IL1-β, and IL-6 in response to LPS stimulation. The plasma membrane location and increased levels of RAC1-GTP could potentially explain this finding. Knockdown of RAC1, but not RAC2, expression in *Pggt1*-deficient macrophages reduced cytokine production. These results suggest that GGTase-I deficiency results in hyperactivation of macrophages partly or fully through a gain-of-function of nonprenylated RAC1. The hyperactivated macrophages can likely explain the *in vivo* phenotype: *Pggt1b*$^{fl/fl}$*L* mice developed chronic progressive joint inflammation with bone erosions and high levels of serum autoantibodies including rheumatoid factor. The phenotypes were both transplantable and reversible in bone marrow transplantation experiments. Thus, inactivating GGTase-I in macrophages induces erosive arthritis in mice.

This result might throw cold water on the concept of inhibiting GGTase-I in the treatment of inflammatory disorders and should be informative for ongoing clinical trials with GGTIs in cancer patients. On the other hand, the study has resulted in a simple genetic mouse model of human chronic rheumatoid arthritis which can be used to study mechanisms of disease development and for testing new therapies. The results should stimulate interest in further understanding the biochemical and physiological role of *CAAX* protein geranylgeranylation.

VI. Simultaneous Inactivation of FTase and GGTase-I

The inactivation of *Fntb* and *Pggt1b* reduced tumor growth and improved survival to a similar extent in mice with K-RAS-induced lung cancer. In both cases, K-RAS^{G12D} remains prenylated, membrane associated, and fully functional. FTI/GGTI combinations and dual-prenylation inhibitors were previously shown to block K-RAS prenylation *in vivo*, but the doses required were toxic [49,50]. We bred *Fntb*$^{fl/\Delta}$*Pggt1b*$^{fl/\Delta}$*KL* and littermate control mice to test the hypothesis that inactivating both FTase and GGTase-I would block prenylation of K-RAS and further reduce tumor growth. A significant proportion of K-RAS was found in the cytosolic fraction in lung lysates of $Fntb^{fl/\Delta}Pggt1b^{fl/\Delta}KL$ mice and the mice survived far longer than mice lacking either enzyme alone ($P<0.0001$). Remarkably, the lungs of 1- to 3-month-old $Fntb^{fl/\Delta}Pggt1b^{fl/\Delta}KL$ mice showed normal histology despite widespread expression of K-RAS^{G12D} and the mice were fertile and healthy. Moreover, $Fntb^{fl/\Delta}Pggt1b^{fl/\Delta}L$ mice (harboring the L allele but lacking the K oncogene) were healthy, had a normal life span, and showed no evidence of apoptosis in the lung. The simultaneous inactivation of $Fntb$ and $Pggt1b$ inhibited tumorigenesis in a second K-RAS-induced lung tumor model. Cre-adenovirus was administered intranasally to $Fntb^{fl/\Delta}Pggt1b^{fl/\Delta}K$ and littermate control mice (K mice), and 8 weeks later, tumor number and size were determined. Control K mice had many large tumors that were visible on the lung surface (Figure 14.3). In contrast, lungs of $Fntb^{fl/\Delta}Pggt1b^{fl/\Delta}K$ mice were smooth and showed a 76% reduction in tumor number and a 79% reduction in tumor size (Figure 14.3). Thus, inactivating both $Fntb$ and $Pggt1b$ in the lung reduces K-RAS prenylation and membrane association, efficiently reduces K-RAS-induced tumor growth, and is compatible with cell survival.

Inactivating both $Fntb$ and $Pggt1b$ induced apoptosis in fibroblasts [31]. Similarly, E17.5 $Fntb^{fl/fl}Pggt1b^{fl/fl}KCre$ embryos had increased amounts of apoptotic keratinocytes in the skin, compared with embryos lacking $Fntb$ alone or $Pggt1b$ alone [33]. In contrast, lungs of $Fntb^{fl/\Delta}Pggt1b^{fl/\Delta}L$ mice showed no evidence of increased apoptosis despite the presence of many cells lacking both FTase and GGTase-I activity. Thus, some cells are clearly viable in the absence of both enzymes whereas others undergo apoptosis. Based on these findings, it is difficult to predict the potential toxicity of highly specific dual-prenylation inhibitors.

VII.   Concluding Remarks

The genetic studies with $Fntb$ and $Pggt1b$ conditional knockout alleles support the idea that farnesylation is essential for H-RAS membrane association and HDJ2 and prelamin A prenylation and that both farnesylation and geranylgeranylation is essential for fibroblast and keratinocyte proliferation $in\ vitro$. Knockout of $Fntb$ alone and $Pggt1b$ alone significantly impairs K-RAS-induced tumor growth $in\ vivo$ despite the fact that K-RAS remains prenylated under both conditions. The results also support the idea that simultaneous inhibition of FTase and GGTase-I blocks K-RAS prenylation and could be therapeutically useful. However, our

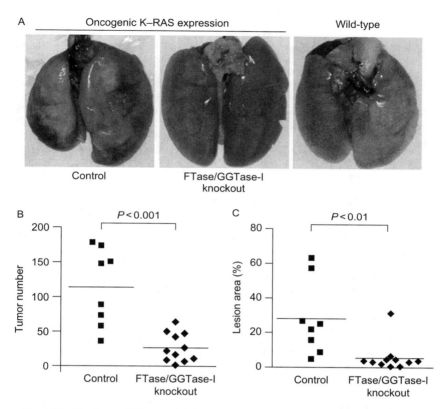

FIG. 14.3. Knockout of FTase and GGTase-I reduces K-RAS-induced lung tumor development. (A) Photographs of lungs from $Fntb^{fl/+}Pggt1b^{fl/+}K$ (left), $Fntb^{fl/\Delta}Pggt1b^{fl/\Delta}K$ (middle), and wild-type (right) mice 8 weeks after inhalation of a Cre-adenovirus. Note, gross tumors visible as white streaks on the surface of lung of control mouse at left. No tumors are visible on the surface of the $Fntb^{fl/\Delta}Pggt1b^{fl/\Delta}K$ lung. (B) Quantification of lung tumor number. (C) Quantification of lung tumor area. (See color plate section in the back of the book.)

results also indicate that the absence of the prenyltransferases could be harmful: the inactivation of $Pggt1b$ in macrophages causes severe chronic erosive arthritis and the absence of $Pggt1b$ and $Fntb$ in keratinocytes results in embryonic lethality and alopecia, respectively. We have yet to determine the consequences of widespread inactivation of $Fntb$ and $Pggt1b$ in adult mice and to define which $CAAX$ protein substrates are involved in the antitumor effects and disease phenotypes triggered by the absence of FTase and GGTase-I activity *in vivo*.

REFERENCES

1. Reid, T.S., Terry, K.L., Casey, P.J., and Beese, L.S. (2004). Crystallographic analysis of CaaX prenyltransferases complexed with substrates defines rules of protein substrate selectivity. *J Mol Biol* 343:417–433.
2. Basso, A.D., Kirschmeier, P., and Bishop, W.R. (2006). Thematic review series: lipid posttranslational modifications. Farnesyl transferase inhibitors. *J Lipid Res* 47:15–31.
3. Sahai, E., Olson, M.F., and Marshall, C.J. (2001). Cross-talk between Ras and Rho signalling pathways in transformation favours proliferation and increased motility. *EMBO J* 20:755–766.
4. Vega, F.M., and Ridley, A.J. (2008). Rho GTPases in cancer cell biology. *FEBS Lett* 582:2093–2101.
5. Qiu, R.G., Abo, A., McCormick, F., and Symons, M. (1997). Cdc42 regulates anchorage-independent growth and is necessary for Ras transformation. *Mol Cell Biol* 17:3449–3458.
6. Lim, K.H., Baines, A.T., Fiordalisi, J.J., Shipitsin, M., Feig, L.A., Cox, A.D., Der, C.J., and Counter, C.M. (2005). Activation of RalA is critical for Ras-induced tumorigenesis of human cells. *Cancer Cell* 7:533–545.
7. Bodemann, B.O., and White, M.A. (2008). Ral GTPases and cancer: linchpin support of the tumorigenic platform. *Nat Rev Cancer* 8:133–140.
8. Clark, E.A., Golub, T.R., Lander, E.S., and Hynes, R.O. (2000). Genomic analysis of metastasis reveals an essential role for RhoC. *Nature* 406:532–535.
9. Omer, C.A., Chen, Z., Diehl, R.E., Conner, M.W., Chen, H.Y., Trumbauer, M.E., Gopal-Truter, S., Seeburger, G., Bhimnathwala, H., Abrams, M.T., *et al.* (2000). Mouse mammary tumor virus-Ki-*ras*B transgenic mice develop mammary carcinomas that can be growth-inhibited by a farnesyl:protein transferase inhibitor. *Cancer Res* 60:2680–2688.
10. Whyte, D.B., Kirschmeier, P., Hockenberry, T.N., Nunez-Oliva, I., James, L., Catino, J.J., Bishop, W.R., and Pai, J.-K. (1997). K- and N-Ras are geranylgeranylated in cells treated with farnesyl protein transferase inhibitors. *J Biol Chem* 272:14459–14464.
11. James, G., Goldstein, J.L., and Brown, M.S. (1996). Resistance of K-RasB^{V12} proteins to farnesyltransferase inhibitors in Rat1 cells. *Proc Natl Acad Sci USA* 93:4454–4458.
12. Fong, L.G., Frost, D., Meta, M., Qiao, X., Yang, S.H., Coffinier, C., and Young, S.G. (2006). A protein farnesyltransferase inhibitor ameliorates disease in a mouse model of progeria. *Science* 311:1621–1623.
13. Bergo, M.O., Wahlstrom, A.M., Fong, L.G., and Young, S.G. (2008). Genetic analyses of the role of RCE1 in RAS membrane association and transformation. *Methods Enzymol* 438:367–389.
14. Svensson, A.W., Casey, P.J., Young, S.G., and Bergo, M.O. (2006). Genetic and pharmacologic analyses of the role of Icmt in Ras membrane association and function. *Methods Enzymol* 407:142–159.
15. He, B., Chen, P., Chen, S.Y., Vancura, K.L., Michaelis, S., and Powers, S. (1991). RAM2, an essential gene of yeast, and RAM1 encode the two polypeptide components of the farnesyltransferase that prenylates a-factor and Ras proteins. *Proc Natl Acad Sci USA* 88:11373–11377.
16. Song, J.L., and White, T.C. (2003). RAM2: an essential gene in the prenylation pathway of Candida albicans. *Microbiology* 149:249–259.
17. Powers, S., Michaelis, S., Broek, D., Anna-A, S.S., Field, J., Herskowitz, I., and Wigler, M. (1986). *RAM*, a gene of yeast required for a functional modification of *RAS* proteins and for production of mating pheromone **a**-factor. *Cell* 47:413–422.

18. Ohya, Y., Goebl, M., Goodman, L.E., Petersen-Bjorn, S., Friesen, J.D., Tamanoi, F., and Anraku, Y. (1991). Yeast CAL1 is a structural and functional homologue to the DPR1 (RAM) gene involved in ras processing. *J Biol Chem* 266:12356–12360.

19. Kelly, R., Card, D., Register, E., Mazur, P., Kelly, T., Tanaka, K.I., Onishi, J., Williamson, J.M., Fan, H., Satoh, T., *et al.* (2000). Geranylgeranyltransferase I of Candida albicans: null mutants or enzyme inhibitors produce unexpected phenotypes. *J Bacteriol* 182:704–713.

20. Therrien, M., Chang, H.C., Solomon, N.M., Karim, F.D., Wassarman, D.A., and Rubin, G.M. (1995). KSR, a novel protein kinase required for RAS signal transduction. *Cell* 83:879–888.

21. Running, M.P., Lavy, M., Sternberg, H., Galichet, A., Gruissem, W., Hake, S., Ori, N., and Yalovsky, S. (2004). Enlarged meristems and delayed growth in plp mutants result from lack of CaaX prenyltransferases. *Proc Natl Acad Sci USA* 101:7815–7820.

22. Reiss, Y., Goldstein, J.L., Seabra, M.C., Casey, P.J., and Brown, M.S. (1990). Inhibition of purified p21[ras] farnesyl:protein transferase by Cys-AAX tetrapeptides. *Cell* 62:81–88.

23. Seabra, M.C., Reiss, Y., Casey, P.J., Brown, M.S., and Goldstein, J.L. (1991). Protein farnesyltransferase and geranylgeranyltransferase share a common alpha subunit. *Cell* 65:429–434.

24. Moores, S.L., Schaber, M.D., Mosser, S.D., Rands, E., O'Hara, M.B., Garsky, V.M., Marshall, M.S., Pompliano, D.L., and Gibbs, J.B. (1991). Sequence dependence of protein isoprenylation. *J Biol Chem* 266:14603–14610.

25. Moomaw, J.F., and Casey, P.J. (1992). Mammalian protein geranylgeranyltransferase. Subunit composition and metal requirements. *J Biol Chem* 267:17438–17443.

26. Yokoyama, K., Goodwin, G.W., Ghomashchi, F., Glomset, J.A., and Gelb, M.H. (1991). A protein geranylgeranyltransferase from bovine brain: implications for protein prenylation specificity. *Proc Natl Acad Sci USA* 88:5302–5306.

27. Mijimolle, N., Velasco, J., Dubus, P., Guerra, C., Weinbaum, C.A., Casey, P.J., Campuzano, V., and Barbacid, M. (2005). Protein farnesyltransferase in embryogenesis, adult homeostasis, and tumor development. *Cancer Cell* 7:313–324.

28. Lerner, E.C., Qian, Y., Blaskovich, M.A., Fossum, R.D., Vogt, A., Sun, J., Cox, A.D., Der, C.J., Hamilton, A.D., and Sebti, S.M. (1995). Ras CAAX peptidomimetic FTI-277 selectively blocks oncogenic Ras signaling by inducing cytoplasmic accumulation of inactive Ras–Raf complexes. *J Biol Chem* 270:26802–26806.

29. Willumsen, B.M., Christensen, A., Hubbert, N.L., Papageorge, A.G., and Lowy, D.R. (1984). The p21 ras C-terminus is required for transformation and membrane association. *Nature* 310:583–586.

30. Yang, S., Bergo, M., Farber, E., Qiao, X., Fong, L., and Young, S. (2009). Caution! Analyze transcripts from conditional knockout alleles. *Transgenic Res* 18:483–489.

31. Liu, M., Sjogren, A.K., Karlsson, C., Ibrahim, M.X., Andersson, K.M., Olofsson, F.J., Wahlstrom, A.M., Dalin, M., Yu, H., Chen, Z., *et al.* (2010). Targeting the protein prenyltransferases efficiently reduces tumor development in mice with K-RAS-induced lung cancer. *Proc Natl Acad Sci USA* 107:6471–6476.

32. Sjogren, A.K., Andersson, K.M., Liu, M., Cutts, B.A., Karlsson, C., Wahlstrom, A.M., Dalin, M., Weinbaum, C., Casey, P.J., Tarkowski, A., *et al.* (2007). GGTase-I deficiency reduces tumor formation and improves survival in mice with K-RAS-induced lung cancer. *J Clin Invest* 117:1294–1304.

33. Lee, R., Chang, S.Y., Trinh, H., Tu, Y., White, A.C., Davies, B.S., Bergo, M.O., Fong, L.G., Lowry, W.E., and Young, S.G. (2010). Genetic studies on the functional relevance of the protein prenyltransferases in skin keratinocytes. *Hum Mol Genet* 19:1603–1617.

34. Zhang, F.L., Moomaw, J.F., and Casey, P.J. (1994). Properties and kinetic mechanism of recombinant mammalian protein geranylgeranyltransferase type I. *J Biol Chem* 269:23465–23470.

35. Ohya, Y., Qadota, H., Anraku, Y., Pringle, J.R., and Botstein, D. (1993). Suppression of yeast geranylgeranyl transferase I defect by alternative prenylation of two target GTPases, Rho1p and Cdc42p. *Mol Biol Cell* 4:1017–1025.

36. Braun, B.S., Tuveson, D.A., Kong, N., Le, D.T., Kogan, S.C., Rozmus, J., Le Beau, M.M., Jacks, T.E., and Shannon, K.M. (2004). Somatic activation of oncogenic Kras in hematopoietic cells initiates a rapidly fatal myeloproliferative disorder. *Proc Natl Acad Sci USA* 101:597–602.

37. Chan, I.T., Kutok, J.L., Williams, I.R., Cohen, S., Kelly, L., Shigematsu, H., Johnson, L., Akashi, K., Tuveson, D.A., Jacks, T., *et al.* (2004). Conditional expression of oncogenic K-ras from its endogenous promoter induces a myeloproliferative disease. *J Clin Invest* 113:528–538.

38. Wahlstrom, A.M., Cutts, B.A., Karlsson, C., Andersson, K.M., Liu, M., Sjogren, A.K., Swolin, B., Young, S.G., and Bergo, M.O. (2007). Rce1 deficiency accelerates the development of K-RAS-induced myeloproliferative disease. *Blood* 109:763–768.

39. Wahlstrom, A.M., Cutts, B.A., Liu, M., Lindskog, A., Karlsson, C., Sjogren, A.K., Andersson, K.M., Young, S.G., and Bergo, M.O. (2008). Inactivating Icmt ameliorates K-RAS-induced myeloproliferative disease. *Blood* 112:1357–1365.

40. Kogan, S.C., Ward, J.M., Anver, M.R., Berman, J.J., Brayton, C., Cardiff, R.D., Carter, J.S., de Coronado, S., Downing, J.R., Fredrickson, T.N., *et al.* (2002). Bethesda proposals for classification of nonlymphoid hematopoietic neoplasms in mice. *Blood* 100:238–245.

41. Hall, A. (1998). G proteins and small GTPases: distant relatives keep in touch. *Science* 280:2074–2075.

42. Heasman, S.J., and Ridley, A.J. (2008). Mammalian Rho GTPases: new insights into their functions from in vivo studies. *Nat Rev Mol Cell Biol* 9:690–701.

43. Connor, A.M., Berger, S., Narendran, A., and Keystone, E.C. (2006). Inhibition of protein geranylgeranylation induces apoptosis in synovial fibroblasts. *Arthritis Res Ther* 8:R94.

44. Nagashima, T., Okazaki, H., Yudoh, K., Matsuno, H., and Minota, S. (2006). Apoptosis of rheumatoid synovial cells by statins through the blocking of protein geranylgeranylation: a potential therapeutic approach to rheumatoid arthritis. *Arthritis Rheum* 54:579–586.

45. Greenwood, J., Steinman, L., and Zamvil, S.S. (2006). Statin therapy and autoimmune disease: from protein prenylation to immunomodulation. *Nat Rev Immunol* 6:358–370.

46. Khan, O.M., Ibrahim, M.X., Jonsson, I.M., Karlsson, C., Liu, M., Sjogren, A.K., Olofsson, F.J., Brisslert, M., Andersson, S., Ohlsson, C., *et al.* (2011). Geranylgeranyltransferase type I (GGTase-I) deficiency hyperactivates macrophages and induces erosive arthritis in mice. *J Clin Invest* 121:628–639.

47. Lindholm, M.W., and Nilsson, J. (2007). Simvastatin stimulates macrophage interleukin-1beta secretion through an isoprenylation-dependent mechanism. *Vascul Pharmacol* 46:91–96.

48. Kuijk, L.M., Beekman, J.M., Koster, J., Waterham, H.R., Frenkel, J., and Coffer, P.J. (2008). HMG-CoA reductase inhibition induces IL-1beta release through Rac1/PI3K/PKB-dependent caspase-1 activation. *Blood* 112:3563–3573.

49. deSolms, S.J., Ciccarone, T.M., MacTough, S.C., Shaw, A.W., Buser, C.A., Ellis-Hutchings, M., Fernandes, C., Hamilton, K.A., Huber, H.E., Kohl, N.E., *et al.* (2003). Dual protein farnesyltransferase:geranylgeranyltransferase-I inhibitors as potential cancer chemotherapeutic agents. *J Med Chem* 46:2973–2984.

50. Lobell, R.B., Omer, C.A., Abrams, M.T., Bhimnathwala, H.G., Brucker, M.J., Buser, C.A., Davide, J.P., deSolms, S.J., Dinsmore, C.J., Ellis-Hutchings, M.S., *et al.* (2001). Evaluation of farnesyl:protein transferase and geranylgeranyl, protein transferase inhibitor combinations in preclinical models. *Cancer Res* 61:8758–8768.

15

Farnesyl Transferase Inhibitors: From Targeted Cancer Therapeutic to a Potential Treatment for Progeria

W. ROBERT BISHOP[a] • RONALD DOLL[b] • PAUL KIRSCHMEIER[a]

[a]Oncology Discovery
Merck Research Laboratories, Kenilworth
New Jersey, USA

[b]Charles A. Dana Research Institute
Drew University, Hall of Sciences
Madison, New Jersey, USA

I.   Abstract

Farnesyl transferase inhibitors (FTIs) were initially designed to inhibit the activity of Ras oncoproteins and represent one of the first attempts to develop a targeted cancer therapy. The high prevalence of Ras mutations in human disease and its critical role in proliferative signaling make it an important target for cancer therapeutics. The discovery that the posttranslational prenylation of Ras by farnesyl transferase (FTase) is critical for its function led to a concerted effort to identify inhibitors of this enzyme. A number of structurally distinct FTIs were identified; several of these advanced into clinical trials in cancer patients. Despite some promising activity in early phase clinical studies, no FTI demonstrated robust activity in larger, randomized trials. One factor that likely contributed to this lack of robust efficacy was the inability to select for responsive patient subpopulations. While FTIs effectively inhibit the function of the H-Ras protein, they fail to inhibit K-Ras and N-Ras, the isoforms most often mutated in human cancer. This is due to posttranslational modification of

ISSN NO: 1874-6047
DOI: 10.1016/B978-0-12-381339-8.00015-9

K-Ras and N-Ras by a distinct prenyl transferase when FTase is inhibited. As our knowledge of farnesylated proteins has grown, diseases driven by proteins not subject to efficient alternative prenylation have been identified. These include Hutchinson–Gilford Progeria Syndrome (HGPS), a premature aging disorder caused by a mutation in the *Lmna* gene. One FTI is undergoing clinical evaluation in children with HGPS. Other potential therapeutic opportunities for FTIs, including Costello Syndrome, a genetic disorder caused by a mutation in H-Ras, are worthy of consideration.

II.   Targeting Ras with Farnesyl Transferase Inhibitors

The initial surge in interest in FTIs arose when the posttranslational processing pathway for the Ras oncoproteins was first elucidated [1,2]. There are three *ras* genes encoding four Ras proteins (H-Ras, K-Ras-4A, K-Ras-4B, and N-Ras). Ras proteins are small guanine nucleotide-binding proteins that bind and hydrolyze GTP [3,4]. They can exist in an active (Ras-GTP) or inactive (Ras-GDP) state. Ras-GTP plays a critical role in transmission of proliferative signals in cells.

A high frequency of activating mutations in the *ras* genes is found in human cancers [3,4]. These activating mutations encode Ras proteins with suppressed GTPase activity, causing Ras to remain in the active GTP-bound state independent of upstream activation. This results in constitutive signal transduction to downstream effectors. Activating mutations in K-Ras are prevalent in epithelial cancers including pancreatic ($>90\%$), colorectal (50%), and non-small cell lung cancer (NSCLC) (30%) [5]. Activating mutations in N-Ras occur in melanoma (10–20%) and some hematologic malignancies [5]. H-Ras mutations are rarer in human cancer but have been reported in bladder cancer (15–20%) as well as some thyroid and prostate cancers [5,6]. The high prevalence of Ras mutations in human disease and its critical role in proliferative signaling make it an attractive target for cancer therapeutics; however, targeting Ras with small molecule inhibitors has proven a formidable challenge.

Site-directed mutagenesis studies showed that the carboxy-terminus of Ras is essential for its transforming activity [7,8]. Homology of the C-terminus of the Ras proteins to the C-terminus of yeast mating factors suggested that Ras proteins may be farnesylated [9]. Subsequent experiments demonstrated that both H- and K-Ras incorporated [^{3}H] mevalonic acid in the form of a farnesyl moiety on the carboxy-terminal cysteine [9,10] and that this cysteine is essential for Ras transforming activity [8]. All Ras isoforms are farnesylated *in vitro* and in cells by the enzyme FTase, with K-Ras being a higher-affinity substrate than H-Ras [9,11]. Following farnesylation, Ras proteins undergo additional carboxy-terminal modifications (Figure 15.1) [2,7,12]. The first step is proteolytic removal of the three

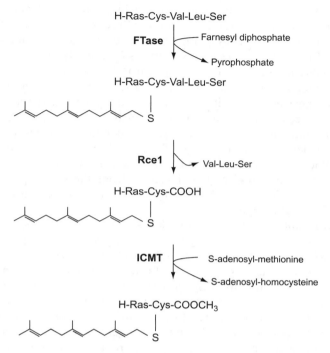

Fig. 15.1. Posttranslational processing pathway of the H-Ras protein. The first posttranslational processing step is prenylation of the carboxy-terminal cysteine with the 15 carbon isoprene, farnesyl, catalyzed by farnesyl transferase (FTase). This is followed by proteolytic removal of the three amino acids following the farnesyl cysteine, catalyzed by Ras converting enzyme (Rce1). The newly formed farnesyl-cysteine carboxy-terminus is then methylated by isoprenyl cysteine methyl transferase (ICMT). Following these events, H-Ras and N-Ras undergo further modification by palmitoylation.

amino acid residues following the farnesyl cysteine, catalyzed by Ras converting enzyme 1 (Rce1) [13]. Subsequently, the farnesylated cysteine is methylated by isoprenyl cysteine methyl transferase (ICMT) [14]. The elucidation of these posttranslational processing steps and the cloning and purification of FTase led to a concerted effort in academic and pharmaceutical laboratories to identify potent, selective inhibitors of this enzyme.

III. Discovery of Farnesyl Transferase Inhibitors

FTase recognizes the CAAX motif on substrate proteins (where C is cysteine, A is usually an aliphatic residue, and X is any other amino acid) and catalyzes the formation of a farnesylthioether on the cysteine residue.

CAAX analogues provided a logical starting point for development of FTIs, since this tetrapeptide is the minimal sequence required to bind to the enzyme [15,16]. CAAX peptides in which X = Ser or Met are good substrates for FTase and are poor substrates for the related prenyl transferase, geranylgeranyltransferase-I (GGTase-I), which transfers a 20-carbon isoprene onto proteins. The opposite is true for peptides where X = Leu [17]. This suggested that inhibitors with selectivity for FTase over GGTase-1 could be discovered. Tetrapeptides in which the A2 of the CAAX motif was an aromatic amino acid and which contained a basic cysteine amino group were identified as FTIs but not FTase substrates [18,19]. For example, Cys-Val-Phe-Met was an FTI, but acetyl-Cys-Val-Phe-Met was a substrate. By removing two amide carbonyl groups from these tetrapeptides, the Merck group devised the CAAX-competitive peptidomimetics L-731,734 and L-731,735 [20] (Figure 15.2).

L-731,734 and L-731,735 inhibit FTase with IC_{50} values of 282 and 18 nM, respectively, and are highly selective against other prenyl transferases. L-731,735 is competitive with respect to the CAAX substrate and noncompetitive with respect to the farnesyl diphosphate prenyl group donor. L-731–734 inhibits Ras farnesylation in cells with an IC_{50} of 100 nM and does not block protein geranylgeranylation. At a concentration of 1 μM, L-731,734 inhibits growth and induces a morphology change in Rat1 fibroblasts transformed with a *ras* oncogene, but not fibroblasts transformed with oncogenes that do not require farnesylation. In contrast, L-731,735 has little effect on cells. Presumably, L-731,734 acts as a prodrug, which readily penetrates cells and is subsequently hydrolyzed to L-731,735. These initial compounds showed that selective FTase inhibition is possible and can inhibit the growth of Ras-transformed tumor cells. This led to the discovery of other peptidomimetic FTIs, some lacking the mercapto group found in cysteine [21–24] and the identification of L-778,123 (Figure 15.3) that was progressed into human clinical trials by Merck.

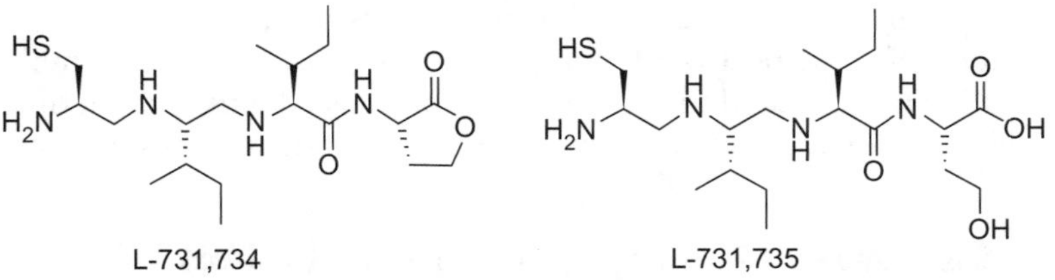

FIG. 15.2. Early peptidomimetic inhibitors of farnesyl transferase. This figure shows the structure of two of the early FTIs based on modifications of the CAAX sequence. These compounds, identified by Kohl *et al.* at Merck Research Labs were shown to be potent and selective inhibitors of H-Ras-induced cellular transformation.

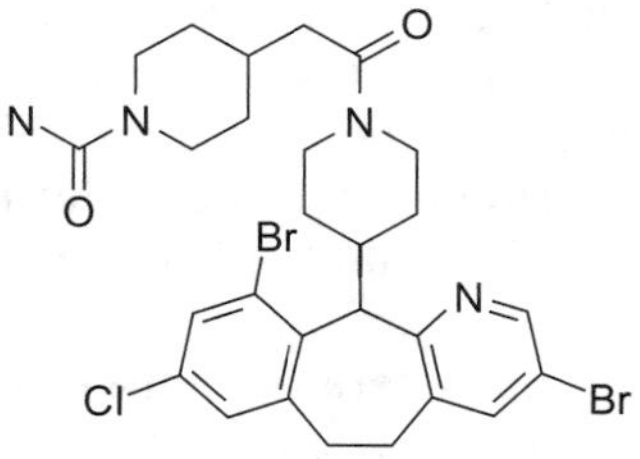

Lonafarnib 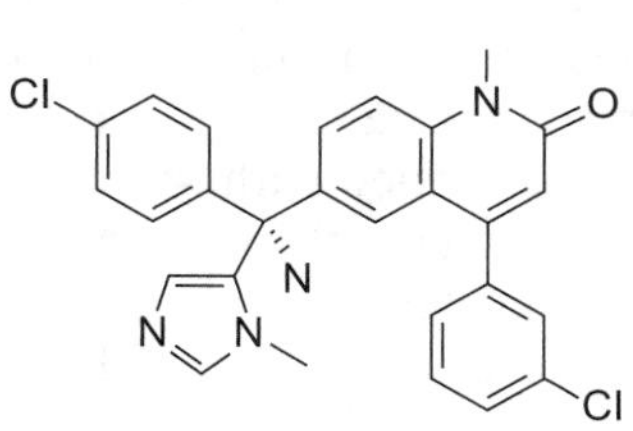   Tipifarnib

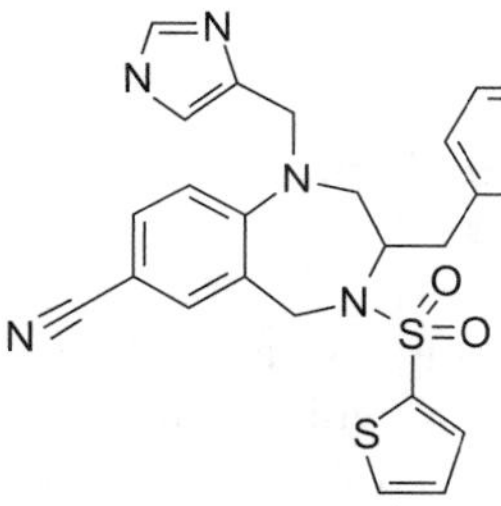

BMS-214662 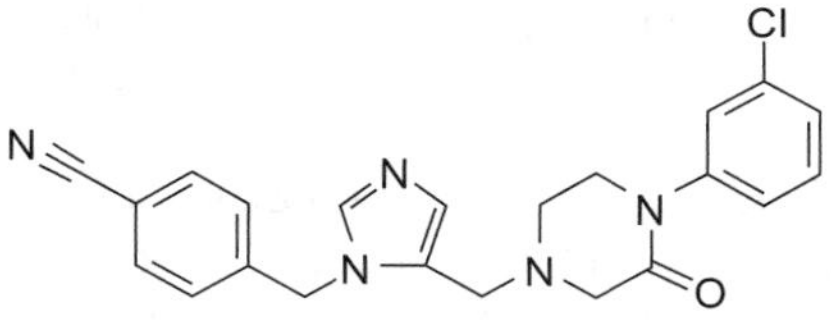   L-778,123

FIG. 15.3. FTIs that advanced into clinical studies in cancer. This figure shows the structures of four of the FTIs that were advanced into human studies in cancer patients. These compounds show a high degree of structural diversity. Tipifarnib and lonafarnib were advanced into later stage clinical trials in cancer patients where they failed to demonstrate robust efficacy despite encouraging signs of activity in smaller studies. Lonafarnib is currently being evaluated for the treatment of children with Hutchinson–Gilford Progeria Syndrome. This figure was originally published in Ref. [25]. © The American Society for Biochemistry and Molecular Biology.

By screening compound collections, followed by medicinal chemistry efforts, other groups identified nonpeptidomimetic FTIs of diverse structures that were competitive with respect to CAAX peptide binding. These include R-11577 (tipifarnib) from Janssen Pharmaceuticals, SCH 66336 (lonafarnib) from Schering-Plough Research Institute, and BMS-214662 from Bristol–Meyers Squibb. Their structures are shown in Figure 15.3 [26]. These compounds were progressed into clinical trials with tipifarnib and lonafarnib advancing into later stage trials in cancer patients. Potent and selective FTIs that are competitive with farnesyl pyrophosphate have also been reported but not have entered clinical studies [21,22].

X-ray crystallographic studies have been very useful in elucidating the binding modes of FTIs and aiding in the design of higher-affinity molecules. FTase has a large active site containing a zinc ion that is involved in catalysis. Substrates and inhibitors containing a mercapto group, such as a cysteine, bind to FTase with the sulfur coordinating to the zinc [27]. Other zinc coordinating groups have been incorporated into FTIs. The imidazole moiety of tipifarnib and BMS-214662 coordinate to the zinc while other interactions are formed with the remaining functionality in the molecules [28]. It is interesting that lonafarnib is a potent FT inhibitor but does not coordinate to zinc [29].

Recent publications on novel FTIs indicate that there continues to be interest in targeting this enzyme [30,31]. Some of these compounds have selectivity for inhibiting microbial FTase over mammalian FTase [32,33], suggesting potential use as anti-infective agents.

IV.   Activity of FTIs in Preclinical Cancer Models and Alternative Prenylation of Ras

There have been numerous reports describing the activity of various FTIs in preclinical cancer models. This review will briefly summarize these reports, with a focus on lonafarnib. In general, structurally diverse FTIs have demonstrated similar activity across a variety of models. One exception is BMS-214662 which is reported to induce a robust apoptotic response in cells in culture [34]. The reason for the distinct profile of this compound is not clear, but likely suggests inhibition of another cellular target in addition to FTase, possibly GGTase-II which modifies the Rab family of small GTPases [35].

The initial preclinical characterization of FTIs was performed primarily using H-Ras-driven cancer models. The results in these models were very promising and suggested that FTIs would be effective cancer therapeutics.

A.   IN VITRO ACTIVITY

In vitro studies showed that FTIs prevented the farnesylation and membrane localization of H-Ras and reversed H-Ras-induced morphological transformation. In addition, FTIs potently inhibited proliferation on plastic and in soft agar of H-Ras-transformed cells [20,36]. In contrast to the dramatic effects on H-Ras function, fibroblasts transformed by K-Ras show a reduced sensitivity to FTIs. An additional study showed that cells transformed by a chimeric Ras protein comprising the N-terminal 164 amino acids of H-Ras and the C-terminal 24 amino acids of K-Ras also displayed reduced sensitivity to FTI treatment [8]. Differential effects of FTIs in H-Ras- and K-Ras-driven models were also observed in animal models as described below.

Studies with human tumor cell lines showed that the presence or absence of a *ras* mutation did not predict FTI sensitivity [37]. Lonafarnib inhibits the anchorage-independent growth of a variety of human cancer cell lines [38]. This growth inhibition is independent of *ras* mutational status and is observed in cells with activated H-*ras*, K-*ras,* or N-*ras* and in cells that lack an activated *ras* oncogene. Lonafarnib slows growth of anchorage-dependent human tumor cell lines on plastic; in many lines, cells accumulate in the G2/M phase of the cell cycle [38]. Growth inhibition is enhanced by combining FTIs with other chemotherapeutic agents; in particular, FTIs have demonstrated a marked synergistic interaction when combined with antimicrotubule agents, including paclitaxel and docetaxel [39–41].

Lonafarnib potently inhibits the growth of BCR/ABL-transformed cells *in vitro* [42], including primary human chronic myeloid leukemia cells. It synergizes with the BCR/ABL kinase inhibitor STI-571 (Gleevec®) to inhibit proliferation and induce apoptosis in BCR/ABL-positive leukemic cell lines, including lines that are Gleevec-resistant [42–44]. Enhanced antiproliferative activity is also seen when lonafarnib is combined with antiestrogens to treat estrogen-receptor-positive breast cancer cells *in vitro* [45].

B.   ALTERNATIVE PRENYLATION OF K-RAS AND N-RAS

The differential sensitivity of H- and K-Ras-transformed cells to FTIs and the Ras-independent activity observed in human cancer cell lines led to a reevaluation of the effects of FTIs on the individual Ras proteins [46]. Using highly purified enzyme preparations of FTase and GGTase-I, kinetic studies were conducted using purified Ras proteins. All four Ras proteins were efficient substrates for FTase with K-Ras having the highest catalytic efficiency. N-, K-4A-, and K-4B-Ras, but not H-Ras, were also found to be GGTase-I substrates *in vitro* [11]. The catalytic efficiency of the reaction with FTase is much higher than that with GGTase-I, accounting for the fact that these proteins normally undergo farnesylation in cells. These results suggested the possibility that K- and N-Ras might be geranylgeranylated in FTI-treated cells, providing a mechanistic explanation for the observation that cells transformed by K-Ras proteins are generally more resistant to FTI treatment.

Experiments in cell culture showed that K- and N-Ras were farnesylated when overexpressed in COS monkey kidney cells, however, following FTI treatment both K- and N-Ras became geranylgeranylated. In contrast, H-Ras prenylation was completely blocked by FTI treatment. Further, upon treatment of DLD-1 human colorectal cancer cells with FTI,

282   W. ROBERT BISHOP, ET AL.

endogenously expressed mutant K-Ras and wild-type N-Ras underwent a switch from farnesylation to geranylgeranylation [47,48].

This "alternative prenylation" of K-Ras with a 20-carbon geranylgeranyl isoprene supports plasma membrane localization (Figure 15.4) and supports proliferative signaling. In addition, geranylgeranylated mutant Ras proteins are capable of transforming cells, similar to their farnesylated counterparts [50].

Overall, the effects of FTI treatment on Ras proteins are isoform specific. The prenylation of H-Ras is completely abrogated, the nonprenylated H-Ras accumulates in the cytoplasm and its physiologic functions are blocked. FTI treatment of H-Ras-transformed fibroblasts or human tumor cell lines harboring a mutant H-Ras allele results in reversion of the oncogenic phenotype. This is accompanied by G1 arrest and apoptosis under certain conditions (e.g., when cells are deprived substratum attachment)

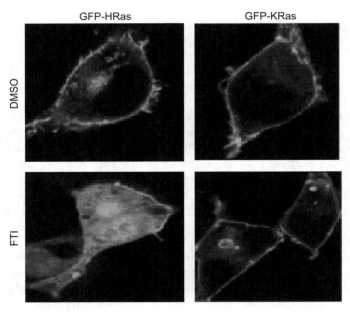

FIG. 15.4. FTIs block the plasma membrane association of H-Ras, but not K-Ras. This figure shows fluorescence micrographs of fibroblasts expressing H-Ras or K-Ras tagged with green fluorescent protein at their amino-termini. In DMSO-treated cells, both proteins localize to the plasma membrane. In cells treated with lonafarnib, H-Ras becomes cytosolic. In contrast, K-Ras remains membrane-associated in the presence of FTI due to its alternative prenylation by geranylgeranyl transferase-I. This figure was originally published in Ref. [49]. © The American Society for Biochemistry and Molecular Biology. (See color plate section in the back of the book.)

[51]. The presence of a mutant H-Ras allele in a tumor or transformed cell correlates with sensitivity to FTIs. In contrast, FTIs do not abrogate the prenylation of K-Ras or block its membrane association. The presence of a mutant K-Ras allele in human tumor cells does not predict sensitivity or resistance to FTIs.

C. Activity in Animal Models of Cancer

As would be predicted based on the observations described above, cancer models driven by mutant H-Ras are very sensitive to FTIs. Lona-farnib treatment of mice inoculated with H-Ras-transformed NIH-3T3 cells resulted in complete tumor growth inhibition; all mice in the treatment group were tumor free at the end of the experiment [51a]. This result was extended to a human H-Ras tumor xenograft model, EJ bladder cancer, where complete tumor growth inhibition was also observed [51a]. The sensitivity of H-Ras dependent tumors to lonafarnib extended to transgenic animal models. Mice engineered to express a mutated H-Ras under the control of the Whey-acidic protein promoter spontaneously develop both salivary and mammary gland tumors. Treatment of these mice with lona-farnib after tumor development resulted in significant tumor regression and near tumor eradication [52]. Similar results were reported by Kohl *et al.* [53] using a distinct FTI in an MMTV-v-*Ha-ras* transgenic mouse model.

The *in vivo* results with K- or N-Ras mutant tumors were not as striking. Activity in K- and N-Ras transgenic mouse models was less robust than that observed in H-Ras models, with tumor stasis, but not regressions, typically observed [54,55]. Some human tumor xenografts expressing mutant K- or N-Ras remain sensitive to lonafarnib, however, neither mutation predicted sensitivity to FTI treatment [52,56]. These *in vitro* and *in vivo* results, taken together, suggest that FTIs will not be effective inhibitors of Ras signaling in patients with K- or N-Ras mutant tumors. Despite this, FTI treatment often resulted in significant tumor growth inhibition in a variety of human tumor xenografts, including some with activating mutations in K-Ras.

Lonafarnib treatment also resulted in long-term, disease-free survival in mice injected with BCR/ABL-transformed cells and in transgenic mice harboring a BCR/ABL oncogene (e.g., Refs. [42,56a]). In addition, FTIs demonstrate enhanced *in vivo* efficacy in animal models when combined with various cytotoxic chemotherapeutic agents including taxanes. Syner-gistic tumor growth inhibition or regression was demonstrated in breast, ovarian, and prostate cancer xenograft models when lonafarnib was admi-nistered in combination with docetaxel or paclitaxel [57,58]. Enhanced *in vivo* activity was also demonstrated when lonafarnib was combined with hormonal therapies to treat estrogen-receptor-positive breast cancer xenografts [45,59].

This broad antitumor activity in preclinical models both as single agent and in combination led to an intensive search for other farnesylated proteins that may contribute to this antitumor activity.

V.   Impact of FTIs on Other Farnesylated Proteins

A number of other cellular proteins are isoprenylated by addition of either farnesyl or the 20-carbon geranylgeranyl isoprene. Geranylgeranylation is catalyzed by two distinct transferases (GGTase-I and GGTase-II). Lonafarnib and tipifarnib are selective inhibitors of FTase and do not inhibit protein geranylgeranylation [36,56], while L-778,123 also has modest activity against GGTase-I [60].

Farnesylated proteins fall into two classes with respect to their response to FTase inhibition. One class of proteins, including H-Ras, accumulates in their unprenylated state. The other class of proteins, including K- and N-Ras, becomes alternatively prenylated by GGTase-I in FTI-treated cells. A number of proteins have been analyzed for their ability to be alternatively prenylated. Since alternative prenylation supports at least some of the same functions as farnesylation, we hypothesize that farnesylated proteins that are not alternatively prenylated are more likely to be targets responsible for the antitumor effects of FTIs (although it cannot be ruled out that switching the isoprene content of some proteins may result in subtle changes in function). Additional consideration was given to proteins that participate in oncogenic signaling pathways or other pathways that may, in part, explain some of the biological responses to FTIs. Candidate proteins that meet these criteria include the centromere-associated proteins, CENP-E and CENP-F, and the small GTPase Rheb (Ras homolog enriched in brain) [61,62]. Some of these candidate proteins are discussed below. It is likely that inhibition of farnesylation of a single protein does not explain the complex and pleiotropic effects observed following FTase inhibition.

A.   SMALL GTPASE SUBSTRATES OF FTASE

In addition to the Ras proteins, a number of other small GTPases including RhoD, Rnd1, Rnd2, Rnd3, and TC10 are substrates for FTase. Like K- and N-Ras, these proteins are also alternatively prenylated in FTI-treated cells [63]. Interestingly, despite the alternative prenylation of the Rnd family of small GTPases, they become mislocalized in FTI-treated cells; this is accompanied by changes in cellular phenotype suggestive of inhibition of cell rounding. Interestingly, Rnd3 is implicated in melanoma migration and the data presented by Roberts *et al.* [63] suggest that FTIs may inhibit this process.

1. RhoB

The small GTPase RhoB is a substrate for both FTase and GGTase-I in untreated cells [64]. In cells treated with FTIs, RhoB becomes exclusively geranylgeranylated. Some studies have suggested functional differences between farnesylated RhoB (RhoB-F) and geranylgeranylated RhoB (RhoB-GG) and that FTI treatment results in a gain of growth inhibitory function for RhoB-GG [65]. In support of a growth inhibitory role for RhoB-GG, overexpression of a form of RhoB engineered to be only geranylgeranylated reverses the transformed phenotype of Ras-transformed Rat1 fibroblasts. Further, in Ras-transformed NIH-3T3 cells, RhoB-F enhanced and RhoB-GG suppressed cell growth [64,66]. In addition, RhoB-GG, but not RhoB-F, was found to inhibit Ras-induced activation of AKT and NFκB. Although these results are intriguing, expression of wild-type RhoB, RhoB-F, or RhoB-GG were all reported to inhibit cell focus formation and growth in soft agar of the human pancreatic tumor cell line Panc-1 [67]. In addition, FTIs were found to inhibit anchorage-independent growth of RhoB null cells [68]. RhoB that become exclusively geranylgeranylated in the presence of FTIs and Rnd1, 2, and 3, whose cellular localization is altered by FTIs may contribute to some of the cellular changes these molecules induce. Whether these changes are critical to the anticancer effects of FTIs remains to be determined.

2. Rheb

Rheb1 and 2 are farnesylated small GTPases that are ubiquitously expressed in human tissue and upregulated in transformed cells and human tumor cell lines [69–71]. Neither isoform is alternatively prenylated in cells treated with FTIs. Rheb is a component of the mTOR signaling pathway. The tuberous sclerosis complex TSC1/TSC2 (hamartin/tuberin) serves as a GTPase activating protein for Rheb, accelerating its rate of GTP hydrolysis and promoting the formation of the inactive, GDP-bound form [72]. TSC1/TSC2 therefore regulates Rheb function and reduces the level of activated Rheb-GTP.

The TSC1/TSC2 complex is negatively regulated by AKT-mediated phosphorylation of tuberin, and as a result, Rheb activity is increased [73]. Rheb directly interacts with mTOR and positively regulates mTOR signaling [74]. Overexpression of Rheb induces phosphorylation of mTOR substrates, including S6 kinase [75]. Disruption of Rheb function results in inhibition of phosphorylation of the mTOR substrates S6 kinase and 4EBP-1 [70,71,75]. Recent studies demonstrated that the antitumor activity of an FTI in a PTEN-deficient lymphoma model is due to its ability to inhibit Rheb farnesylation [76]. These data suggest that FTIs may have activity in tumors where the mTOR pathway is activated and, potentially, in genetic disorders such as tuberous sclerosis, where Rheb activity is upregulated. Suppression of Rheb function may also contribute to the enhanced activity observed when FTIs are combined with other cancer therapies [70].

B. Cellular Phosphatases as Substrates of FTase

The three members of the PTP-CAAX or PRL family of protein tyrosine phosphatases play a role in regulating cell growth and mitosis [77]. These proteins are farnesylated and are weak substrates for GGTase-I (Refs. [78,79]; W. Robert Bishop and Paul Kirschmeier, unpublished). Ectopic expression of a form of PRL1 engineered to be unprenylated results in defects in mitosis and cytokinesis characterized by chromosome bridges and lagging chromosomes [80]. PRL1 mRNA is elevated in numerous human tumor cell lines and a role for PRL in tumorigenesis has been demonstrated. Overexpression of PRL1 or PRL2 in epithelial cells results in a transformed phenotype in culture and tumor growth in nude mice [78]. PRL3 is also overexpressed in a number of tumors and has been implicated in tumor metastases [81]. Cells expressing mutant PRL1 properly arrest in mitosis following nocodazole treatment suggesting that farnesylated PRL1 is required for proper spindle dynamics but not for the spindle checkpoint [80]. Inhibition of PRL farnesylation may, in part, account for FTI-induced accumulation of cells in the G2/M phase of the cell cycle. Further studies are warranted on the impact of FTIs on this family of phosphatases.

C. Nuclear substrates of FTase

Several nuclear proteins are FTase substrates, including the kinesin motor proteins CENP-E and F and nuclear lamins A and B.

1. CENP-E and -F

CENP-E is a kinesin motor protein that is involved in the alignment of chromosomes on the metaphase plate during mitosis [82,83]. CENP-F is a centromere-associated protein which is thought to play a role in the mitotic spindle checkpoint [25]. Both CENP-E and CENP-F are farnesylated proteins not subject to alternative prenylation [61].

In cells treated with FTIs, CENP-E and CENP-F still localize to kine-tochores during prometaphase; however, it has been reported that FTIs reduce the levels of CENP-F at the kinetochore. FTI treatment also blocks CENP-E association with spindle microtubules [61]. Ectopic expression of the C-terminal kinetochore-binding domain of CENP-F delayed progres-sion through G2/M in a farnesylation-dependent manner, suggesting that farnesylated CENP-F is required for efficient G2/M progression [84]. Lona-farnib also inhibits CENP-F farnesylation [61]. Disruption of CENP-E and CENP-F function by FTIs may contribute to the failure of chromosomes to properly align in metaphase and to the subsequent accumulation of some human tumor cells in G2/M observed in response to FTI treatment [61,85,86]. The relative contribution of inhibition of CENP and PRL func-tion to the observed cell cycle effects remains to be determined. Effects on CENP function may also contribute to the synergistic antitumor activity of FTIs in combination with taxanes.

2. Nuclear Lamins

The nuclear lamina consists of lamin proteins that are required for nuclear envelope assembly [87,88]. Lamin B was one of the first proteins shown to be modified by prenylation [89]. Lamin A is also farnesylated, but it is unique among farnesylated human proteins [90]. Following modifica-tions by FTase, Rce1, and ICMT, prelamin A undergoes a farnesylation-dependent proteolytic cleavage catalyzed by the zinc metalloprotease ZMPSTE24. This cleavage removes the carboxy-terminal 15 amino acids, including the farnesylated cysteine [49,91]. As a consequence, mature lamin A lacks a farnesyl modification. Defects in the nuclear envelope have not been reported in cells treated with FTIs, suggesting that the farnesylated intermediate is not required for normal nuclear envelope assembly. The inhibition of lamin A farnesylation provides the rationale for studying FTIs in the premature aging disorder HGPS as described in Section VI.

VI. Clinical Studies of FTIs in Cancer

Several FTIs were advanced into clinical trials. The most extensive clinical experience has been with lonafarnib and tipifarnib; this review will focus primarily on more recent clinical results with these two mole-cules. Earlier studies are discussed in more detail in the 2006 review by Basso et al. [92]. While both lonafarnib and tipifarnib demonstrated some promising signs of efficacy in early clinical studies, both compounds failed to display robust efficacy in later stage, larger, randomized studies.

A. PHASE 1 STUDIES: SAFETY/TOLERABILITY

Lonafarnib was generally well tolerated at doses of up to 200 mg when dosed orally on a continuous twice-daily schedule. Higher doses of lonafarnib were associated with dose-limiting gastrointestinal toxicities (e.g., Ref. [93]). Myelosuppression was also observed using higher twice-daily doses. Tipifarnib is dosed orally at 300–600 mg twice daily for 21 consecutive days followed by 1 week off. Dose-limiting toxicity is myelosuppression manifested typically as neutropenia [94]. Some single-agent clinical activity was reported in Phase 1 studies of lonafarnib (e.g., Ref. [95]). Both tipifarnib and lonafarnib were advanced into Phase 2 studies as single agent and in combination with chemotherapy. A brief summary of these studies is given below.

B. HEMATOLOGIC MALIGNANCIES

FTIs have been evaluated in a number of hematologic cancers. Some promising results were seen in Phase 2 studies in myelodysplastic syndrome (MDS)/chronic myelomonocytic leukemia (CMML) and acute myelogenous leukemia (AML).

A Phase 2 study of lonafarnib in patients with MDS/CMML was conducted by Feldman et al. [96]. An overall response rate of 24% was observed, with two patients achieving complete remission and one a partial response. A number of patients demonstrated hematological improvement and 5 of 19 subjects who were platelet transfusion dependent at baseline became transfusion-free for a median duration of 185 days. Based on these encouraging results, a Phase 3 study of lonafarnib versus placebo was initiated in subjects with MDS/CMML who were platelet transfusion dependent. This study was terminated early based on a treatment-blinded review of data that indicated that it was unlikely to meet the protocol-defined response criteria (http://www.clinicaltrials.gov/ct2/results?term=lonafarnib).

Tipifarnib was tested as monotherapy in several Phase 1 and 2 trials in patients with AML. The results of these trials suggested clinical activity in elderly patients with newly diagnosed AML. In one study, clinical responses (complete and partial remissions) were observed in 23% of patients (e.g., Ref. [97]). Based on these data, a Phase 3 trial of tipifarnib, administered at 600 mg twice daily for 21 of 28 days, was conducted in 457 elderly newly diagnosed AML patients. The study was randomized to best supportive care and the primary endpoint was overall survival. The complete response rate in the tipifarnib arm was 8% lower than in previous studies, and the overall survival was essentially unchanged between the treatment arm and the control arm [98].

Tipifarnib has also been evaluated in combination with other agents for the treatment of AML. A recent study of tipifarnib (300 mg BID) in combination with idarubicin and cytarabine was conduced in 95 patients with previously untreated acute AML or high-risk MDS. Complete responses were reported in 64% of patients [99]. The results of these combination studies appear encouraging but additional clinical trials are needed to confirm the benefit of adding tipifarnib to approved therapies in AML.

C. SOLID TUMORS

Clinical activity has also been observed with FTIs in breast cancer. Tipifarnib administered as a single agent by either continuous or intermittent oral dosing demonstrated objective response rates of 10% and 14%, respectively, in patients with advanced breast cancer [100]. When evaluated in combination with hormonal agents, however, FTIs have not demonstrated adequate efficacy to warrant Phase 3 evaluation in breast cancer (e.g., Ref. [101]).

Encouraging early clinical data was also generated with the combination of lonafarnib and paclitaxel in NSCLC patients who failed prior taxane therapy [102,103]. Based on this, a randomized Phase 3 study of paclitaxel and carboplatin with either lonafarnib or placebo as first-line treatment of subjects with NSCLC was initiated. This study was terminated early after an interim analysis determined that there was insufficient evidence of efficacy to warrant further enrollment (http://www.clinicaltrials.gov/ct2/results?term=lonafarnib). It is unclear why the activity observed in Phase 2 did not carry over into Phase 3 studies, although this may reflect differences in patient population and/or trial design.

A number of other solid tumor indications have been explored with tipifarnib and lonafarnib including pancreatic cancer [104], squamous cell carcinoma of the head and neck [105], and brain tumors. Most of these studies have not demonstrated sufficient activity to support further clinical trials. While there are still several ongoing clinical studies with FTIs in cancer (e.g., http://www.clinicaltrials.gov/ct2/results?term=tipifarnib), the data to date have been disappointing and underscore the high failure rate of cancer therapeutics in later stage trials.

D. PHARMACODYNAMIC ACTIVITY

Various proteins have been evaluated as potential pharmacodynamic markers of FTI activity. Due to alternative prenylation, K-Ras and N-Ras cannot be used in this way, since their farnesylated and geranylgeranylated forms are indistinguishable based on electrophoretic or immunohistochemical properties. Attempts to utilize H-Ras as a marker have been limited by the low level of expression of this protein in many tissues. It is important to note that CAAX-competitive FTIs display different IC_{50} values when tested against different FTase substrates. Farnesylation of low-affinity substrates such as H-Ras is inhibited at lower FTI concentrations than is farnesylation of higher-affinity substrates such as K-Ras and HDJ-2. This is an important consideration when surrogate proteins are used to track pharmacodynamic activity. Most efforts to demonstrate target inhibition in the clinic with FTIs have focused on the nuclear lamin, prelamin A, and the cochaperone protein HDJ-2 (DNAJ).

As described above, during the production of mature lamin A, the carboxy-terminal 15 amino acids, including the farnesylated cysteine, are removed by proteolytic cleavage. In FTI-treated cells, there is an accumulation of prelamin A which retains this 15-amino acid epitope. An antibody directed against this region can be used to detect accumulation of unfarnesylated prelamin A by immunoblot analysis or immunohistochemistry. An electrophoretic mobility shift assay is used to detect inhibition of HDJ-2 farnesylation. The unfarnesy-lated protein displays a slower electrophoretic mobility.

When tumor cells in culture are treated with lonafarnib, unfarnesylated prelamin A and HDJ-2 accumulate [106]. Similar experiments have been performed evaluating HDJ-2 gel mobility shifts from xenograft tumors derived from mice treated with lonafarnib or vehicle. In general, greater antitumor activity is observed in tumors that displayed greater accumulation of unfarnesylated HDJ-2 (unpublished data), however a robust model defining the relationship between antitumor efficacy and the level of unfarnesylated HDJ-2 is lacking.

The levels of unfarnesylated HDJ2 and, to a lesser extent, prelamin A have been evaluated in patient-derived samples from a number of clinical studies. Clinical lonafarnib doses of greater than 100 mg BID resulted in inhibition of protein farnesylation. Accumulation of prelamin A and/or unfarnesylated HDJ-2 has been observed in tumor biopsies (melanoma, head, and neck), buccal mucosa, and peripheral blood mononuclear cells derived from subjects treated with lonafarnib and tipifarnib (Figure 15.5) (e.g., Refs. [95,104]).

E.   IDENTIFICATION OF FTI-RESPONSIVE PATIENTS

Failure to elucidate the critical FTI target proteins has posed a significant challenge for FTI clinical development in oncology as responsive patient subpopulations are not able to be identified. Based on retrospective gene expression profiling, Raponi et al. [107] reported a two gene classifier that correlated with clinical response to tipifarnib in patients with relapsed

Prelamin A Lamin A

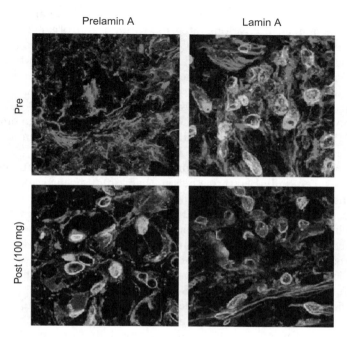

FIG. 15.5. Accumulation of unfarnesylated prelamin A in a tumor biopsy from a patient with squamous cell carcinoma of the head and neck following treatment with lonafarnib. Tumor biopsies were derived from a patient with squamous cell carcinoma of the head and neck both before treatment and following 2 weeks of treatment with lonafarnib at a dose of 100 mg, twice daily. Cells were stained with an antibody that specifically recognizes the 15-amino acid carboxy-terminal epitope that is retained when lamin A farnesylation is inhibited (prelamin A) or with an antibody that recognizes an epitope nearer the amino-terminus which is present in both processed and unprocessed lamin A. While lamin A is detected in both sets of samples, prelamin A is only seen in the posttreatment sample, consistent with farnesyl transferase inhibition. (See color plate section in the back of the book.)

or refractory AML. Tools such as this have not been validated prospectively in a clinical study. The availability of a clear genetic profile that could serve as a patient selection biomarker would have greatly improved the probability of success for the clinical studies of FTIs.

VII. Hutchinson–Gilford Progeria Syndrome: A Disease of Farnesylation?

HGPS is a rare (frequency 1 in 4 million), fatal, premature aging disorder [108]. In 2003, the mutation responsible for HGPS was identified as an autosomal dominant C to T transition at nucleotide 1824 in the gene encoding the nuclear lamin A protein (*LMNA*) [109,110]. Lamin A plays a critical role in nuclear envelope integrity, chromosome organization, DNA replication, and gene transcription [111]. Lamin A undergoes a complex series of posttranslational processing steps involving a farnesylated intermediate and subsequent proteolytic removal of the farnesylated carboxy-terminus. The mutation in HGPS activates a cryptic mRNA splice site, resulting in synthesis of a lamin A protein with an internal deletion of 50 amino acids. This deletion leaves the CAAX motif intact, but eliminates the ZMPSTE24 cleavage site, resulting in accumulation of an abnormal, farnesylated form of lamin A, termed "progerin" [111,112].

Children with HGPS undergo normal fetal and early postnatal development, but severe failure to thrive becomes apparent during the first year of life. Other clinical manifestations of HGPS include sclerodermatous skin changes, osteoporosis, alopecia, and joint contractures. The children exhibit no deficit in intellectual development [113]. Death at an average age of 12.6 years occurs from myocardial infarction or stroke, as a consequence of widespread occlusive vascular disease [114]. A more severe, perinatal-lethal progeroid disorder termed restrictive dermopathy results from homozygous loss of function mutations in ZMPSTE24 [111].

The discovery that the genetic defect in HGPS results in the accumulation of a farnesylated protein led to great interest in evaluating the effects of FTIs in preclinical models of this disease. Initial *in vitro* studies demonstrated that, when grown in the presence of FTIs, the nuclei of fibroblasts derived from HGPS patients lost their misshapen appearance and their characteristic nuclear blebbing [115–117]. Several genetically engineered mouse models of HGPS have been used to evaluate FTIs. In a *Zmpste24*$^{-/-}$ mouse model, Fong *et al.* [118] demonstrated that FTI treatment improved survival, improved grip strength performance, and reduced the number of rib fractures, although animals still succumbed to progeroid disease. In gene-targeted mice bearing a mutant *Lmna* allele designed to produce only progerin, FTI treatment increased adipose tissue mass, improved body weight curves, reduced the number of rib fractures, and improved bone mineralization and bone cortical thickness [119]. Transgenic mice carrying a copy of the G608G *LMNA* mutation displays a phenotype characterized by a progressive loss of vascular smooth muscle cells in the media of large arteries, a phenotype that is similar to that observed in autopsy studies of HGPS patients. Importantly, tipifarnib was shown to prevent both the onset and progression of cardiovascular disease in this model [120].

A clinical study of lonafarnib in children with HGPS was initiated in May 2007. Twenty-seven children from around the world participated in this study which was conducted at Children's Hospital, Boston. The trial is a single-agent, open-label, historically controlled study where all children receive lonafarnib treatment for 2 years. One factor that will be critical to the successful interpretation of the results of this study was the significant effort that was made prior to study initiation to establish the detailed phenotype of HGPS [113] and to establish baseline growth parameters for each individual child who entered the study. Clinical results from this study are expected shortly. These children are now participating in a second study where they are continuing on lonafarnib in combination with a statin and a bisphosphonate. A total of 45 children are enrolled in this triple combination trial.

VIII. Conclusions and Other Therapeutic Opportunities for FTIs

FTIs have experienced a number of challenges during their development as cancer therapeutics. Despite alternative prenylation of K- and N-Ras, they demonstrated promising activity in preclinical cancer models and early evidence of activity in Phase 1 and 2 cancer studies was observed. This early activity failed to translate into robust clinical activity in randomized, controlled trials. Clearly, one of the issues this class of molecules faced was the inability to prospectively identify a responsive patient population based on a specific mutation or gene expression profile. Based on the knowledge we have gained with respect to Ras processing pathways, and the demonstration that FTIs fully block the transforming activity of H-Ras, it remains possible that FTIs could provide benefit to patients with tumors driven by an activating H-Ras mutation. This hypothesis has not been effectively tested clinically, but in the age of personalized therapy and efficient sequencing technologies, it probably deserves additional examination. The role of FTI inhibition of the Rheb/mTOR signaling pathway is also worthy of further investigation and could potentially lead to a distinct responder hypothesis based on activation of this pathway.

Beyond cancer, there are several infectious diseases where FTI therapy has been considered. Hepatitis delta virus (HDV) is a satellite virus of hepatitis B virus that can increase the incidence and severity of liver disease in coinfected individuals. Farnesylation of the large antigen of HDV is required for viral particle assembly. Geranylgeranylation of this protein has not been observed, raising the possibility that FTIs may be useful agents for treatment of HDV infection [121]. FTIs have demonstrated efficacy in an animal model of HDV infection [122].

There are also several parasitic diseases caused by protozoan pathogens where species-specific FTIs may have promise, including malaria, caused by *Plasmodium falciparum,* and African sleeping sickness, caused by

Trypanosoma brucei [123]. Gelb and colleagues have identified a number of potent, selective FTIs that have potential in this area [32,33]. FTI selectivity for the parasitic enzymes is being guided by structural studies [124]. Hopefully, approaches such as this will be advanced through Product Development Partnerships such as the Medicines for Malaria Venture to meet the enormous unmet medical need.

In addition to HGPS, there are other rare genetic disorders where FTIs and other signal transduction inhibitors should be considered. There is a family of related disorders termed "Rasopathies" which are driven by activating mutations in the Ras-MAP kinase signaling pathway [125]. Among these is Costello syndrome, where the underlying molecular defect is an activating mutation in the H-*ras* gene [126,127]. The mutation found in Costello patients is typically Gly^{12}Ser (or another substitution at this position), rather than the Gly^{12}Val mutation more often found in human cancer. The evaluation of FTIs in preclinical models of Costello Syndrome is warranted. The ideal FTI for evaluation in Costello may be a brain-penetrant molecule with potential to address both the peripheral and central features of this disease.

Since the discovery of Ras farnesylation in 1989, the preclinical and clinical development of FTIs has faced many challenges. Despite the setbacks in providing a novel therapeutic option for the treatment of major epithelial cancers driven by K-Ras mutation, there remain a number of opportunities for targeting FTase to have an impact on human disease. These range from opportunities in H-Ras driven cancers to extremely rare genetic disorders such as HGPS to diseases with huge global impact and unmet medical need such as malaria.

ACKNOWLEDGMENTS

The authors would like to acknowledge our collaborators, Dr. Merrill Kies (MD Anderson Cancer Center) and Dr. Fadlo Khuri (Emory University Medical Center), who provided the tumor samples for immunohistochemical analysis.

REFERENCES

1. Casey, P.J., Solski, P.A., Der, C.J., and Buss, J.E. (1989). p21ras is modified by a farnesyl isoprenoid. *Proc Natl Acad Sci USA* 86:8323–8327.
2. Gibbs, J.B. (1991). Ras C-terminal processing enzymes—new drug targets. *Cell* 65:1–4.
3. Barbacid, M. (1987). *Ras* genes. *Annu Rev Biochem* 56:779–827.
4. Bos, J.L. (1989). *Ras* oncogenes in human cancer: a review. *Cancer Res* 49:4682–4689.
5. Bos, J.L. (1989). Detection of ras oncogenes using PCR. In PCR Technology, (H.A. Erlich, Ed.), p. 228. . PCR Technology, Stockton Press, New York, NY.

6. Rivera, M., Ricarte-Filho, J., Knauf, J., Shaha, A., Tuttle, M., Fagin, J.A., and Ghossein, R.A. (2010). Molecular genotyping of papillary thyroid carcinoma follicular variant according to its histological subtypes (encapsulated vs. infiltrative) reveals distinct BRAF and RAS mutation patterns. *Mod Pathol* 23:1191–1200.

7. Willumsen, B., Norris, K., Papageorge, A., Hubbert, N.L., and Lowy, D.R. (1984). Harvey murine sarcoma virus p21 ras protein: biological and biochemical significance of the cysteine nearest the carboxy terminus. *EMBO J* 3:2581–2585.

8. Willumsen, B.M., Christensen, A., Hubbert, N.L., Papageorge, A.G., and Lowy, D.R. (1984). The p21 ras C-terminus is required for transformation and membrane association. *Nature* 310:583–586.

9. Hancock, J., Magee, A., Childs, J., and Marshall, C. (1989). All ras proteins are poly-isoprenylated but only some are palmitoylated. *Cell* 57:1167–1177.

10. Casey, P.J., Solski, P.A., Der, C.J., and Buss, J.E. (1989). p21ras is modified by a farnesyl isoprenoid. *Proc Natl Acad Sci USA* 86:8323–8327.

11. Zhang, F.L., Kirschmeier, P., Carr, D., James, L., Bond, R.W., Wang, L., Patton, R., Windsor, W.T., Syto, R., Zhang, R., and Bishop, W.R. (1997). Characterization of H, K and N-ras as in vitro substrates for isoprenyl protein transferases. *J Biol Chem* 272:10232–10239.

12. Lowe, P.N., Sydenham, M., and Page, M.J. (1990). The Ha-ras protein, p21, is modified by a derivative of mevalonate and methyl-esterified when expressed in the insect/baculovirus system. *Oncogene* 5:1045–1048.

13. Otto, J.C., Kim, E., Young, S.G., and Casey, P.J. (1999). Cloning and characterization of a mammalian prenyl protein-specific protease. *J Biol Chem* 274:8379–8382.

14. Dai, Q., Choy, E., Chiu, V., Romano, J., Slivka, S.R., Steitz, S.A., Michaelis, S., and Philips, M.R. (1998). Mammalian prenylcysteine carboxyl methyltransferase is in the endoplasmic reticulum. *J Biol Chem* 273:15030–15034.

15. Reiss, Y., Goldstein, J.L., Serabra, M.C., Casey, P.J., and Brown, M.S. (1990). Inhibition of purified p21ras farnesyl: proteintransferase by Cys-AAX tetrapeptides. *Cell* 62:81–88.

16. Schaber, M.D., O'Hara, M.B., Garsky, V.M., Mosser, S.C., Bergstorm, D.J., Moores, S.L., Marshall, M.S., Friedman, P.A., Dixon, R.A., and Gibbs, J.B. (1990). Polyisoprenylation of Ras in vitro by a farnesyl-protein transferase. *J Biol Chem* 265:14701–14704.

17. Moores, S.L., Schaber, M.D., Mosser, S.D., Rands, E., O'Hara, M.B., Garsky, V.M., Mardhall, M.S., Pompliano, D.L., and Gibbs, J.B. (1991). Sequence dependence of protein isoprenylation. *J Biol Chem* 266:14603–14610.

18. Goldstein, J.L., Brown, M.S., Stradley, S.J., Reiss, Y., and Gierasch, L.M. (1991). Nonfarnesylated tetrapeptide inhibition of protein farnesyltransferase. *J Biol Chem* 266:15575–15578.

19. Brown, M.S., Goldstein, J.L., Paris, K.J., Burnier, J.P., and Masters, J.C. (1992). Tetrapeptide inhibitors of protein farnesyltransferase: amino-terminal substitution in phenylalanine-containing tetrapeptides restores farnesylation. *Proc Natl Acad Sci USA* 89:8313–8316.

20. Kohl, N.E., Mosser, S.D., deSolms, J., Giuliani, D.L., Pompliano, D.L., Grahm, S.L., Smith, R.L., Scolnick, E.M., and Oliff, A. (1993). Selective inhibition of ras-dependent transformation by a farnesyltransferase inhibitor. *Science* 260:1934–1937.

21. Graham, S. (1995). Inhibitors of protein farnesylation: a new approach to cancer chemo-therapy. *Expert Opin Ther Pat* 5:1269–1285.

22. Ayral-Kaloustain, S., and Skotnicki, J.S. (1996). Ras farnesyltransferase inhibitors. In Annual Reports in Medicinal Chemistry, (J. Bristol, Ed.), Vol. 31, pp. 171–180. Academic Press, San Diego.

23. Leonard, D.M. (1997). Ras farnesyltransferase: a new therapeutic target. *J Med Chem* 40:2971–2990.
24. Williams, T.M., and Dinsmore, C.J. (1999). Farnesyl transferase inhibitors: design of a new class of cancer chemotherapeutic agents. *Adv Med Chem* 4:273–314.
25. Liao, H., Winkfein, R.J., Mack, G., Rattner, J.B., and Yen, T.J. (1995). CENP-F is a protein of the nuclear matrix that assembles onto kinetochores at late G2 and is rapidly degraded after mitosis. *J Cell Biol* 130:507–518.
26. Doll, R.J. (2001). Inhibition of farnesyl protein transferase—a review of the recent patent literature. *IDrugs* 4:1382–1388.
27. Strickland, C.L., and Weber, P.C. (1999). Farnesyl protein transferase: a review of structural studies. *Curr Opin Drug Discov Devel* 2:475–483.
28. Reid, T.S., and Beese, L.S. (2004). Crystal structures of the anticancer clinical candidates R11577 (Tipifarnib) and BMS-214662 complexed with protein farnesyltransferase suggest a mechanism of FTI selectivity. *Biochemistry* 43:6877–6884.
29. Strickland, C.L., Weber, P.C., Windsor, W.T., Wu, Z., Le, H.V., Albanese, M.M., Alvarez, C.S., Cesarz, D., del Rosario, J., Deskus, J., Mallams, A.K., Njoroge, F.G., Piwinski, J.J., Remiszewski, S., Rossman, R.R., Taveras, A.G., Vibulbhan, B., Doll, R. J., Girijavallabhan, V.M., and Ganguly, A.K. (1999). Tricyclic farnesyl protein transferase inhibitors: crystallographic and calorimetric studies of structure-activity relationships. *J Med Chem* 42:2125–2135.
30. Zhu, H.Y., Cooper, A.B., Desai, J., Njoroge, G., Kirschmeier, P., Bishop, W.R., Strickland, C., Hruza, A., Doll, R.J., and Girijavallabhan, V.M. (2010). Discovery of C-imidazole azaheptapyridine FPT inhibitors. *Bioorg Med Chem Lett* 20:1134–1136.
31. Fletcher, S., Keaney, E.P., Cummings, C.G., Blaskovich, M.A., Hast, M.A., Glenn, M.P., Chang, S.-Y., Bucher, C.J., Floyd, R.J., Katt, W.P., Gelb, M.H., Van Voorhis, W.C., Beese, L.S., Sebti, S.M., and Hamilton, A.D. (2010). Structure-based design and synthesis of potent, ethylenediamine-based, mammalian farnesyltransferase inhibitors as anticancer agents. *J Med Chem* 53:6867–6888.
32. Bendale, P., Olepu, S., Suryadevara, P.K., Bulbule, V., Rivas, K., Nallan, L., Smart, B., Yokoyama, K., Ankala, S., Pendyala, R.P., Floyd, D., Lombardo, L.J., Williams, D.K., Buckner, F.S., Chakrabarti, D., Verlinde, C.L.M.J., Van Voorhis, W.C., and Gelb, M.H. (2007). Second generation tetrahydroquinoline-based protein farnesyltransferase inhibitors as antimalarials. *J Med Chem* 50:4585–4605.
33. Bulbule, V.J., Rivas, K., Verlinde, C.L.M.J., Van Voorhis, W.C., and Gelb, M.H. (2008). 2-Oxotetrahydroquinoline-based antimalarials with high potency and metabolic stability. *J Med Chem* 51:384–387.
34. Manne, V., Lee, F.Y., Bol, D.K., Gullo-Brown, J., Fairchild, C.R., Lombardo, L.J., Smykla, R.A., Vite, G.D., Wen, M.L., Yu, C., Wong, T.W., and Hunt, J.T. (2004). Apoptotic and cytostatic farnesyltransferase inhibitors have distinct pharmacology and efficacy profiles in tumor models. *Cancer Res* 64:3974–3980.
35. Lackner, M.R., Kindt, R.M., Carroll, P.M., Brown, K., Cancilla, M.R., Chen, C., de Silva, H., Franke, Y., Guan, B., Heuer, T., Hung, T., Keegan, K., Lee, J.M., Manne, V., O'Brien, C., Parry, D., Perez-Villar, J.J., Reddy, R.K., Xiao, H., Zhan, H., Cockett, M., Plowman, G., Fitzgerald, K., Costa, M., and Ross-Macdonald, P. (2005). Chemical genetics identifies Rab geranylgeranyl transferase as an apoptotic target of farnesyl transferase inhibitors. *Cancer Cell* 7:325–336.
36. Bishop, W.R., Bond, R., Petrin, J., Wang, L., Patton, R., Doll, R., Njoroge, G., Catino, J., Schwartz, J., Windsor, W., Syto, R., Schwartz, J., Carr, D., James, L., and Kirschmeier, P. (1995). Novel tricyclic inhibitors of farnesyl protein transferase: biochemical

characterization and inhibition of ras modification in transfected Cos cells. *J Biol Chem* 270:30611–30618.

37. Sepp-Lorenzino, L., Ma, Z., Rands, E., Kohl, N.E., Gibbs, J.B., Oliff, A., and Rosen, N. (1995). A peptidomimetic inhibitor of farnesyl:protein transferase blocks the anchorage-dependent and -independent growth of human tumor cell lines. *Cancer Res* 55:5302–5309.

38. Ashar, H.R., James, L., Gray, K., Carr, D., McGuirk, M., Maxwell, E., Black, S., Armstrong, L., Doll, R.J., Taveras, A.G., Bishop, W.R., and Kirschmeier, P. (2000). The farnesyl transferase inhibitor SCH 66336 induces a G2/M or G1 phase pause in sensitive human tumor cell lines. *Exp Cell Res* 262:17–27.

39. Moasser, M.M., Sepp-Lorenzino, L., Kohl, N.E., Oliff, A., Balog, A., Su, D.S., Danishefsky, S.J., and Rosen, N. (1998). Farnesyl transferase inhibitors cause enhanced mitotic sensitivity to taxol and epothilones. *Proc Natl Acad Sci USA* 95:1369–1374.

40. Shi, B., Yaremko, B., Hajian, G., Terracina, G., Bishop, W.R., Liu, M., and Nielsen, L.L. (2000). The farnesyl protein transferase inhibitor lonafarnib synergizes with taxanes in vitro and enhances their antitumor activity in vivo. *Cancer Chemother Pharmacol* 46:387–393.

41. Nakamura, K., Yamaguchi, A., Namiki, M., Ishihara, H., Nagasu, T., Kowalczyk, J.J., Garcia, A.M., Lewis, M.D., and Yoshimatsu, K. (2001). Antitumor activity of ER-51785, a new peptidomimetic inhibitor of farnesyl transferase: synergistic effect in combination with paclitaxel. *Oncol Res* 12:477–484.

42. Peters, D.G., Hoover, R.R., Gerlach, M.J., Koh, E.Y., Zhang, H., Choe, K., Kirschmeier, P., Bishop, W.R., and Daley, G.Q. (2001). Activity of the farnesyl protein transferase inhibitor SCH 66336 against BCR/ABL-induced murine leukemia and primary cells from patients with chronic myeloid leukemia. *Blood* 97:1404–1412.

43. Nakajima, A., Tauchi, T., Sumi, M., Shimamoto, T., Bishop, W.R., and Ohyashiki, K. (2003). Efficacy of SCH 66336, a farnesyl transferase inhibitor, in conjunction with imatinib against BCR-ABL positive cells. *Mol Cancer Ther* 2:219–224.

44. Hoover, R.R., Mahon, F.X., Melo, J.V., and Daley, G.Q. (2002). Overcoming STI571-resistance with the farnesyl transferase inhibitor SCH 66336. *Blood* 100:1068–1071.

45. Martin, L.A., Head, J.E., Pancholi, S., Salter, J., Quinn, E., Detre, S., Kaye, S., Howes, A., Dowsett, M., and Johnston, S.R. (2007). The farnesyltransferase inhibitor R115777 (tipifarnib) in combination with tamoxifen acts synergistically to inhibit MCF-7 breast cancer cell proliferation and cell cycle progression *in vitro* and *in vivo*. Mol Cancer Ther 6:2458–2467.

46. James, G.L., Goldstein, J.L., and Brown, M.S. (1995). Polylysine and CVIM sequences of K-RasB dictate specificity of prenylation and confer resistance to benzodiazepine peptidomimetic in vitro. *J Biol Chem* 270:6221–6226.

47. Whyte, D.B., Kirschmeier, P., Hockenberry, T.N., Nunez-Oliva, I., James, L., Catino, J.J., Bishop, W.R., and Pai, J.K. (1997). K- and N-Ras are geranylgeranylated in cells treated with farnesyl protein transferase inhibitors. *J Biol Chem* 272:14459–14464.

48. Rowell, C.A., Kowalczyk, J.J., Lewis, M.D., and Garcia, A.M. (1997). Direct demonstration of geranylgeranylation and farnesylation of Ki-Ras in vivo. *J Biol Chem* 272:14093–14097.

49. Corrigan, D.P., Kuszczak, D., Rusinol, A.E., Thewke, D.P., Hrycyna, C.A., Michaelis, S., and Sinensky, M.S. (2005). Prelamin A endoproteolytic processing in vitro by recombinant Zmpste24. *Biochem J* 387:129–138.

50. Hancock, J.F., Cadwallader, K., Paterson, H., and Marshall, C.J. (1991). A CAAX or a CAAL motif and a second signal are sufficient for plasma membrane targeting of ras proteins. *EMBO J* 10:4033–4039.

51. Lebowitz, P.F., Sakamuro, D., and Prendergast, G.C. (1997). Farnesyltransferase inhibitors induce apoptosis in Ras-transformed cells denied substratum attachment. *Cancer Res* 57:708–713. (a) Liu, M., Bishop, W.R., Nielsen, L.L., Bryant, M.S., and Kirschmeier, P. (2001). Orally bioavailable farnesyltransferase inhibitors as anticancer agents in transgenic and xenograft models. *Methods Enzymol* 333:306–318.

52. Liu, M., Bryant, M.S., Chen, J., Lee, S., Yaremko, B., Lipari, P., Malkowski, M., Ferrari, E., Nielsen, L., Prioli, N., Dell, J., Sinha, D., Syed, J., Korfmacher, W.A., Nomeir, A.A., Lin, C.C., Taveras, A.G., Doll, R.J., Njoroge, F.G., Mallams, A.K., Remiszewski, S., Catino, J.J., Girijavallabhan, V.M., Kirschmeier, P., and Bishop, W.R. (1998). Antitumor activity of SCH 66336, an orally bioavailable tricyclic inhibitor of farnesyl protein transferase, in human tumor xenograft models and wap-ras transgenic mice. *Cancer Res* 58:4947–4956.

53. Kohl, N.E., Omer, C.A., Conner, M.W., Anthony, N.J., Davide, J.P., deSolms, S.J., Giuliani, E.A., Gomez, R.P., Graham, S.L., and Hamilton, K. (1995). Inhibition of farnesyltransferase induces regression of mammary and salivary carcinomas in ras transgenic mice. *Nat Med* 1:792–797.

54. Mangues, R., Corral, T., Kohl, N.E., Symmans, W.F., Lu, S., Malumbres, M., Gibbs, J.B., Oliff, A., and Pellicer, A. (1998). Antitumor effect of a farnesyl protein transferase inhibitor in mammary and lymphoid tumors overexpressing N-ras in transgenic mice. *Cancer Res* 58:1253–1259.

55. Omer, C.A., Chen, Z., Diehl, R.E., Conner, M.W., Chen, H.Y., Trumbauer, M.E., Gopal-Truter, S., Seeburger, G., Bhimnathwala, H., Abrams, M.T., Davide, J.P., Ellis, M.S., Gibbs, J.B., Greenberg, I., Koblan, K.S., Kral, A.M., Liu, D., Lobell, R.B., Miller, P.J., Mosser, S.D., O'Neill, T.J., Rands, E., Schaber, M.D., Senderak, E.T., Oliff, A., and Kohl, N.E. (2000). Mouse mammary tumor virus-Ki-rasB transgenic mice develop mammary carcinomas that can be growth-inhibited by a farnesyl:protein transferase inhibitor. *Cancer Res* 60:2680–2688.

56. End, D.W., Smets, G., Todd, A.V., Applegate, T.L., Fuery, C.J., Angibaud, P., Venet, M., Sanz, G., Poignet, H., Skrzat, S., Devine, A., Wouters, W., and Bowden, C. (2001). Characterization of the antitumor effects of the selective farnesyl protein transferase inhibitor R115777 in vivo and in vitro. *Cancer Res* 61:131–137. (a) Reichert, A., Heisterkamp, N., Daley, G.Q., and Groffen, J. (2001). Treatment of BCR/ABLpositive ALL in P190 transgenic mice with the farnesyl transferase inhibitor SCH 66336. *Blood* 97:1399–1403.

57. Taylor, S.A., Marrinan, C.H., Liu, G., Nale, L., Bishop, W.R., Kirschmeier, P., Liu, M., and Long, B.J. (2008). Combining the farnesyltransferase inhibitor lonafarnib with paclitaxel results in enhanced growth inhibitory effects on human ovarian cancer models in vitro and in vivo. *Gynecol Oncol* 109:97–106.

58. Liu, G., Taylor, S.A., Marrinan, C.H., Hsieh, Y., Bishop, W.R., Kirschmeier, P., and Long, B.J. (2009). Continuous and intermittent dosing of lonafarnib potentiates the therapeutic efficacy of docetaxel on preclinical human prostate cancer models. *Int J Cancer* 125:2711–2720.

59. Liu, G., Marrinan, C.H., Taylor, S.A., Black, S., Basso, A.D., Kirschmeier, P., Bishop, W.R., Liu, M., and Long, B.J. (2007). Enhancement of the anti-tumor activity of tamoxifen and anastrozole by the farnesyltransferase inhibitor lonafarnib (SCH 66336). *Anticancer Drugs* 18:923–931.

60. Huber, H.E., Robinson, R.G., Watkins, A., Nahas, D.D., Abrams, M.T., Buser, C.A., Lobell, R.B., Patrick, D., Anthony, N.J., Dinsmore, C.J., Graham, S.L., Hartman, G.D., Lumma, W.C., Williams, T.M., and Heimbrook, D.C. (2001). Anions modulate the potency of geranylgeranyl-protein transferase I inhibitors. *J Biol Chem* 276:24457–24465.

61. Ashar, H.R., James, L., Gray, K., Carr, D., Black, S., Armstrong, L., Bishop, W.R., and Kirschmeier, P. (2000). Farnesyl transferase inhibitors block the farnesylation of CENP-E and CENP-F and alter the association of CENP-E with microtubules. *J Biol Chem* 275:30451–30457.

62. Basso, A.D., Mirza, A., Liu, G., Long, B.J., Bishop, W.R., and Kirschmeier, P. (2005). The farnesyl transferase inhibitor (FTI) SCH 66336 (lonafarnib) inhibits Rheb farnesylation and mTOR signaling. Role in FTI enhancement of taxane and tamoxifen anti-tumor activity. *J Biol Chem* 280:31101–31108.

63. Roberts, P.J., Mitin, N., Keller, P.J., Chenette, E.J., Madigan, J.P., Currin, R.O., Cox, A.D., Wilson, O., Kirschmeier, P., and Der, C.J. (2008). Rho family GTPase modification and dependence on CAAX motif-signaled posttranslational modification. *J Biol Chem* 283:25150–25163.

64. Lebowitz, P.F., Casey, P.J., Prendergast, G.C., and Thissen, J.A. (1997). Farnesyltransferase inhibitors alter the prenylation and growth-stimulating function of RhoB. *J Biol Chem* 272:15591–15594.

65. Lebowitz, P.F., Davide, J.P., and Prendergast, G.C. (1995). Evidence that farnesyltransferase inhibitors suppress Ras transformation by interfering with Rho activity. *Mol Cell Biol* 15:6613–6622.

66. Du, W., and Prendergast, G.C. (1999). Geranylgeranylated RhoB mediates suppression of human tumor cell growth by farnesyltransferase inhibitors. *Cancer Res* 59:5492–5496.

67. Chen, Z., Sun, J., Pradines, A., Favre, G., Adnane, J., and Sebti, S.M. (2000). Both farnesylated and geranylgeranylated RhoB inhibit malignant transformation and suppress human tumor growth in nude mice. *J Biol Chem* 275:17974–17978.

68. Liu, A., Du, W., Liu, J.P., Jessell, T.M., and Prendergast, G.C. (2000). RhoB alteration is necessary for apoptotic and antineoplastic responses to farnesyltransferase inhibitors. *Mol Cell Biol* 20:6105–6113.

69. Gromov, P.S., Madsen, P., Tomerup, N., and Celis, J.E. (1995). A novel approach for expression cloning of small GTPases: identification, tissue distribution and chromosome mapping of the human homolog of rheb. *FEBS Lett* 377:221–226.

70. Basso, A.D., Mirza, A., Liu, G., Long, B.J., Bishop, W.R., and Kirschmeier, P. (2005). The farnesyl transferase inhibitor (FTI) SCH66336 (lonafarnib) inhibits Rheb farnesylation and mTOR signaling. Role in FTI enhancement of taxane and tamoxifen anti-tumor activity. *J Biol Chem* 280:31101–31108.

71. Clark, G.J., Kinch, M.S., Rogers-Graham, K., Sebti, S.M., Hamilton, A.D., and Der, C.J. (1997). The Ras-related protein Rheb is farnesylated and antagonizes Ras signaling and transformation. *J Biol Chem* 272:10608–10615.

72. Zhang, Y., Gao, X., Saucedo, L.J., Ru, B., Edgar, B.A., and Pan, D. (2003). Rheb is a direct target of the tuberous sclerosis tumour suppressor proteins. *Nat Cell Biol* 5:578–581.

73. Inoki, K., Li, Y., Zhu, T., Wu, J., and Guan, K.L. (2002). TSC2 is phosphorylated and inhibited by Akt and suppresses mTOR signalling. *Nat Cell Biol* 4:648–657.

74. Long, X., Lin, Y., Ortiz-Vega, S., Yonezawa, K., and Avruch, J. (2005). Rheb binds and regulates the mTOR kinase. *Curr Biol* 15:702–713.

75. Castro, A.F., Rebhun, J.F., Clark, G.J., and Quilliam, L.A. (2003). Rheb binds tuberous sclerosis complex 2 (TSC2) and promotes S6 kinase activation in a rapamycin- and farnesylation-dependent manner. *J Biol Chem* 278:32493–32496.

76. Mavrakis, K.J., Zhu, H., Silva, R.L., Mills, J.R., Teruya-Feldstein, J., Lowe, S.W., Tam, W., Pelletier, J., and Wendel, H.G. (2008). Tumorigenic activity and therapeutic inhibition of Rheb GTPase. *Genes Dev* 22:2178–2188.

77. Bessette, D.C., Qiu, D., and Pallen, C.J. (2008). PRL PTPs: mediators and markers of cancer progression. *Cancer Metastasis Rev* 27:231–252.

78. Cates, C.A., Michael, R.L., Stayrook, K.R., Harvey, K.A., Burke, Y.D., Randall, S.K., Crowell, P.L., and Crowell, D.N. (1996). Prenylation of oncogenic human PTP(CAAX) protein tyrosine phosphatases. *Cancer Lett* 110:49–55.

79. Zeng, Q., Si, X., Horstmann, H., Xu, Y., Hong, W., and Pallen, C.J. (2000). Prenylation-dependent association of protein-tyrosine phosphatases PRL-1, -2, and -3 with the plasma membrane and the early endosome. *J Biol Chem* 275:21444–21452.

80. Wang, J., Kirby, C.E., and Herbst, R. (2002). The tyrosine phosphatase PRL-1 localizes to the endoplasmic reticulum and the mitotic spindle and is required for normal mitosis. *J Biol Chem* 277:46659–46668.

81. Liang, F., Liang, J., Wang, W.Q., Sun, J.P., Udho, E., and Zhang, Z.Y. (2007). PRL3 promotes cell invasion and proliferation by down-regulation of Csk leading to Src activation. *J Biol Chem* 282:5413–5419.

82. Yao, X., Anderson, K.L., and Cleveland, D.W. (1997). The microtubule-dependent motor centromere-associated protein E (CENP-E) is an integral component of kinetochore corona fibers that link centromeres to spindle microtubules. *J Cell Biol* 139:435–447.

83. Wood, K.W., Sakowicz, R., Goldstein, L.S., and Cleveland, D.W. (1997). CENP-E is a plus end-directed kinetochore motor required for metaphase chromosome alignment. *Cell* 91:357–366.

84. Hussein, D., and Taylor, S.S. (2002). Farnesylation of CENP-F is required for G2/M progression and degradation after mitosis. *J Cell Sci* 115:3403–3414.

85. Crespo, N.C., Ohkanda, J., Yen, T.J., Hamilton, A.D., and Sebti, S.M. (2001). The farnesyltransferase inhibitor FTI02153 blocks bipolar spindle formation and chromosome alignment and causes prometaphase accumulation during mitosis of human lung cancer cells. *J Biol Chem* 276:16161–16167.

86. Schafer-Hales, K., Iaconelli, J., Snyder, J.P., Prussia, A., Nettles, J.H., El-Naggar, A., Khuri, F.R., Giannakakou, P., and Marcus, A.I. (2007). Farnesyl transferase inhibitors impair chromosomal maintenance in cell lines and human tumors by compromising CENP-E and CENP-F function. *Mol Cancer Ther* 6:1317–1328.

87. Aebi, U., Cohn, J., Buhle, L., and Gerace, L. (1986). The nuclear lamina is a meshwork of intermediate-type filaments. *Nature* 323:560–564.

88. Ulitzur, N., Harel, A., Feinstein, N., and Gruenbaum, Y. (1992). Lamin activity is essential for nuclear envelope assembly in a Drosophila embryo cell-free extract. *J Cell Biol* 119:17–25.

89. Wolda, S.L., and Glomset, J.A. (1998). Evidence for modification of lamin B by a product of mevalonic acid. *J Biol Chem* 263:5997–6000.

90. Beck, L.A., Hosick, T.J., and Sinensky, M. (1990). Isoprenylation is required for the processing of the lamin A precursor. *J Cell Biol* 110:1489–1499.

91. Weber, K., Plessmann, U., and Traub, P. (1989). Maturation of nuclear lamin A involves a specific carboxy-terminal trimming, which removes the polyisoprenylation site from the precursor; implications for the structure of the nuclear lamina. *FEBS Lett* 257:411–414.

92. Basso, A., Kirschmeier, P., and Bishop, W.R. (2006). Farnesyl transferase inhibitors. *J Lipid Res* 47:15–31.

93. Eskens, F.A., Awada, A., Cutler, D.L., de Jonge, M.J., Luyten, G.P., Faber, M.N., Statkevich, P., Sparreboom, A., Vewweij, J., Hanauske, A.R., and Piccart, M. (2001). Phase I and pharmacokinetic study of the oral farnesyl transferase inhibitor SCH 66336 given twice daily to patients with advanced solid tumors. *J Clin Oncol* 19:1167–1175.

94. Punt, C.J., van Maanen, L., Bol, C.J., Seifert, W.F., and Wagener, D.J. (2001). Phase I and pharmacokinetic study of the orally administered farnesyl transferase inhibitor R115777 in patients with advanced solid tumors. *Anticancer Drugs* 12:193–197.
95. Adjei, A.A., Erlichman, C., Davis, J.N., Cutler, D.L., Sloan, J.A., Marks, R.S., Hanson, L.J., Svingen, P.A., Atherton, P., Bishop, W.R., Kirschmeier, P., and Kaufmann, S.H. (2000). A phase I trial of the farnesyl transferase inhibitor SCH 66336: evidence for biological and clinical activity. *Cancer Res* 60:1871–1877.
96. Feldman, E.J., Cortes, J., DeAngelo, D.J., Holyoake, T., Simonsson, B., O'Brien, S.G., Reiffers, J., Turner, A.R., Roboz, G.J., Lipton, J.H., Maloisel, F., Colombat, P., Martinelli, G., Nielsen, J.L., Petersdorf, S., Guilhot, F., Barker, J., Kirschmeier, P., Frank, E., Statkevich, P., Zhu, Y., Loechner, S., and List, A. (2008). On the use of lonafarnib in myelodysplastic syndrome and chronic myelomonocytic leukemia. *Leukemia* 22:1707–1711.
97. Lancet, J.E., Gojo, I., Gotlib, J., Feldman, E.J., Greer, J., Liesveld, J.L., Bruzek, L.M., Morris, L., Park, Y., Adjei, A.A., Kaufmann, S.H., Garret-Mayer, E., Greenberg, P.L., Wright, J.J., and Karp, J.E. (2007). A phase 2 study of the farnesyltransferase inhibitor tipifarnib in poor-risk and elderly patients with previously untreated acute myelogenous leukemia. *Blood* 109:1387–1394.
98. Harousseau, J.L., Martinelli, G., Jedrzejczak, W.W., Brandwein, J.M., Bordessoule, D., Masszi, T., Ossenkoppele, G.J., Alexeeva, J.A., Beutel, G., Maertens, J., Vidriales, M.B., Dombret, H., Thomas, X., Burnett, A.K., Robak, T., Khuageva, N.K., Golenkov, A.K., Tothova, E., Mollgard, L., Park, Y.C., Bessems, A., De Porre, P., and Howes, A.J. FIGHT-AML-301 Investigators (2009). A randomized phase 3 study of tipifarnib compared with best supportive care, including hydroxyurea, in the treatment of newly diagnosed acute myeloid leukemia in patients 70 years or older. *Blood* 114:1166–1173.
99. Jabbour, E., Kantarjian, H., Ravandi, F., Garcia-Manero, G., Estrov, Z., Verstovsek, S., O'Brien, S., Faderl, S., Thomas, D.A., Wright, J.J., and Cortes, J. (2010). A phase 1–2 study of a farnesyltransferase inhibitor, tipifarnib, combined with idarubicin and cytarabine for patients with newly diagnosed acute myeloid leukemia and high-risk myelodysplastic syndrome. *Cancer* 2010 Oct 19. [Epub ahead of print].
100. Johnston, S.R., Hickish, T., Ellis, P., Houston, S., Kelland, L., Dowsett, M., Salter, J., Perez-Ruixo, J.J., Palmer, P., and Howes, A. (2003). Phase II study of the efficacy and tolerability of two dosing regimens of the farnesyl transferase inhibitor, R115777, in advanced breast cancer. *J Clin Oncol* 21:2492–2499.
101. Dalenc, F., Doisneau-Sixou, S.F., Allal, B.C., Marsili, S., Lauwers-Cances, V., Chaoui, K., Schiltz, O., Monsarrat, B., Filleron, T., Renée, N., Malissein, E., Meunier, E., Favre, G., and Roché, H. (2010). Tipifarnib plus tamoxifen in tamoxifen-resistant metastatic breast cancer: a negative phase II and screening of potential therapeutic markers by proteomic analysis. *Clin Cancer Res* 16:1264–1271.
102. Khuri, F.R., Glisson, B.S., Kim, E.S., Statkevich, P., Thall, P.F., Meyers, M.L., Herbst, R. S., Munden, R.F., Tendler, C., Zhu, Y., Bangert, S., Thompson, E., Lu, C., Wang, X.M., Shin, D.M., Kies, M.S., Papadimitrakopoulou, V., Fossella, F.V., Kirschmeier, P., Bishop, W.R., and Hong, W.K. (2004). Phase I study of the farnesyltransferase inhibitor lonafarnib with paclitaxel in solid tumors. *Clin Cancer Res* 10:2968–2976.
103. Kim, E.S., Kies, M.S., Fossella, F.V., Glisson, B.S., Zaknoen, S., Statkevich, P., Munden, R.F., Summey, C., Pisters, K.M., Papadimitrakopoulou, V., Tighiouart, M., Rogatko, A., and Khuri, F.R. (2005). Phase II study of the farnesyltransferase inhibitor lonafarnib with paclitaxel in patients with taxane-refractory/resistant nonsmall cell lung carcinoma. *Cancer* 104:561–569.

104. Cohen, S.J., Ho, L., Ranganathan, S., Abbruzzese, J.L., Alpaugh, R.K., Beard, M.,
     Lewis, N.L., McLaughlin, S., Rogatko, A., Perez-Ruixo, J.J., Thistle, A.M.,
     Verhaeghe, T., Wang, H., Weiner, L.M., Wright, J.J., Hudes, G.R., and Meropol, N.J.
     (2003). Phase II and pharmacodynamic study of the farnesyltransferase inhibitor R115777
     as initial therapy in patients with metastatic pancreatic adenocarcinoma. *J Clin Oncol*
     21:1301–1306.
105. Hanrahan, E.O., Kies, M.S., Glisson, B.S., Khuri, F.R., Feng, L., Tran, H.T., Ginsberg, L.E.,
     Truong, M.T., Hong, W.K., and Kim, E.S. (2009). A phase II study of Lonafarnib
     (SCH66336) in patients with chemorefractory, advanced squamous cell carcinoma of the
     head and neck. *Am J Clin Oncol* 32:274–279.
106. Adjei, A.A., Davis, N., Erlichman, C., Svingen, P.A., and Kaufmann, S.H. (2000). Com-
     parison of potential markers of farnesyltransferase inhibition. *Clin Cancer Res*
     6:2318–2325.
107. Raponi, M., Lancet, J.E., Fan, H., Dossey, L., Lee, G., Gojo, I., Feldman, E.J., Gotlib, J.,
     Morris, L.E., Greenberg, P.L., Wright, J.J., Harousseau, J.L., Löwenberg, B., Stone, R.M.,
     De Porre, P., Wang, Y., and Karp, J.E. (2008). A 2-gene classifier for predicting response
     to the farnesyltransferase inhibitor tipifarnib in acute myeloid leukemia. *Blood*
     111:2589–2596.
108. Kieran, M.W., Gordon, L., and Kleinman, M. (2007). New approaches to progeria.
     Pediatrics 120:834–841.
109. De Sandre-Giovannoli, A., Bernard, R., Cau, P., Navarro, C., Amiel, J., Boccaccio, I.,
     Lyonnet, S., Stewart, C.L., Munnich, A., Le Merrer, M., and Lévy, N. (2003). Lamin a
     truncation in Hutchinson–Gilford progeria. *Science* 300:2055.
110. Eriksson, M., Brown, W.T., Gordon, L.B., Glynn, M.W., Singer, J., Scott, L., Erdos, M.R.,
     Robbins, C.M., Moses, T.Y., Beglund, P., Dutra, A., Pak, E., Durkin, S., Csoka, A.B.,
     Boehnke, M., Glover, T.W., and Collins, F.S. (2003). Recurrent de novo point mutations
     in lamin A cause Hutchinson–Gilford progeria syndrome. *Nature* 423:293–298.
111. Worman, H.J., Fong, L.G., Muchir, A., and Young, S.G. (2009). Laminopathies and the
     long strange trip from basic cell biology to therapy. *J Clin Invest* 119:1825–1836.
112. Goldman, R.D., Shumaker, D.K., Erdos, M.R., Eriksson, M., Goldman, A.E., Gordon, L.B.,
     Gruenbaum, Y., Khuon, S., Mendez, M., Varga, R., and Collins, F.S. (2004). Accumulation
     of mutant lamin A causes progressive changes in nuclear architecture in Hutchinson–Gilford
     progeria syndrome. *Proc Natl Acad Sci USA* 101:8963–8968.
113. Merideth, M.A., Gordon, L.B., Clauss, S., Sachdev, V., Smith, A.C., Perry, M.B.,
     Brewer, C.C., Zalewski, C., Kim, H.J., Solomon, B., Brooks, B.P., Gerber, L.H.,
     Turner, M.L., Domingo, D.L., Hart, T.C., Graf, J., Reynolds, J.C., Gropman, A.,
     Yanovski, J.A., Gerhard-Herman, M., Collins, F.S., Nabel, E.G., Cannon, R.O.,
     Gahl, W.A., and Introne, W.J. (2008). Phenotype and course of Hutchinson–Gilford
     progeria syndrome. *N Engl J Med* 358:592–604.
114. Olive, M., Harten, I., Mitchell, R., Beer, J.K., Djabali, K., Cao, K., Erdos, M.R., Blair, C.,
     Funke, B., Smoot, L., Gerhard-Herman, M., Machan, J.T., Kutys, R., Virmani, R.,
     Collins, F.S., Wight, T.N., Nabel, E.G., and Gordon, L.B. (2010). Cardiovascular pathol-
     ogy in Hutchinson–Gilford progeria: correlation with the vascular pathology of aging.
     Arterioscler Thromb Vasc Biol 30:2301–2309.
115. Davies, B.S., Fong, L.G., Yang, S.H., Coffinier, C., and Young, S.G. (2009). The post-
     translational processing of prelamin A and disease. *Annu Rev Genomics Hum Genet*
     10:153–174.
116. Yang, S.H., Bergo, M.O., Toth, J.I., Qiao, X., Hu, Y., Sandoval, S., Meta, M., Bendale, P.,
     Gelb, M.H., Young, S.G., and Fong, L.G. (2005). Blocking protein farnesyltransferase

improves nuclear blebbing in mouse fibroblasts with a targeted Hutchinson–Gilford progeria syndrome mutation. *Proc Natl Acad Sci USA* 102:10291–10296.

117. Capell, B.C., Erdos, M.R., Madigan, J.P., Fiordalisi, J.J., Varga, R., Conneely, K.N., Gordon, L.B., Der, C.J., Cox, A.D., and Collins, F.S. (2005). Inhibiting farnesylation of progerin prevents the characteristic nuclear blebbing of Hutchinson–Gilford progeria syndrome. *Proc Natl Acad Sci USA* 102:12879–12884.

118. Fong, L.G., Frost, D., Meta, M., Qiao, X., Yang, S.H., Coffinier, C., and Young, S.G. (2006). A protein farnesyltransferase inhibitor ameliorates disease in a mouse model of progeria. *Science* 311:1621–1623.

119. Yang, S.H., Meta, M., Qiao, X., Frost, D., Bauch, J., Coffnier, C., Majumdar, S., Bergo, M.O., Young, S.G., and Fong, L.G. (2006). A farnesyltransferase inhibitor improves disease phenotypes in mice with a Hutchinson–Gilford progeria syndrome mutation. *J Clin Invest* 116:2115–2121.

120. Capell, B.C., Olive, M., Erdos, M.R., Cao, K., Faddah, D.A., Tavarez, U.L., Conneely, K. N., Qu, X., San, H., Ganesh, S.K., Chen, X., Avallone, H., Kolodgie, F.D., Virmani, R., Nabel, E.G., and Collins, F.S. (2008). A farnesyltransferase inhibitor prevents both the onset and late progression of cardiovascular disease in a progeria mouse model. *Proc Natl Acad Sci USA* 105:15902–15907.

121. Otto, J.C., and Casey, P.J. (1996). The hepatitis delta virus large antigen is farnesylated both *in vitro* and in animal cells. *J Biol Chem* 271:4569–4572.

122. Bordier, B.B., Ohkanda, J., Liu, P., Lee, S.Y., Salazar, F.H., Marion, P.L., Ohashi, K., Meuse, L., Kay, M.A., Casey, J.L., Sebti, S.M., Hamilton, A.D., and Glenn, J.S. (2003). In vivo antiviral efficacy of prenylation inhibitors against hepatitis delta virus. *J Clin Invest* 112:407–414.

123. Eastman, R.T., Buckner, F.S., Yokoyama, K., Gelb, M.H., and Van Voorhis, W.C. (2006). Thematic review series: lipid posttranslational modifications. Fighting parasitic disease by blocking protein farnesylation. *J Lipid Res* 47:233–240.

124. Hast, M.A., Fletcher, S., Cummings, C.G., Pusateri, E.E., Blaskovich, M.A., Rivas, K., Gelb, M.H., Van Voorhis, W.C., Sebti, S.M., Hamilton, A.D., and Beese, L.S. (2009). Structural basis for binding and selectivity of antimalarial and anticancer ethylenediamine inhibitors to protein farnesyltransferase. *Chem Biol* 16:181–192.

125. Tidyman, W.E., and Rauen, K.A. (2009). The RASopathies: developmental syndromes of Ras/MAPK pathway dysregulation. *Curr Opin Genet Dev* 19:230–236.

126. Rauen, K.A. (2007). HRAS and the Costello syndrome. *Clin Genet* 71:101–108.

127. Aoki, Y., Niihori, T., Kawame, H., Kurosawa, K., Ohashi, H., Tanaka, Y., Filocamo, M., Kato, K., Suzuki, Y., Kure, S., and Matsubara, Y. (2005). Germline mutations in HRAS proto-oncogene cause Costello syndrome. *Nat Genet* 37:1038–1040.

Author Index

Numbers in regular font are reference numbers and indicate that an author's work is referred to although the name is not cited in the text. Numbers in italics refer to the page numbers on which the complete reference appears.

A

Abagyan, R., 150–151, *160*
Abbruzzese, J. L., 80, 82–83, 92–94, 289, 290, *302*
Abe, K., 187, 190–191, *193*
Abo, A., 75, 78, *91*, 260, *272*
Abrams, M. T., 260–261, 262, 269–270, *272*, 274, 283, 284, *298*
Abu-Abied, M., 165, 168–171, 173, *178*
Abu Kwaik, Y., 210, 227, *230*, *234*
Adam, S. A., 13–14, *19*
Adamson, P., 107, *122*
Ada-Nguema, A., 73, *89*
Aditya, A. V., 107, *122*, *132*, 216–217, 218, 244, 245, *255*
Adjei, A. A., 288, 290, *301*, *302*
Adnane, J., 285, *299*
Aebersold, R., 98, 100, *118*
Aebi, U., 287, *300*
Afar, D., 67, 73, *90*
Agarwal, R., 83, *94*
Agell, N., 70–71, *87*
Agnew, B. J., 196, 199, *205*
Ahearn, I. M., 70–71, *87*
Aitchison, J. D., 52, *58*
Aivazian, D., 156, *161*
Akashi, K., 266–267, *274*
Akgoz, M., 103, 110–113, *120*, *121*
Akino, T., 18, *21*, *23*, 98, 100, 103, 108, 109, 126, 129–132, *142*, *143*
Albanese, M. M., 279–280, *296*
Albert, I., 71, *87*
Alexandrov, K., 69, 78–79, *86*, *92*, 148–155, 156–159, *160*, *161*, 196, 202–203, *206*, 211, *231*

Alexeeva, J. A., 288, *301*
Ali, B. R., 78–79, *92*, 156, *161*
Ali, W., 175, 176–177, *182*
Al-Khodor, S., 227, *234*
Allal, B. C., 289, *301*
Allal, C. M., 252, *256*
Allen, G. J., 172–173, *181*
Alory, C., 166, *179*
Alpaugh, R. K., 80, *92*, 289, 290, *302*
Al-Quadan, T., 210, 227, *230*, *234*
Alspaugh, J. A., 252–254, *257*
Alvarez, C. S., 279–280, *296*
Ambroziak, P., 23, *37*, 83, *94*
Amiel, J., 291–292, *302*
Anant, J. S., 150, 158–159, *160*, *162*
Anderson, J. C., 190, *194*
Anderson, K. L., 286, *300*
Anderson, R. E., 135, *144*
Anderson, R. G. W., 46, *54*
Andersson, K. M., 82–83, *93*, *94*, 262–263, 265, 266–267, 270, *273*, *274*
Andersson, L. O., 109–110, *123*
Andersson, S., 267–268, *274*
Andres, D. A., 13–14, *19*, 26, 28, 29, *39*, *40*, 67, 69, 73–74, *86*, *90*, 196, 201–203, *206*, 211–214, 219, 223, *231–233*
Andre, T., 80, *93*
Andrews, D. W., 45–46, *53*
Angelovici, R., 164–165, *178*
Angibaud, P., 283, 284, *298*
Ankala, S., 280, 293–294, *296*
Anna-A, S. S., 261, *272*
Anraku, Y., 51–52, *58*, 166, *178*, 226, *234*, 261, 265, *273*, 274
Ansaldi, M., 185–187, 190, 191, *193*

315

Gruenberg, J., 156, *161*
Gruissem, W., 164–171, 173–175, *178–182*,
    203, *206*, 261, *273*
Grunau, S., 45–46, *53*
Grzybek, M., 14, *19*
Guan, B., 83, *94*, 280, *296*
Guan, K. L., 285–286, *299*
Guerra, C., 262, *273*
Guilhot, F., 288, *301*
Guillen, N., 185, *193*
Guilluy, C., 76, *91*
Gullo-Brown, J., 280, *296*
Guo, L., 196, 199, *205*
Guo, Z., 69, 78–79, *86*, 150–153, *160*, *161*, 196,
    202–203, *206*
Gutierrez, L., 70–71, *87*
Gutman, O., 165, 168–171, 173, *178*, *180*
Gutowski, S., 100, *118*

H

Haan, G. J., 49, *56*
Habenicht, A. J., 3, *4*, 6, *16*
Habyarimana, F., 227, *234*
Hach, A., 70, *86*
Hadden, D. A., 46, *55*
Haga, T., 129, *143*
Hajara, A. K., 43–44, *53*
Hajian, G., 281, *297*
Hajjar, D. P., 100–102, *119*
Hake, S., 170, *180*, 173, 261, *273*
Haklai, R., 70–71, *87*
Hala, M., 164, 166, 174, *178*
Halbach, A., 46, 49, 51, *55–57*
Hall, A., 70–71, *87*, 107, *122*, 267–268, *274*
Hamblet, C., 10, *17*
Hamilton, A. D., 67, 69–72, *86*, *88*, 244, *255*,
    262, *273*, 280, 285–287, 293–294, *296*, *299*,
    300, *303*
Hamilton, K. A., 82–83, *94*, 269–270, *274*,
    283, *298*
Hamilton, M. H., 106–107, *122*
Hamm, H. E., 98, 103–104, 108, *117*, *121*, *123*
Hamoen, L. W., 185, *192*
Hanauske, A. R., 288, *300*
Hanazaki, M., 209, 211, *229*
Hancock, J. F., 15, *19*, 60–63, 69–71, 82, *84*, *86*,
    87, 98, 100–102, *118*, *119*, 126, *142*, 169,
    170–171, *180*, 196, *205*, 211, *230*, 276–277,
    282, *295*, *297*

Hancock, P. J., 219, 220–223, *233*, 247–248,
    249, *247–249*, *256*
Hang, H. C., 196, 197, 202–203, *206*, 210, 213,
    227, *230*, *231*
Hanker, A. B., 72, 83, *89*, 114, *124*
Hannah, V. C., 107, *122*
Hanrahan, E. O., 289, *302*
Hanson, L. J., 288, 290, *301*
Hanzal-Bayer, M., 70–71, *87*
Harada, K., 78, *91*
Harden, T. K., 103, *121*, 131–132, *144*
Harder, W., 51, *57*
Hardie, R., 137–139, *144*
Harding, A., 71, *87*
Harel, A., 287, *300*
Harker, L., 2, *3*
Harlow, E., 13–14, *19*
Haro, D., 6, *16*
Harousseau, J. L., 288, 290–291, *301*, *302*
Hart, C., 196, 199, *205*
Harten, I., 292, *302*
Hartig, A., 45–46, *53*
Hartley, N. M., 173–174, *181*
Hartman, G. D., 284, *298*
Hartman, H. L., 208–210, 214–224, 227–228,
    229, *230*, 232, 237, 243–245, *254*, *255*
Hartmann, M. A., 171, *181*
Hart, M. J., 76, *90*
Hart, T. C., 292–293, *302*
Hartwig, J., 8–9, 13, *16*
Harvey, K. A., 286, *300*
Hashimoto, Y., 129, *143*
Hast, M. A., 225, *233*, 245–247, 252–254, *256*,
    280, 293–294, *296*, *303*
Hatten, M. E., 24, *38*
Hattori, M., 67, 71, *87*
Håvarstein, L. S., 184, 189, *192*
Haworth, K. B., 213–214, *232*
Hazama-Shimada, Y., 2–3, *4*
Head, J. E., 281, 283–284, *297*
Heal, W. P., 202, *206*
Heasman, S. J., 267–268, *274*
He, B., 226, *234*, 261, *272*
Hegardt, F. G., 6, *16*
He, G.-P., 114, *124*
Heilmeyer, L. M. G., 111, *123*
Heimbrook, D. C., 284, *298*
Heintz, D., 171, 181
Heldt, H. W., 43–44, *53*
Hellinga, H. W., 219–223, *233*, 247–249, *256*

I

Iaconelli, J., 287, *300*
Iakovenko, A., 150, 151, 153–154, *160*
Ibrahim, M. X., 82–83, *93*, 262–263, 267–268, 270, *273, 274*
Ichijo, T., 108, *123*
Ignatev, A., 153–154, *161*
Ihalmo, P., 73, *89*
Ikawa, Y., 71, *87*
Illenberger, D., 170, *180*
Illes, A., 80, *93*
Imanaka, T., 46, 48–49, *54, 55*
Imoto, M., 247, *256*
Im, S. Y., 67, 73, *90*
Ingmundson, A., 156–158, *162*
Iñiguez-Lluhi, J. A., 103, *120*, 129, *143*
Inoki, K., 285–286, *299*
Inoue, H., 62, *94*
Inoue, T., 60–63, 67, 73–74, *85*, 169, *179*
Introne, W. J., 292–293, *302*
Iqbal, S., 80, *93*
Isberg, R. R., 156–158, *162*
Ishibashi, Y., 187, 190–191, *194*
Ishihara, H., 281, *297*
Ishiyama, K., 173–174, *182*
Islinger, M., 43–44, *52*
Isogai, A., 187, 189, 190–191, *194*
Itoh, H., 173–174, *182*
Itoh, Y., 186–187, 190, 191, *193*
Ito, M., 49, *56*
Itzen, A., 156–158, *162*
Ivanov, I. E., 70, *86*
Ivanov, S. S., 210, 227, *230*
Ivaska, J., 73, *89*
Iveson, T., 80, *93*
Iwata, H., 187, 188–189, *193*
Izumi, M., 23–24, *38*

J

Jabbour, E., 289, *301*
Jackson, C. L., 62, *95*
Jackson, J. H., 60–63, 69–70, *84, 85*, 131, *143*
Jacks, T. E., 266–267, *274*
Jahnke, B., 184–185, *192*
James, G. L., 14, *19*, 46, *54*, 80, 82, *92, 93*, 107, *122*, 165, 169, *178*, 260–261, *272*, 281, *297*
James, L., 15, *19*, 67, 69–70, *86*, 170–171, *180*, 260–261, *272*, 276–277, 280, 281–282, 284, 286, 287, *295–297, 299*

Jameson, S. A., 13–14, *19*
Jaunbergs, J., 196, 197–198, 203, *205*
Jedd, G., 43–44, *52*
Jedrzejczak, W. W., 288, *301*
Jenkins, S. M., 164–165, *178*
Jensen, E. C., 184–185, *192*
Jensen, G. J., 141, *145*
Jensen, O. N., 170, *180*
Jensen, P., 67, 73, *90*
Jeong, J., 14, *19*
Jessell, T. M., 285, *299*
Jiang, C., 196, 197–198, 203, *205*
Jiang, H., 72, *88*
Jiang, S. Y., 62, *94*
Jian, X., 103–104, *120*
Joachimiak, A., 190, *194*
Joberty, G., 66, 78–79, *85*
Johnsborg, O., 184, 189, *192*
Johnson, B. R., 13–14, *19*
Johnson, C. D., 165, 173, *178*
Johnson, D. A., 25, *38*
Johnson, L., 266–267, *274*
Johnson, R., 129, *143*
Johnson, R. L.II., 66–69, 70, *85, 86*, 244, *255*
Johnston, S. R., 281, 283–284, 289, *297, 301*
Jomaa, H., 252, *257*
Jones, J. M., 46–47, 48–49, 50, *55, 56*
Jones, S. C., 210, 227, *230*
Jones, T., 129, *142*
Jonsson, I. M., 267–268, *274*
Judd, S. R., 3, *4*
Jung, H. J., 13, *18*
Jurgens, G., 169–171, *180*
Jurisica, I., 83, *94*
Just, W. W., 50, *57*

K

Kadono-Okuda, K., 67, 73–74, *90*
Kahms, M., 70–71, *87*
Kahn, C. R., 67, 73, *90*
Kahn, R. A., 62, 79, *92, 94*
Kaibuchi, K., 76, *90*
Kakesova, H., 170, *180*
Kalia, A., 227, *234*
Kalinin, A., 150, 153, *160*
Kalish, J. E., 46, *54*
Kalyanaraman, V., 98, 99, 102–103, 110–113, *116, 120, 121*
Kaminek, M., 167, 170–171, *179*

Index

A

Alkyne-modified isoprenoids tagging, 202–203, 202*f*

Anilinogeraniol tagging, 202–203, 202*f*

8-Anilinogeranyl diphosphate (AGPP) substrate, 211–213, 212*f*

Annexin A2, 203

APETALA1 (AP1) protein, 168

Arabidopsis prenyltransferases. *See* Plant protein prenyltransferases

Arabidopsis REP (AtREP), 166

Arf family small GTPases, 79

A-type lamins, 25–27

Azide tagging method
applications, 201–202
basic idea, 197
farnesylated proteins detection
mass spectrometry analysis, 199
protocol, 197–198, 198*f*
SDS polyacrylamide gel, 199
geranylgeranylated proteins detection
azido-geranylgeranyl alcohol, 200
pH fractions, isoelectric focusing, 200–201
protocol, 200*f*
improvements, 201
reason, azide usage, 197

Azido-farnesyl diphosphate (FPP-azide), 197

B

Bioorthogonal ligation methods, 213

Biotin-geranyl diphosphate (BGPP)
affinity tag approach, 213
structure, 212*f*

Biotin geranylpyrophosphate tagging, 202–203, 202*f*

BMS-214663 inhibitors, 250–251, 251*f*

Brain G proteins. *See also* Heterotrimeric G proteins
carboxymethylation
C-terminal cysteinylation, 109–110
unmethylated Gγ, 108–109
HPLC, 105*f*, 106–107
MALDI spectra, 104–105, 105*f*
modifications, 106*f*
prenyl moiety variation
geranylgeranylated *versus* farnesylated Gγ2, 107
unprenylated C-terminus, 107–108
proteolytic processing, Rce1
Gγ5 and Gγ7, 111–113
sequence determinants, Gγ5, 111–113
unprocessed CaaX sequences, 114–115
study analysis, 105*f*

B-type lamins, 23–25

C

CaaX motifs
alternate prenylation, prelamin A, 32–33
Gγ subunit prenylation, 100–104
prenylated Ras and Rho proteins, 66

CaaX protein prenyltransferases
cancer chemotherapeutics, 247–251
FTase
crystal structure, 236–241, 236*f*
genetic analyses, 262–264
inhibitors, malaria treatment, 252–254
reaction cycle, 236–241, 240*f*
simultaneous inactivation, 269–270
GGTase-I
Candida albicans, 245–247
crystal structure, 241–244, 242*f*
genetic analyses, 265–269
reaction cycle, 241–244, 243*f*
simultaneous inactivation, 269–270

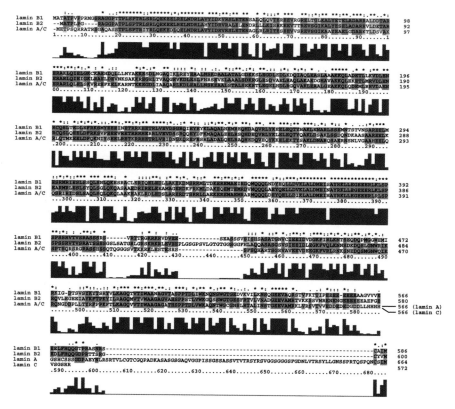

Chapter 3. Figure 1. (See legend in text)

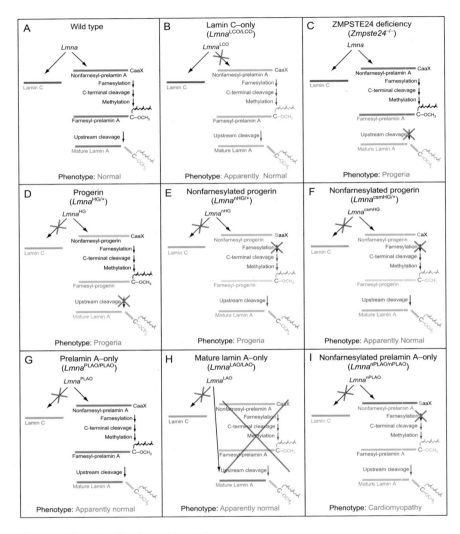

Chapter 3. Figure 2. (See legend in text.)

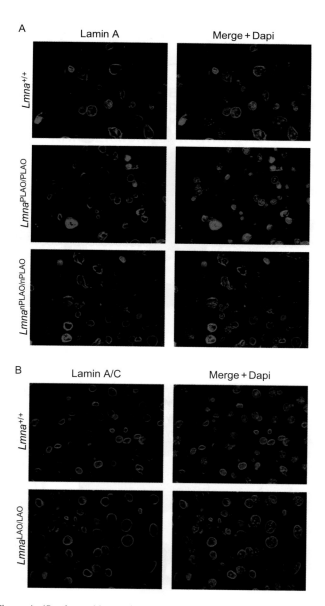

Chapter 3. Figure 4. (See legend in text.)

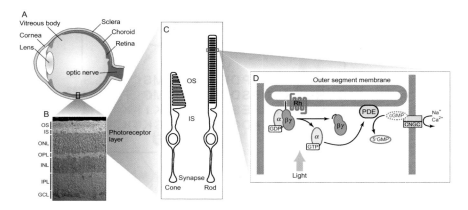

Chapter 7. Figure 1. (See legend in text)

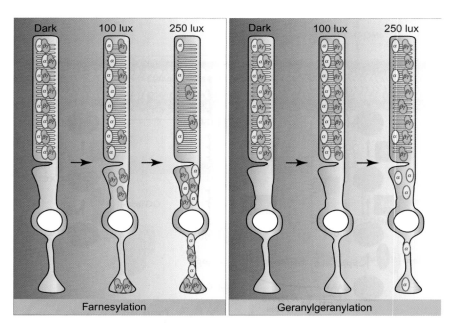

Chapter 7. Figure 5. (See legend in text.)

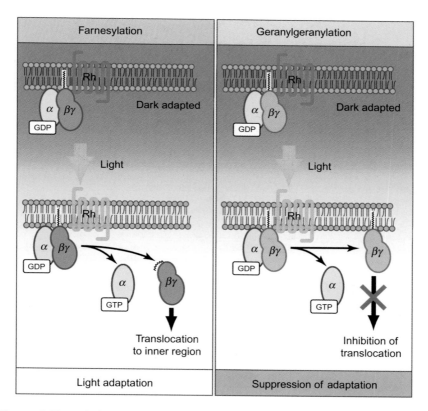

Chapter 7. Figure 6. (See legend in text.)

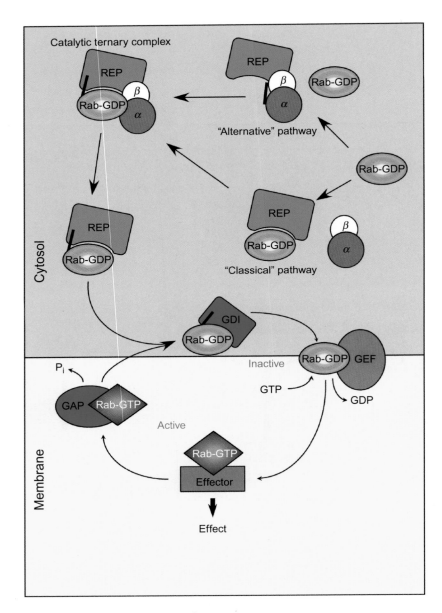

Chapter 8. Figure 1. (See legend in text)

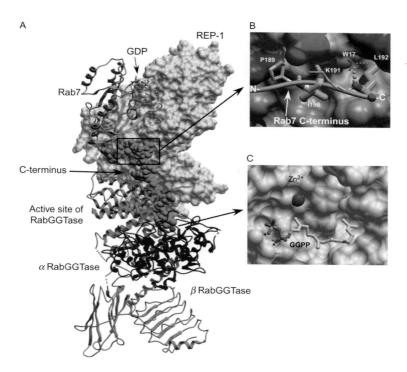

Chapter 8. Figure 2. (See legend in text.)

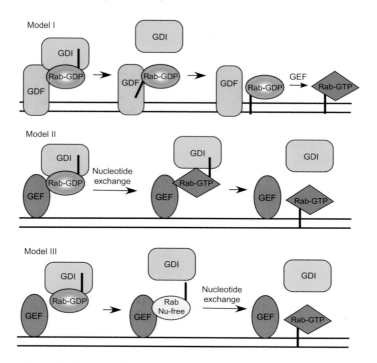

Chapter 8. Figure 3. (See legend in text.)

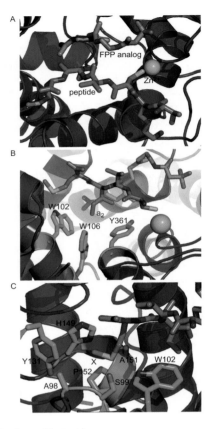

Chapter 12. Figure 3. (See legend in text.)

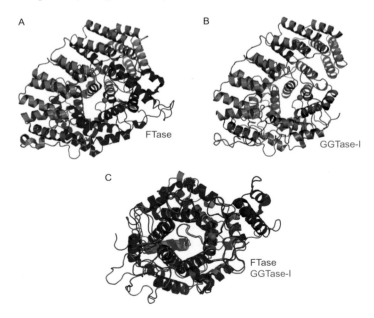

Chapter 13. Figure 1. (See legend in text.)

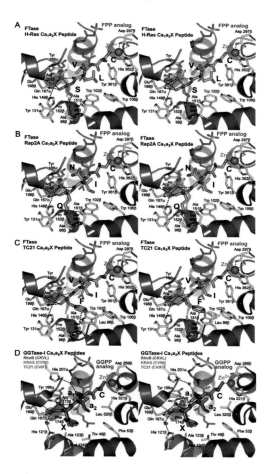

Chapter 13. Figure 2. (See legend in text.)

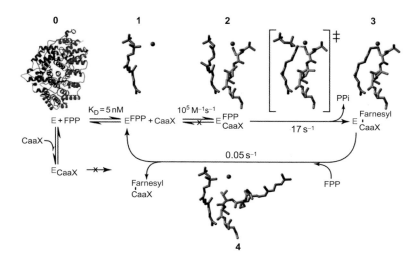

Chapter 13. Figure 3. (See legend in text.)

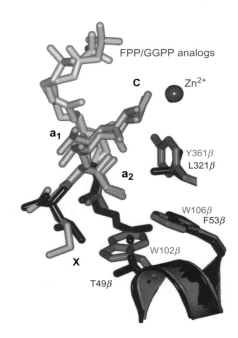

Chapter 13. Figure 4. (See legend in text.)

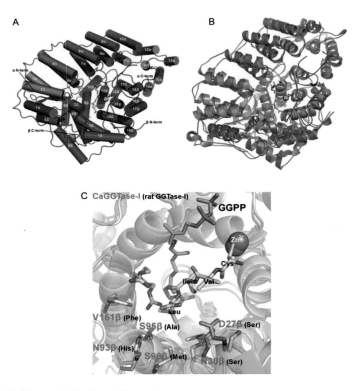

Chapter 13. Figure 6. (See legend in text.)

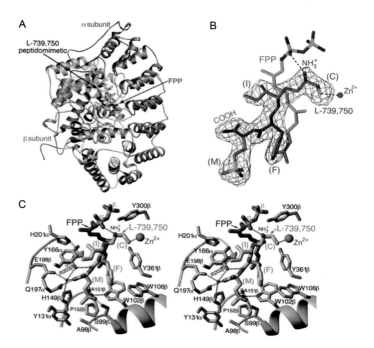

Chapter 13. Figure 7. (See legend in text.)

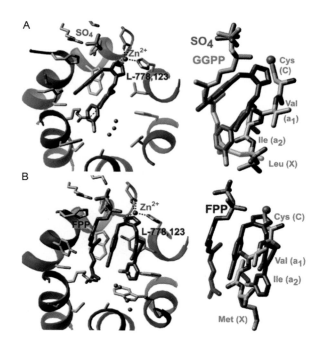

Chapter 13. Figure 8. (See legend in text.)

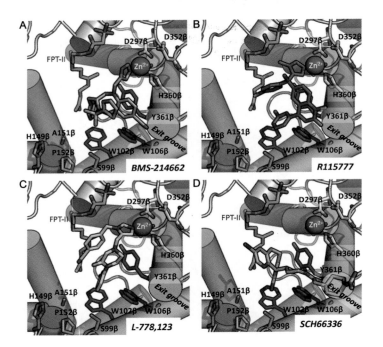

Chapter 13. Figure 9. (See legend in text.)

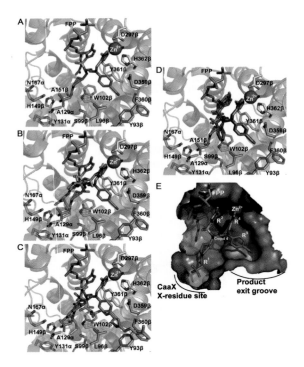

Chapter 13. Figure 10. (See legend in text.)

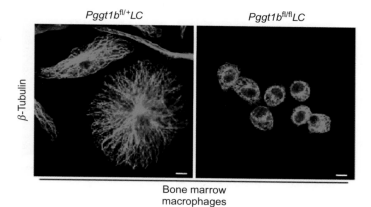

Bone marrow
macrophages

Chapter 14. Figure 2. (See legend in text.)

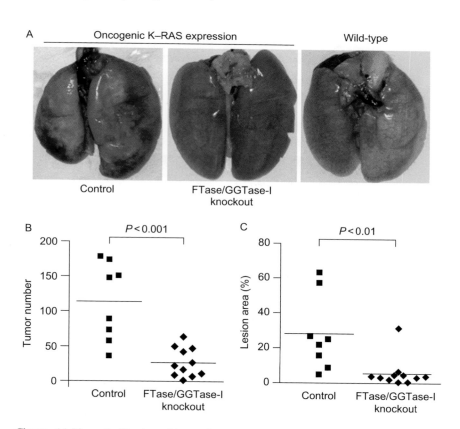

Chapter 14. Figure 3. (See legend in text.)

GFP-HRas GFP-KRas

DMSO

FTI

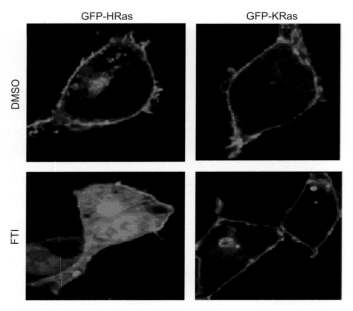

Chapter 15. Figure 4. (See legend in text.)

Prelamin A Lamin A

Pre

Post (100 mg)

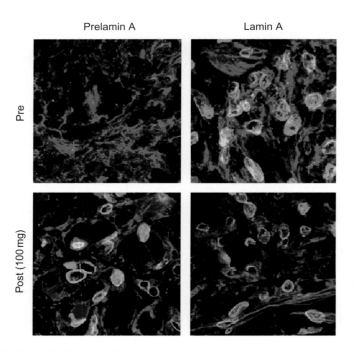

Chapter 15. Figure 5. (See legend in text.)